Praise for *Biomolecular Archaeology*

"This book is a perfect introduction into biomolecular archaeology not only for students interested in the field but also for experienced archaeologists, palaeontologists and archaeobiologists who engage in interdisciplinary research involving the analysis of biomolecules. It is written by one of the most prominent genomic textbook authors, Terry Brown, a pioneer in ancient DNA research and the origins of plant domestication. In this book, his qualities as both an excellent textbook writer as well as a brilliant molecular biologist merge to explain even the most advanced sequencing methods used in palaeogenomics in a way that is understandable for non-experts. The contribution of Keri Brown ensures that the book is relevant to researchers working in the field. *Biomolecular Archaeology* makes for an ideal manual for archaeologists and students eager to exploit the newest scientific developments to answer typical archaeological questions and better interpret the information buried in the archaeological sites they are working on."

Eva-Maria Geigl, Université Paris Diderot

"The study of ancient and extant biomolecules has revolutionized archaeological methodologies. This textbook is an excellent, user-friendly introduction to biomolecular techniques and applications for beginning students in archaeology and physical anthropology."

Linda Stone, Professor Emeritus of Anthropology, Washington State University

"This is a timely and welcome contribution to the rapidly developing field of biomolecular archaeology, covering the basic science as well as an introduction to the applications. It will become essential reading."

A.M. Pollard, University of Oxford

"There are fewer and fewer areas of archaeology which are immune to biomolecular analysis. Technological innovation combined with a greater understanding of molecular survival has increased reliability of analyses and interpretation, making biomolecular research amongst the fastest moving and most exciting areas in modern archaeology. This book, helped by its easy and accessible style, leads the reader in a logical progression from the molecules themselves to their application in the study of demography, diet, innovation and migration; it should be recommended reading for all new students of archaeology."

Matthew Collins, University of York

Biomolecular Archaeology
An Introduction

Terry Brown and Keri Brown

A John Wiley & Sons, Ltd., Publication

This edition first published 2011 © 2011 Terry Brown & Keri Brown

Blackwell Publishing was acquired by John Wiley & Sons in February 2007. Blackwell's publishing program has been merged with Wiley's global Scientific, Technical, and Medical business to form Wiley-Blackwell.

Registered Office
John Wiley & Sons Ltd, The Atrium, Southern Gate, Chichester, West Sussex, PO19 8SQ, United Kingdom

Editorial Offices
350 Main Street, Malden, MA 02148-5020, USA
9600 Garsington Road, Oxford, OX4 2DQ, UK
The Atrium, Southern Gate, Chichester, West Sussex, PO19 8SQ, UK

For details of our global editorial offices, for customer services, and for information about how to apply for permission to reuse the copyright material in this book please see our website at www.wiley.com/wiley-blackwell.

The right of Terry Brown & Keri Brown to be identified as the authors of this work has been asserted in accordance with the UK Copyright, Designs and Patents Act 1988.

All rights reserved. No part of this publication may be reproduced, stored in a retrieval system, or transmitted, in any form or by any means, electronic, mechanical, photocopying, recording or otherwise, except as permitted by the UK Copyright, Designs and Patents Act 1988, without the prior permission of the publisher.

Wiley also publishes its books in a variety of electronic formats. Some content that appears in print may not be available in electronic books.

Designations used by companies to distinguish their products are often claimed as trademarks. All brand names and product names used in this book are trade names, service marks, trademarks or registered trademarks of their respective owners. The publisher is not associated with any product or vendor mentioned in this book. This publication is designed to provide accurate and authoritative information in regard to the subject matter covered. It is sold on the understanding that the publisher is not engaged in rendering professional services. If professional advice or other expert assistance is required, the services of a competent professional should be sought.

Library of Congress Cataloging-in-Publication Data

Brown, Keri.
Biomolecular archaeology / Keri Brown, Terry Brown.
p. cm.
Includes bibliographical references and index.
ISBN 978-1-4051-7960-7 (pbk. : alk. paper)
1. Biomolecular archaeology. I. Brown, T. A. (Terence A.) II. Title.
CC79.B56B76 2011
930.1–dc22

2010035177

A catalogue record for this book is available from the British Library.

This book is published in the following electronic formats: ePDFs 9781444392425;
Wiley Online Library 9781444392449; ePub 9781444392432

Set in 10/13pt Minion by SPi Publisher Services, Pondicherry, India

1 2011

Brief Contents

Contents

List of Figures

List of Tables

Preface

This book originated in an MSc course in Biomolecular Archaeology that was taught jointly by the University of Manchester and University of Sheffield for 10 years up to 2006. Our experience as teachers was that there are many students, from both biology and archaeology backgrounds, who are interested in biomolecular archaeology, and that the great challenge is bringing together the two sides of the subject so that the student becomes an expert in both. The objective of this book is therefore to teach the fundamentals of biomolecular archaeology within a context that emphasizes both the biomolecules and the archaeology.

Deciding to write the book was the easy part. Much more difficult was the actual writing. It became clear early on that we could not cover every biomolecular archaeology project that has ever been published, nor even all the important ones. Those that we describe are chosen because they illustrate key themes and important scientific approaches. We decided not to name the researchers responsible for individual pieces of work in the text, as that would be abnormal in a textbook, but instead to cite the papers describing these projects in the "Further Reading" sections. We were also conscious that as most of our own work has been with DNA we might give this part of biomolecular archaeology a greater emphasis than it deserves. We have tried hard to avoid a DNA bias, and believe that in the book as a whole the different types of biomolecule are given degrees of coverage consistent with their relative importance in biomolecular archaeology. One message that we hope comes across is that the most informative projects are those that use a range of different biomolecular techniques to address the question being asked.

A number of people have helped us in various ways, such as providing figures or permissions to use published or unpublished work, and in suggesting improvements to drafts. We would therefore like to thank Abigail Bouwman, Angela Thomas, Bettina Stoll-Tucker, Charlotte Roberts, Chris Dudar, Claudia Grimaldo, David Beresford-Jones, Diane Lister, Eva-Maria Giegl, Glynis Jones, Ingrid Mainland, John Prag, Julie Wilson,

Martin Richards, Matthew Collins, Mike Richards, Mike Taylor, Peter Rowley-Conwy, Richard Allen, Richard Evershed, Robert Hedges, Robert Tykot, Rosalie David, Sandra Bunning, Susan McCouch, and Wolfgang Haak. We would also like to thank all of our research students, postdocs, and technicians, past and present, for making the last few years so interesting.

Keri Brown
Terry Brown

PART I

BIOMOLECULES AND HOW THEY ARE STUDIED

1

What is Biomolecular Archaeology?

A curiosity about our past is one of the things that makes us human. Over the last century **archaeology** has developed into a sophisticated discipline that interprets the past through examination of the physical remains of human life, those remains often but not exclusively recovered by excavation of archaeological sites. Science has always played an important role in archaeology, increasingly so since the 1950s when techniques invented by nuclear physicists for measuring the decay of radioactive atoms were first used by scientific archaeologists to date artifacts. Biological methods have become equally important. Knowledge of human anatomy and pathology enables **osteoarchaeologists** to use skeletal features to identify the sex of a person, to work out an approximate age at the time of death, and to determine if the person had been suffering from diseases such as tuberculosis or anemia. **Archaeobotanists** are similarly adept at studying seeds and other plant remains, and from these identifying the types of plants that were grown and consumed by people in the past. By combining information from different kinds of analysis, we can address broader issues such as the development of agriculture in particular parts of the world, and how agriculture and the concomitant changes such as increases in population density affected human diet and health.

Since 1985 the way in which biological remains have been studied by scientific archaeologists has undergone a remarkable revolution. Osteology, archaeobotany, and other approaches that involve examination of the physical structure of remains are still vitally important, but they have been supplemented with techniques in which the biomolecular content of the artifact is analyzed. This is called **biomolecular archaeology** and the first thing we must do is understand what this term means.

Biomolecular Archaeology: An Introduction, by Terry Brown and Keri Brown © 2011 Terry Brown & Keri Brown

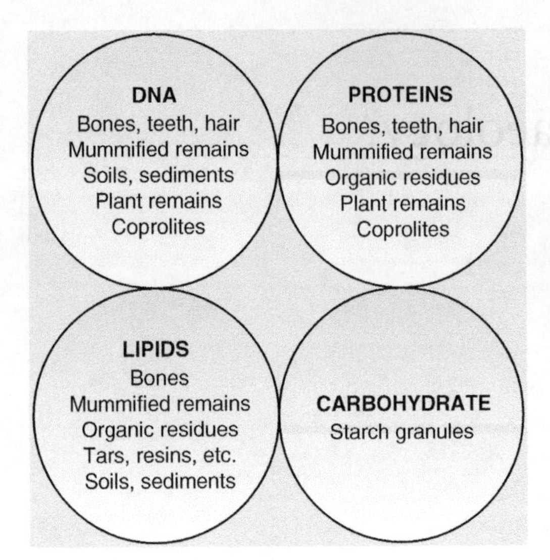

Figure 1.1 The four types of ancient biomolecule studied in biomolecular archaeology, with the main types of archaeological material from which each one can be obtained.

1.1 The Scope of Biomolecular Archaeology

The biomolecules studied by biomolecular archaeologists are the large organic compounds found in living organisms and sometimes present, usually in a partly degraded state, in the remains of those organisms after their death (Figure 1.1).

There are four categories of these **macromolecules:**

- **Nucleic acids** (Chapter 2), of which there are two types, **deoxyribonucleic acid (DNA)** and **ribonucleic acid (RNA)**. Three features of DNA make this molecule valuable in biomolecular archaeology. First, in the cell, DNA acts a store of biological information, which means that DNA can be used to identify at least some of the biological characteristics of an archaeological specimen, such as the sex of a human skeleton (Chapter 10). Second, the DNA of different species can be distinguished, enabling DNA from a pathogen such as *Mycobacterium tuberculosis* to be detected in human bones (Chapter 15). Third, DNA is a record of ancestry, and so can be used to deduce if two human skeletons could be related (Chapter 11), and to map the evolutionary relationships between domesticated animals and their wild progenitors (Chapter 13). RNA molecules are copies of parts of the cell's DNA, and could, theoretically, be used in a similar way to DNA, but RNA has not been extensively studied in biomolecular archaeology. This is because RNA molecules are relatively unstable, and it has been assumed (perhaps incorrectly) that RNA is rarely present in human, animal or plant remains (Section 7.3).
- **Proteins** (Chapter 3) play structural and functional roles in living organisms. Structural proteins, such as collagen and osteocalcin, which are found in all vertebrate bones, are relatively stable and can often be identified in preserved material. Other proteins, usually ones that are less stable, have more limited distributions. Casein, for example, is found only in milk, and can therefore be used as a **marker** for the presence of milk residues in cooking or storage vessels. By showing that certain vessels once contained milk products, the development of dairying in prehistory can be followed (Section 12.3). The blood protein, hemoglobin, has a slightly different structure in different species, and with modern material these differences can be used to identify the origin of a bloodstain. One of the most controversial areas of biomolecular archaeology concerns the analysis of possible blood residues on stone tools to identify the species that were butchered using those implements (Section 9.2).
- **Lipids** (Chapter 4) are a diverse group of macromolecules, the major biochemical classes being fatty acids and their derivatives (which include substances commonly referred to as fats and oils), waxes, steroids, and terpenes. These compounds have various biological functions, both structural (some fatty acids are important components of cell membranes) and functional (some lipids are hormones). Lipids are so hugely diverse that many are species specific – they are found only in a single or small group of species and so can be used as

markers for those species. Analysis of lipids in organic residues from cooking vessels can therefore identify the type of vegetable or meat that was being prepared, and similar studies with storage vessels can show if they were used to hold, for example, a particular type of oil (Section 12.2). Identification of the terpenes in the adhesives used to attach flint arrow heads to wooden shafts can reveal which trees were exploited as sources of tar and pitch, taking biomolecular archaeology into the area of prehistoric technology (Chapter 14).

- **Carbohydrates** (Chapter 5) are important structural and storage compounds in living organisms, and include starch and cellulose in plants, and glycogen in animals. Of the four types of macromolecule, carbohydrates are the least studied by biomolecular archaeologists because, although they are stable over long periods, it is difficult to obtain useful information from them. One exception is the examination of starch grains in archaeological deposits, which can indicate the types of plants that were present at a particular site (Section 13.3).

In studying these four types of macromolecule, biomolecular archaeologists use a variety of methods and analytical techniques, as will be described in Chapters 2–5. Most of these techniques are applicable to just a single type of macromolecule, but one method has greater breadth and is of such general importance that it is often looked on as a distinct area of biomolecular archaeology. This is **stable isotope analysis**, in which ratios of different isotopes of certain elements (primarily carbon and nitrogen) are analyzed in proteins and lipids (Chapter 6). The natural ratios of these elements are constant, but variations can be introduced by biological and environmental processes. These variations can be exploited in studies of diet, as the stable isotope ratios present in bone proteins, or in hair, reflect the types of organism consumed by that individual (Section 12.2). A diet rich in marine resources can be distinguished from, for example, a diet largely made up of terrestrial animal protein, and the presence of maize in the diet can also be detected because this plant has a different isotope ratio to many other cereals and vegetables. In a particularly clever application of the technique, stable isotope analysis has been used to identify lipids derived from dairy products (Section 12.3).

1.2 Ancient and Modern Biomolecules

It will already be clear that research in biomolecular archaeology largely involves the analysis of the compounds that are preserved in archaeological remains. We call these **ancient biomolecules** and archaeologists are not the only scientists interested in their study. Forensic scientists are increasingly using information from preserved biomolecules, especially **ancient DNA**, in samples such as hair, bloodstains, and other bodily fluids collected at crime scenes years or decades ago in order to solve what are popularly called "cold cases." Zoologists also use ancient DNA from animal fossils to study extinct species such as mammoths and moas, and to follow changes in genetic diversity over time in populations of animals such as bison, whose numbers have been affected by climate change and human predation. DNA is rarely preserved for more than a few tens of thousands of years, but other biomolecules are present in much older materials. Proteins have been extracted from dinosaur bones and lipids and carbohydrates from the remains of plants and insects in sediments that are tens of millions of years in age.

The breadth of ancient biomolecules research is important because it means that biomolecular archaeologists have scientific colleagues who have very different interests but who use the same techniques and face the same challenges in planning and interpreting their experiments. Over the years there has been a large amount of cross-fertilization of ideas between researchers working with ancient biomolecules in different disciplines, and this has contributed greatly to the development of biomolecular archaeology. Indeed many biomolecular archaeologists also study ancient biomolecules in non-archaeological material, hence making direct contributions to forensic science, zoology, or paleontology, and the boundaries between the these disciplines and biomolecular archaeology often become blurred.

Although most research in biomolecular archaeology involves the study of ancient biomolecules, this is not exclusively the case. Studies of one biomolecule – DNA – in living organisms can contribute greatly to our understanding of certain archaeological issues. This is because DNA contains a record of the ancestry of individuals and the past evolution of populations and species. We can therefore study the relationships between different human populations by typing DNA taken from living representatives of those populations and using techniques from **molecular phylogenetics** and **population genetics** to analyze the data. This approach, sometimes called **archaeogenetics**, has been particularly informative in understanding the timing and trajectories followed by the migrations of modern humans out of Africa into Asia, Europe, Australasia, and the New World (Chapter 16). Using similar approaches with DNA from living crop plants and domestic animals, information is being obtained on the origins and spread of agriculture (Chapter 13). These studies overlap with evolutionary biology and crop and animal genetics, broadening still further the range of researchers who can be looked on as the scientific partners of biomolecular archaeologists.

1.3 The Challenges of Biomolecular Archaeology

All scientists work at the frontiers of their discipline – that is one of the characteristics of research – and all disciplines provide challenges that must be met and overcome if research is to progress. Biomolecular archaeology is no different, the challenges coming in two guises: technical and intellectual.

The technical challenges are posed by the degradation of ancient biomolecules and by contamination of specimens with modern biomolecules. All biomolecules begin to decay when the organism that contains them dies. Some, especially the nucleic acids, are relatively unstable and may completely degrade within a few years. Others, such as carbohydrates, are more stable and their decay products might still be detectable tens of millions of years after death (Chapter 8). These are not precise comparisons, because the environmental conditions, in particular the temperature and water content, greatly affect the rate at which a biomolecule decays, but the outcome is always the same. Almost every biomolecular archaeology project requires analysis of very small quantities of biomolecules that have undergone a greater or lesser degree of chemical degradation. The small quantities of ancient biomolecules present in archaeological specimens mean that detection techniques have to be pushed to their very limits, and often this affects the amount of information that can be obtained by biomolecular analysis of a specimen. Frequently, results are frustratingly incomplete, sometimes tempting the unwary researcher to make speculations that are not entirely warranted by their data, a problem that seems particularly prevalent in some areas of ancient DNA research.

The changes in chemical structure that occur during **diagenesis** can also confuse the detection processes, so that precise identification of an ancient biomolecule becomes difficult. For example, a process specific for the detection of human hemoglobin in modern bloodstains, when applied to archaeological material, might also give positive results with the partially degraded hemoglobins from other animals. Because of these problems, studies of biomolecular degradation form an essential adjunct to biomolecular archaeology, as it is only by understanding the decay processes for particular biomolecules that misidentifications can be avoided.

The small quantities of ancient biomolecules present in even the best preserved archaeological specimens leads to the second major technical problem, the possibility that modern contaminating molecules swamp the detection process, again leading to erroneous results. This issue is most clearly recognized in ancient DNA studies, because the exquisite sensitivity of the **polymerase chain reaction** (**PCR**), the primary detection method for DNA (Section 2.5), enables samples containing just a few hundred ancient DNA molecules to be examined. Similar or greater numbers of modern DNA molecules are present in human sweat, droplets expelled from the mouth and nose by sneezing, and in aerosols derived from previous PCR experiments that adhere to the clothes and skin of laboratory workers. Ancient and modern human DNA are very difficult to tell apart, and it is very easy to mistakenly assign to an archaeological specimen the genetic attributes of one or a mixture of the people who have handled the specimen. The problem is so acute that ancient DNA researchers carry out their experiments in ultraclean laboratories, wearing overalls that cover their entire body and face, a regime more commonly associated with research on deadly virus pathogens. The aim is not, however, to prevent escape of a pathogen from the test tube, but to prevent entry into the test tube of modern DNA from the researcher. Such practices are possible within the confines of a modern laboratory, but less feasible in the field, so it is almost inevitable that human bones become contaminated with DNA from the excavators who first uncover them. Solving these conundrums has greatly exercised not only ancient DNA researchers but all biomolecular archaeologists, as we will see in Chapter 9.

In addition to these technical issues, biomolecular archaeology poses a major intellectual challenge. Biomolecular archaeology is an interdisciplinary subject, and biomolecular research is of no value if it is not carried out within an archaeological context. This may seem obvious, but frequently projects that reach high standards as far as the biomolecular aspect is concerned fail to interest archaeologists because the results are not relevant to the issues that are important in archaeology. The problem arises because, until recently, very few biomolecular scientists possessed anything more than a rudimentary understanding of archaeology, and few archaeologists had a strong training in the biomolecular sciences. Successful biomolecular archaeology therefore requires collaboration between archaeologists and biomolecular scientists, and meaningful collaboration is often difficult to achieve. It is easy to assemble a "team," but much less easy to reach the mutual intellectual understanding that is required for interdisciplinary research to flourish. It is difficult to become an expert in both biomolecular research and archaeology – both are complex subjects with their own languages and ways of thinking – but such dual expertise has to be the goal of anyone who wishes to become a biomolecular archaeologist. The aim of this book is to help you achieve that goal.

Further Reading

Brothwell, D.R. & Pollard, A.M. (eds.) (2001) *Handbook of Archaeological Sciences*. Wiley, Chichester. [Covers all areas of archaeological science.]

Cox, M. & Nelson, D.L. (2008) *Lehninger Principles of Biochemistry*, 4th edn. Palgrave Macmillan, New York. [One of the best student textbooks in biochemistry.]

Jones, M.K. (2003) *The Molecule Hunt: Archaeology and the Search for Ancient DNA*. Arcade, New York. [A popular account of the early days of biomolecular archaeology.]

Renfrew, C. & Bahn, P. (2008) *Archaeology: Theories, Methods and Practice*, 5th edn. Thames and Hudson, London. [One of the best student textbooks in archaeology.]

2

DNA

Although proteins and lipids were the first biomolecules to be studied in an archaeological context, it was the demonstration in the late 1980s that ancient DNA is sometimes preserved in human bones that marks the true beginning of biomolecular archaeology as a discipline in its own right. This was for two reasons. First, DNA is the genetic material of living cells, which means that it contains a vast amount of information that, if accessed in preserved specimens, could be of immense value in addressing archaeological questions. Second, the discovery of ancient DNA in archaeological specimens was made possible by the invention of the polymerase chain reaction (PCR), an extremely sensitive detection method capable of reading the genetic information in very small numbers of DNA molecules – just one under ideal conditions. This means that even if the vast majority of the DNA in a specimen has decayed, it might still be possible to obtain information from the small quantities that remain. PCR of ancient DNA therefore seemed to be a panacea that would bring biomolecular studies to the fore as a new tool for studying archaeology.

In many respects, the archaeological potential of ancient DNA has indeed been realized, as we will see when we examine the applications of biomolecular archaeology in Part III of this book. The road has, however, been rocky, largely because the challenges presented by contamination of specimens with modern DNA were not taken sufficiently seriously until the early years of the 21st century (Chapter 9), and partly because ancient DNA researchers have sometimes neglected to establish productive collaborations with archaeologists, which means that even when contamination problems have been solved, the results of ancient DNA projects have not always been relevant to the mainstream of archaeological research. Those issues are for later; in this chapter we examine the structure and function of DNA and the methods used to read and analyze the genetic information contained in DNA molecules.

 © 2011 Terry Brown & Keri Brown

2.1 The Importance of DNA in Biomolecular Archaeology

DNA is important in biomolecular archaeology for three reasons. First, DNA specifies the biological characteristics of living organisms, which means that some of the biological characteristics of an archaeological specimen can be identified by studying its ancient DNA. The range of characteristics that can be addressed is limited, because molecular biologists have only an incomplete understanding of the link between DNA structure and biological attributes, and for many characteristics the link is complex and not easily unravelled by examining the DNA. Molecular biologists are a long way from being able to describe the physical appearance of an individual, or of understanding any aspect of personality, from analyzing his or her DNA. But whether or not a person is male or female can be identified from their DNA, and **sex identification** is one of the most frequent applications of ancient DNA analysis with archaeological skeletons, not only of humans but of animals also (Chapter 10). With human remains, ancient DNA typing has also been used to assess the frequency of **lactase persistence** in prehistoric populations, this characteristic conferring the ability to digest lactose, the sugar found in milk. Humans who lack this characteristic are lactose intolerant and become ill if they drink unfermented milk, so a high frequency of lactase persistence is thought to indicate a population whose diet includes dairy products. Typing lactase persistence by ancient DNA analysis can therefore help indicate when dairying was first adopted (Section 12.4). Equally interesting work has been done with plants. For crops a key question is how soon the special characteristics of the domesticated version of a species appeared after farmers first started cultivating the plants rather than collecting them from the wild. In maize, these characters include changes in the architecture of the plant and in the way protein and sugar are produced in the kernels. The development of these features can be followed by typing ancient DNA in preserved maize specimens of different ages, helping archaeologists to understand how the crops were used by the first farmers, and also providing crop geneticists with insights into the evolutionary processes occurring during domestication (Section 13.4).

The second reason why DNA is important in biomolecular archaeology is because the species to which an organism belongs can be identified by typing its DNA. This means that uncertainties about the identities of animal bones recovered by excavation can sometimes be settled by ancient DNA typing, and similar uncertainties about the identities of plant remains can be addressed. More importantly, the species specificity of DNA can be used to identify if the remains of a pathogen such as the tuberculosis bacterium are present in a human bone, a positive result indicating that the individual was infected with that pathogen at the time of death. This does not mean that death was necessarily due to the disease caused by the pathogen, but the information is still vital for understanding the prevalence of different diseases in the past, one of the major goals of **paleopathology** (Chapter 15).

Finally, DNA is important because it contains a record of an individual's ancestry. The DNA of every living organism is inherited from its parents and combines features of both the maternal and paternal DNA. This means that DNA typing can be used to establish if two people are likely to be related or, depending on their ages, are parent and offspring. With DNA samples taken from living people this can be done with a high degree of accuracy, and although it is much more difficult to achieve this accuracy with archaeological material, a great deal of progress is being made in using ancient DNA to determine **kinship** between groups of skeletons buried together at archaeological sites (Chapter 11). DNA also contains a record of broader population affinities, enabling humans to be placed into groups, all members of a single group sharing a common ancestry that dates back thousands of years. These ancient population groups are much older than the national and cultural groups that we recognize today, and so DNA typing cannot be used to assign individuals to populations such as "Roman," but their detailed analysis can reveal patterns in human evolution that can be related to the trajectories followed by modern humans as they migrated out of Africa and colonized the rest of the world during the period 90,000–10,000 years ago (Section 16.3). Similar studies can be carried out with animals and plants in order to trace the origins of domesticated species and the trajectories by which farming spread from its areas of origin (Sections 13.2 and 13.3). This type of research increasingly involves typing of ancient DNA in archaeological specimens to understand the population affinities of particular groups of prehistoric people and of their domestic animals or plants. Similar studies are also frequently carried out with DNA from living people, animals, and plants, with the information on their origins deduced by applying analytical techniques developed by evolutionary biologists and population geneticists.

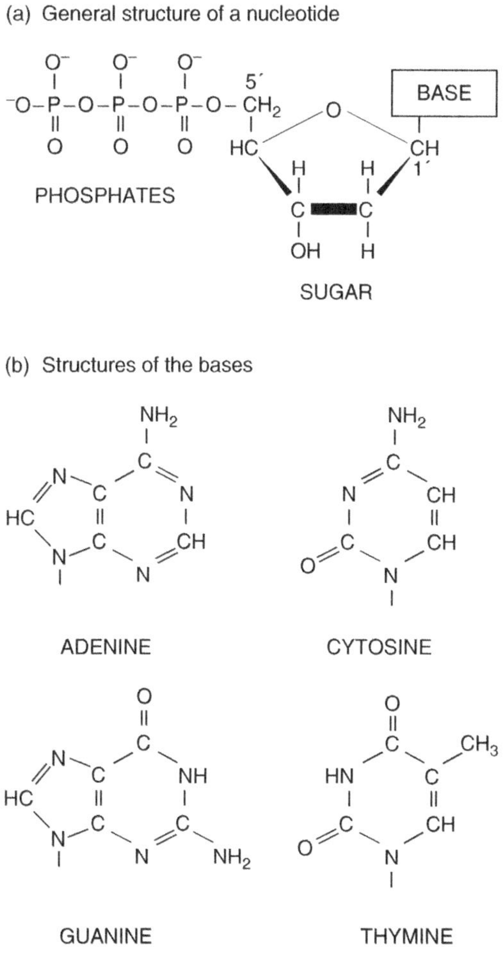

2.2 The Structure of DNA

In chemical terms, DNA (deoxyribonucleic acid) is a relatively simple molecule. It is a linear, unbranched polymer in which the monomeric subunits are four chemically distinct **nucleotides** that can be linked together in any order in **polynucleotide** chains hundreds, thousands, or even millions of nucleotides in length. The biological information contained in a DNA molecule is denoted by its nucleotide sequence, represented as a series of As, Cs, Gs, and Ts, the abbreviations of the full chemical names of the nucleotides. This is the format in which the information is read from ancient DNA or from samples taken from living organisms, and is the only aspect of DNA structure that most biologists consider in their day to day research. It is important, however, that every biomolecular archaeologist be familiar with the underlying details of DNA structure, because of the effects that subtle changes in this structure – such as those occurring during diagenesis – can have on the nucleotide sequence of an ancient DNA molecule (Section 8.1).

First, we consider the structures of the four nucleotides (Figure 2.1). Each of these is made up of three components:

- The first component is **2′-deoxyribose**, which is a **pentose**, a type of sugar composed of five carbon atoms. These five carbons are numbered 1′ (spoken as "one-prime"),

(a) General structure of a nucleotide

(b) Structures of the bases

Figure 2.1 Nucleotide structure.

Table 2.1 The nucleotides found in DNA molecules.

Full name	*Standard abbreviation*	*One-letter abbreviation*
2′-deoxyadenosine 5′-triphosphate	dATP	A
2′-deoxycytidine 5′-triphosphate	dCTP	C
2′-deoxyguanosine 5′-triphosphate	dGTP	G
2′-deoxythymidine 5′-triphosphate	dTTP	T

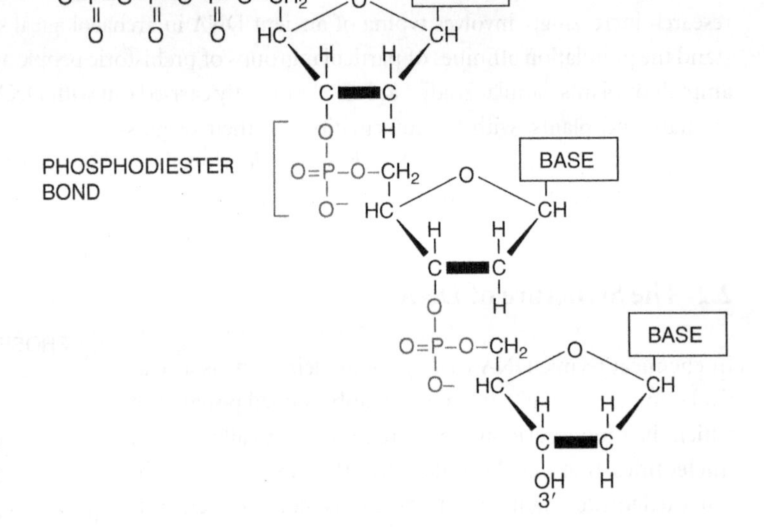

Figure 2.2 A short DNA polynucleotide showing the structure of the phosphodiester bond.

2′, etc., the numbering running clockwise around the sugar ring. The name 2′-deoxyribose indicates that this particular sugar is a derivative of ribose, one in which the hydroxyl (–OH) group attached to the 2′-carbon of ribose has been replaced by a hydrogen (–H) group.

- The second component is the part that distinguishes the four nucleotides. This is a **nitrogenous base**, one of **cytosine, thymine** (single-ring **pyrimidines**), **adenine**, or **guanine** (double-ring **purines**). The base is attached to the 1′-carbon of the sugar by a **β-*N*-glycosidic bond.**
- The final component is a phosphate group, comprising three linked phosphate units attached to the 5′-carbon of the sugar. The phosphates are designated α, β, and γ, with the α-phosphate being the one directly attached to the sugar.

The full chemical names of the four nucleotides that polymerize to make DNA are 2′-deoxyadenosine 5′-triphosphate, 2′-deoxycytidine 5′-triphosphate, 2′-deoxyguanosine 5′-triphosphate, and 2′-deoxythymidine 5′-triphosphate. The abbreviations of these four nucleotides are dATP, dCTP, dGTP, and dTTP, respectively, or, when referring to a DNA sequence, A, C, G, and T, respectively (Table 2.1).

In a polynucleotide, individual nucleotides are linked together by **phosphodiester bonds** between their 5′- and 3′-carbons (Figure 2.2). From the structure of this linkage we can see that the polymerization reaction involves removal of the two outer phosphates (the β- and

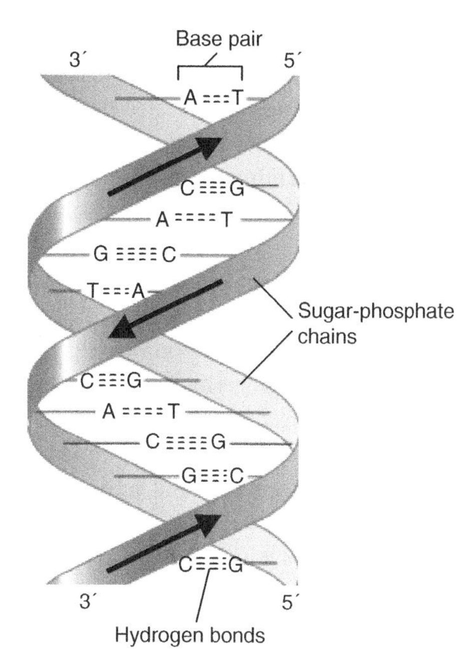

Figure 2.3 The double helix structure for DNA. In this drawing, the sugar-phosphate chain of each polynucleotide is shown as a ribbon, with the nucleotide bases indicated by the letters on the inside of the helix. The two polynucleotides are held together by hydrogen bonds between the bases. The two polynucleotides are antiparallel, as shown by the arrows, which are drawn in the 3′→5′ direction.

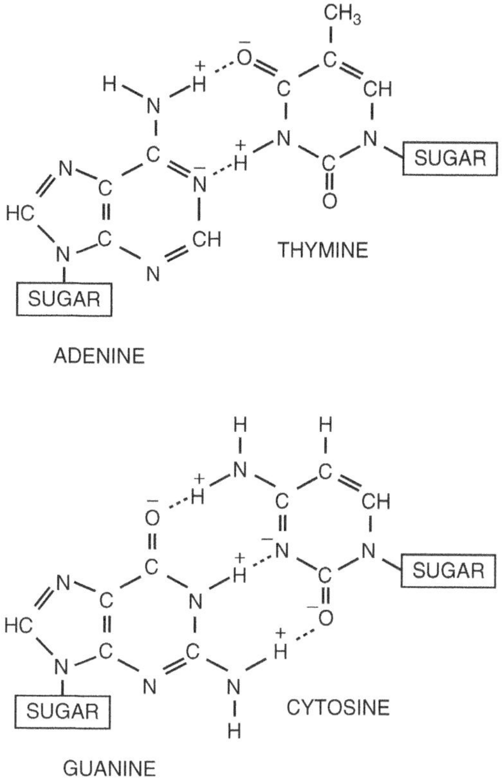

ADENINE

THYMINE

GUANINE

CYTOSINE

Figure 2.4 Structures of the adenine–thymine and guanine–cytosine base pairs. Hydrogen bonds are shown as dotted lines. These form between an electronegative atom such as an oxygen or nitrogen, and a positively charged hydrogen attached to a second electronegative atom.

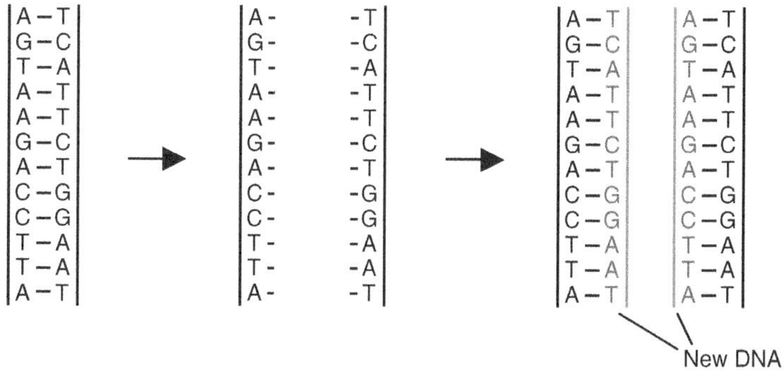

New DNA

Figure 2.5 Two identical copies of a DNA double helix can be made by separating the two strands and using each as a template for synthesis of its complementary polynucleotide.

γ-phosphates) from one nucleotide and replacement of the hydroxyl group attached to the 3′-carbon of the second nucleotide. An important consequence of this reaction is that the two ends of the polynucleotide are chemically distinct, one having an unreacted triphosphate group attached to the 5′-carbon (the **5′** or **5′-P terminus**), and the other having an unreacted hydroxyl attached to the 3′-carbon (the **3′**or **3′-OH terminus**). This means that the polynucleotide has a chemical direction, expressed as either 5′→3′(down in Figure 2.2) or 3′→5′ (up in Figure 2.2).

In living cells, each DNA molecule comprises two polynucleotides wound around one another to form the **double helix**, the iconic symbol of modern biology (Figure 2.3). The double helix is right-handed, which means that if it were a spiral staircase and you were climbing upwards then the rail on the outside of the staircase would be on your right-hand side. The two strands are **antiparallel**, meaning that they run in opposite directions. The helix is stabilized by **base pairing** between the two polynucleotides. This involves the formation of **hydrogen bonds** between an adenine on one strand and a thymine on the other strand, or between a cytosine and a guanine (Figure 2.4). The two base-pair combinations – A base-paired with T, and G base-paired with C – means that the sequences of the two polynucleotides are **complementary**; so once one is known, the other can be predicted. More importantly two identical copies of a double helix can be made by unwinding the polynucleotides and using each one as a **template** for synthesis of a complementary copy (Figure 2.5). This is the basis to DNA replication in living cells, and to test-tube reactions such as PCR, which are used by molecular biologists to make copies of DNA molecules (Section 2.5).

A **base pair** (bp) is the standard unit of length of a double-stranded DNA molecule. Using this convention, 260 million bp is the same as 260 thousand kb (**kilobase pairs**) and 260 Mb (**megabase pairs**).

2.3 Genomes and Genes

Every organism possesses a set of DNA molecules that together contain the biological information needed to construct and maintain a living example of that organism. This set of DNA molecules is called the organism's **genome**, and the information contained within it is packaged into discrete units called **genes**. Understanding how the genes are organized within a genome is one of the major goals of modern molecular biology.

2.3.1 The human genome

The human genome, which is typical of the genomes of all multicellular animals, has two distinct parts:

- The **nuclear genome**, which comprises approximately 3200 Mb of DNA, divided into 24 linear molecules, the shortest 50 Mb in length and the longest 260 Mb, each contained in a different **chromosome**. These 24 chromosomes consist of 22 **autosomes** and the two sex chromosomes, X and Y.
- The **mitochondrial genome**, which is a circular DNA molecule of 16,569 bp, multiple copies of which are located in the energy-generating organelles called mitochondria.

There are approximately 10^{13} cells in the adult human body, and each one has its own copy or copies of the genome, except for a few specialized types, such as red blood cells, which do not have a nucleus. Most cells are **diploid**, meaning that they have two copies of each autosome, and two sex chromosomes, XX for females or XY for males. These are called **somatic cells**, in contrast to **sex cells** or **gametes**, which are **haploid** and have just one copy of each autosome and one sex chromosome. Both types of cell have about 8000 copies of the mitochondrial genome, 10 or so in each mitochondrion.

Chromosomes are not made up entirely of DNA; in fact they comprise roughly equal amounts of DNA and proteins. Some of these proteins control the release of information from the genome to the rest of the cell, but most are involved in **DNA packaging**. A highly organized packaging system is needed to fit a DNA molecule into its chromosome: the average human chromosome is only a few thousandths of a millimetre in length but contains almost 5 cm of DNA. The packaging system was initially visualised by electron microscopy which revealed a "beads-on-a-string" structure, the string being a DNA molecule and the beads being protein complexes called **nucleosomes**, each one made up of eight **histone** proteins, in humans two each of histones H2A, H2B, H3, and H4. Between 140 and 150 bp of DNA (depending on the species) are associated with each nucleosome, and pairs of nucleosomes are separated by 50–70 bp of linker DNA. As well as the proteins of the nucleosome, there is a group of additional histones, all closely related to one another and collectively called **linker histones**, one attached to each nucleosome, possibly forming a clamp preventing the DNA from unwinding.

The "beads-on-a-string" structure is only the first level of DNA packaging and on its own only reduces the length of a DNA molecule by a factor of six. The next level of packaging is the **30 nm fiber** (it is approximately 30 nm in width), which is formed by packing together individual nucleosomes, although the precise arrangement is not known. Although this only reduces the length of the DNA molecule by a further seven times (meaning that a molecule that is 5 cm when fully linear would now be about 1.2 mm in length), it is probably the form taken up by chromosomes in a cell that is not dividing. To achieve the additional size reduction needed to produce the highly condensed chromosomes present in dividing cells, the 30 nm fiber takes up a super-coiled conformation, with loops radiating from a core, called the **scaffold**, comprising non-histone proteins. About 85 kb of DNA (in the form of the 30 nm fiber) is contained in each loop.

Mitochondrial genomes are also associated with their attendant proteins, but we know less about the packaging of these DNA molecules. It is clear, however, that neither in the nucleus of a human cell nor in its mitochondria is DNA ever in a "naked" form: always the DNA is attached to proteins and the associations are so tight that it is unlikely that the proteins simply "fall off" when the cell dies. The breakdown of DNA during diagenesis is therefore closely associated with breakdown of protein, and ancient DNA is probably still attached to protein remnants when it is extracted from archaeological specimens. Although some researchers have recognized this fact and attempted to devise extraction methods that will recover intact DNA-protein complexes from ancient material, it is too frequently assumed that ancient DNA is naked and unprotected within a bone or other type of human remain. This oversight places a question mark against some of the theories that have been put forward about the way in which DNA degrades in archaeological specimens (Section 8.1).

Table 2.2 The nuclear genomes of various organisms relevant to biomolecular archaeology.

Species	*Genome size (Mb)*	*Number of chromosomes*
Humans and other animals		
Human	3200	24
Cattle	2870	31
Sheep	2760	28
Goat	2800	31
Pig	2700	20
Plants		
Maize	2500	10
Rice	466	12
Bread wheat	16,000	21
Disease organisms		
Plasmodium falciparum	22.90	14
Mycobacterium tuberculosis	4.42	1
Mycobacterium leprae	3.27	1
Helicobacter pylori	1.60	1
Treponema pallidum	1.14	1
Yersinia pestis	4.50	1

2.3.2 The genomes of other organisms

The human genome is typical of vertebrate genomes, the only variations between the genomes of humans, cattle, pigs, goats, and other mammals being the number of chromosomes in the nucleus and the exact length of the nuclear genome (Table 2.2). In cattle, there are 31 chromosomes and the genome is 2870 Mb in length (compared with 24 chromosomes and 3200 Mb for humans). The mitochondrial genome is almost exactly the same size in every mammal.

Plants also have nuclear and mitochondrial genomes, the nuclear genomes being more variable in length than those of mammals (e.g. the rice genome is 466 Mb, and that of wheat is 16,000 Mb), and the mitochondrial genomes much larger (e.g. 570 kb for maize). Plants also have separate genomes in their chloroplasts, the organelles within which photosynthesis takes place. These are circular, multicopy molecules, between 120 and 160 kb for most flowering plants. Like the nuclear and mitochondrial genomes, the chloroplast molecules are associated with proteins. It is worth noting that chloroplast genomes are present in all plant tissues, not just those that are actively photosynthesizing. This means that multicopy chloroplast genomes are present in seeds, and fragments of these genomes can be recovered as ancient DNA from archaeological specimens such as preserved grains of wheat.

Much greater differences are seen when we move from mammals and plants to bacteria. Bacteria are **prokaryotes**, their cells lacking a defined nucleus and generally having a less complex structure than the **eukaryotic** cells of mammals (all species from algae, yeast, and fungi up the evolutionary scale to vertebrates and flowering plants). Most bacterial genomes are circular, with the entire genome contained in a single DNA molecule, though there are a few species with linear genomes and/or genomes split into different molecules. The sizes of bacterial genomes range from 0.5 to 10.0 Mb (Table 2.2); that of *Mycobacterium tuberculosis*, the bacterium that causes tuberculosis in humans, is in the mid range at 4.42 Mb. As in eukaryotic nuclei, these molecules are so large that they must be packaged by association with proteins in order to fit into the cell. The resulting structure is called the **nucleoid**, and consists of a protein core from which 40–50 loops radiate, each loop containing approximately 100 kb of super-coiled DNA. The protein component of the nucleoid includes packaging proteins which are structurally different to eukaryotic histones but which act in a similar way, forming beads, each comprising four proteins, around which approximately 60 bp of DNA becomes wound. Again, degradation of the genome after death of the bacterium must be intimately associated with degradation of the attached proteins.

2.3.3 Genes are looked on as the important parts of a genome

Although Gregor Mendel did not use the term "gene," he was the first person to discover their existence, his "unit factors," each one specifying a different biological characteristic, being functionally identical to the entities that we today recognize as "genes." To Mendel, however, unit factors were not physical entities but simply units of inheritance, whereas to us genes are real structures contained within the DNA molecules of an organism's genome.

A gene is simply a segment of a DNA molecule, any length from less than 75 bp to over 2500 kb. The biological information carried by a gene is contained in its nucleotide sequence. In the human nuclear genome, there are 20,000–25,000 genes – the precise number is uncertain because, even though the entire genome sequence is known, identifying the genes by inspecting the sequence is a complex process. Other mammals have similar numbers of genes, but bacteria have far fewer, only 4000 in the *M. tuberculosis* genome, the smaller number reflecting the relative simplicity of a bacterium compared with a mammal. In most bacteria, the genes are fairly evenly spaced within the genome, separated by short **intergenic regions** which might be just a few bp in length and are rarely more than 100 bp. In mammals, the intergenic regions are much longer, and there are variations in gene density in different parts of the genome. In some parts of the human genome, there are stretches of DNA several hundred kb in length that appear to lack any genes at all, whereas in other parts there may be as many as 65 genes per 100 kb. We do not know why this should be the case, neither do we understand for the most part why particular genes are positioned near to one another in the genome. We suspect that the patterns we see today are largely the result of random events such as **recombination** that have occurred in genomes over the millions years during which they have evolved.

One exception to the apparently random distribution of genes is provided by those **multigene families** whose individual members have similar functions and are grouped together at the same position in the genome. The globin genes are an example in mammals. These genes specify the globin proteins that combine to make hemoglobin, found in red blood cells, each molecule of hemoglobin being made up of two α-type and two β-type globins. In humans, the α-type globins are coded by a small multigene family on chromosome 16, and the β-type globins by a second family on chromosome 11 (Figure 2.6A). The genes in each family are similar to one another but sufficiently different for the proteins they code for to have distinctive biochemical properties. These distinctive properties enable the individual globin proteins to play slightly different physiological roles in the bloodstream, each role appropriate for a different stage in mammalian development. For example, in the

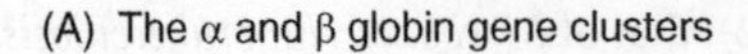

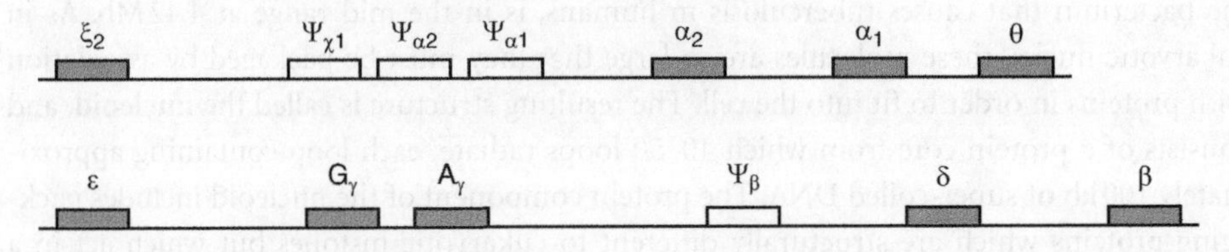

(B) Evolution of the β cluster

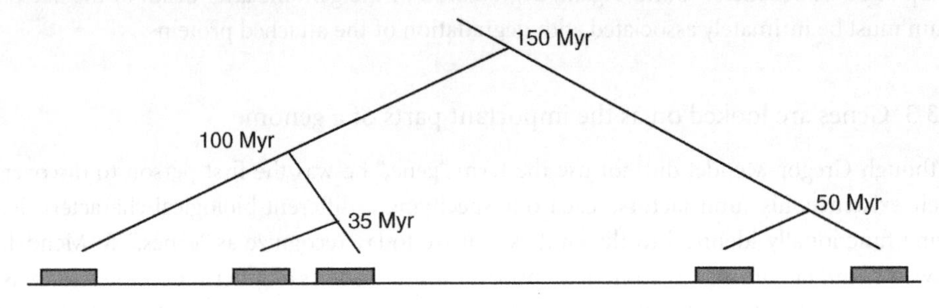

Figure 2.6 The human globin genes. (A) Maps of the α and β clusters. Functional genes are shown as dark boxes and pseudogenes as open boxes. (B) Evolutionary relationships between the functional genes in the β cluster, showing the approximate dates of the ancestral gene duplications. Abbreviation: Myr, million years ago.

β-type cluster ε is active in the early embryo, G_γ and A_γ in the fetus, and δ and β in the adult. We believe that the various globin genes arose by a series of gene duplications, each pair of duplicates initially being identical, but their sequences subsequently diverging because of random **mutations** within each one. By using the **molecular clock** (Section 16.2) to calibrate the amount of sequence difference between each pair of genes we can estimate how many millions of years ago each duplication took place (Figure 2.6B). This analysis indicates that the ancestral globin gene duplicated to give the proto-α and proto-β genes 450 million years ago, and that the duplications within the α and β families took place during the last 200 million years, the most recent of these, giving rise to G_γ and A_γ, occurring just 35 million years ago. This is an example of how DNA sequences can be used to infer evolution, similar to the way in which, albeit on a much shorter timescale, DNA sequence comparisons can be used to chart evolutionary events relevant to archaeological issues such as the origins of humans (Section 16.2) and of domesticated animals and plants (Section 13.2).

The globin multigene families illustrate two other important features of genes. The first of these concerns the genes labelled $\psi_{\chi 1}$, $\psi_{\alpha 1}$ and $\psi_{\alpha 2}$ in the α cluster, and ψ_β in the β group. Each of these is very similar to the other members of its family, but we can tell from their nucleotide sequences that the biological information they contain has become scrambled and is no longer readable. They are evolutionary relics or **pseudogenes**, and more and more of them are being discovered in the genomes of mammals and other organisms – 20,000 have been found so far in the human genome. Indeed it is possible that in mammalian genomes there are more pseudogenes than functional genes.

The second important feature of the globin genes becomes apparent when we examine their nucleotide sequences. Each of these genes is **discontinuous**, meaning that the biological information it contains does not run in an uninterrupted fashion from the beginning of the gene to the end. Instead, the information is split into three segments, called **exons**, separated by non-coding regions called **introns** (Figure 2.7). This discontinuous structure is typical of genes in higher organisms, the numbers of exons in human genes ranging from one (i.e. a non-discontinuous structure) to over 100, with an average of ten. Discontinuous genes are not, however, a common feature of bacterial genomes: virtually all bacterial genes are made up of a single block of uninterrupted biological information.

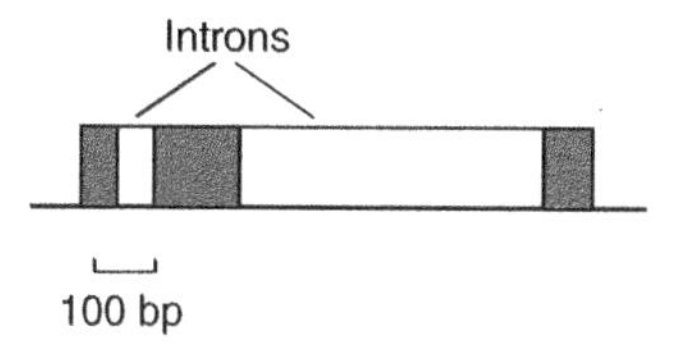

Figure 2.7 The structure of the human β-globin gene, which is a typical discontinuous gene comprising three exons (shown as dark boxes) and two introns (shown as open boxes).

2.3.4 Genes make up only a small part of a mammalian genome

Although the genes are looked on as the important part of the DNA, they make up only a portion of the mammalian genome. All the human genes, for example, if placed end to end would give a DNA molecule only 900 Mb in length, about 28% of the total length of the DNA molecules within which they are found. Even if we include all the pseudogenes we still only account for 38% of the genome. Human genes, and those of other mammals, are separated from one another by tracts of **intergenic DNA** whose function, if any, has so far eluded molecular biologists.

In most organisms, the bulk of the intergenic DNA is made up of repeated sequences of one type or another. Repetitive DNA can be divided into two categories: genome-wide or **interspersed repeats**, whose individual repeat units are distributed around the genome in an apparently random fashion, and **tandemly-repeated DNA**, whose repeat units are placed next to each other in a series. There are many different families of interspersed repeats, each with several thousand copies per genome, with the individual repeat units distributed in an apparently random fashion. For many interspersed repeats, the genome-wide distribution pattern is set up by **transposition**, the process by which a segment of DNA can move from one position to another in a genome. These movable segments are called transposable elements or **transposons**. Transposons are also found in bacteria, but make up a much smaller part of the genome, usually less than 5% of the total length.

There are also several families of tandemly repeated DNA, some of these possibly playing a structural role of some kind within chromosomes. Both the **telomeres**, the specialized DNA-protein structures at the ends of chromosomes, and the **centromeres**, which form the attachment points for the microtubules that draw chromosomes into daughter nuclei when the cells divide, contain tandemly repeated DNA. A different type of tandemly repeated DNA is particularly important in biomolecular archaeology as it is used in **genetic profiling** to establish kinship relationships between individuals (Section 11.2). These repeats are called **short tandem repeats (STRs)** or **microsatellites**. As the name implies, the repeat unit is short, up to 13 bp in length. The commonest type of human STR has dinucleotide repeats, with approximately 140,000 copies in the genome as whole, about half of these being repeats of the motif "CA." Single-nucleotide repeats (e.g. AAAAA) are the next most common (about 120,000 in total). Many STRs are variable, meaning that the number of repeat units in the array is different in different members of a species. This is because "slippage" sometimes

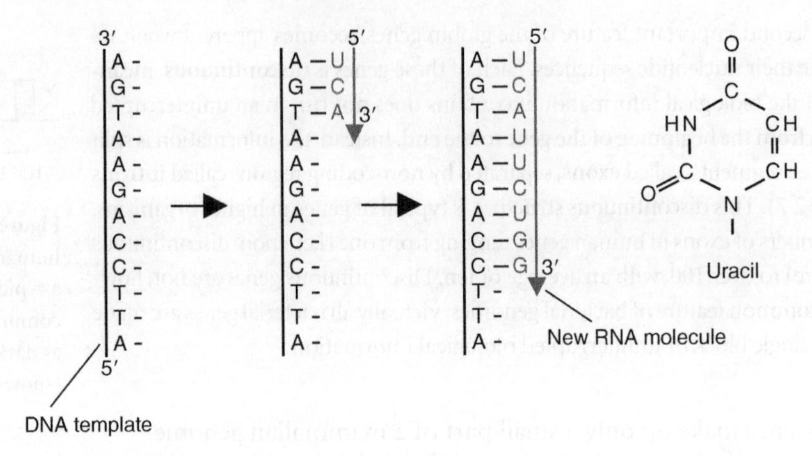

Figure 2.8 Transcription of RNA on a DNA template.

occurs when an STR is copied during DNA replication, leading to insertion or, less frequently, deletion of one or more of the repeat units. No two humans alive today have exactly the same combination of STR length variants: if enough are examined, then a unique profile can be established for every person, except pairs of genetically identical twins.

2.4 From Genomes to Organisms

The genome is a store of biological information, but on its own is unable to release that information to the cell. The information contained within a genome is read by the process called **genome expression**, a complex series of reactions, coordinated by the activity of enzymes and other proteins, which determines the biochemical capabilities of the cell in which the genome is contained. The coordinated activity of different cells, their activities specified by their genomes, results in a living, functioning organism. We will now examine the nature of this link between genome and organism.

2.4.1 There are two major steps in the genome expression pathway

The initial product of genome expression is the **transcriptome**, a collection of RNA molecules derived from those protein-coding genes whose biological information is required by the cell at a particular time. The RNA molecules of the transcriptome, as well as many other RNAs derived from genes that do not code for proteins, are synthesized by the process called **transcription**. The second product of genome expression is the **proteome**, the cell's repertoire of proteins, which specifies the nature of the biochemical reactions that the cell is able to carry out. These proteins are synthesized by **translation** of the mRNA molecules that make up the transcriptome.

Transcription is a simple copying reaction. RNA, like DNA, is a polynucleotide, the only chemical differences being that in RNA the sugar is ribose rather than 5′-deoxyribose, and the base thymine is replaced by uracil (U), which, like thymine, base-pairs with adenine.

Table 2.3 Some of the various functions of proteins.

Function	*Examples*
Structural	Collagen and osteocalcin in mammalian bones; keratin in hair
Catalysis	RNA polymerase; DNA polymerase
Transport	Hemoglobin; serum albumin, which transports fatty acids in the bloodstream
Storage	Ovalbumin, which stores amino acids in egg white; ferritin, which stores iron in the liver
Regulatory	Hormones such as insulin, which regulates glucose metabolism in vertebrates
Protective	Antibodies (immunoglobulins); thrombin, part of the blood clotting mechanism
Mobility	Myosin in muscles

During transcription of a gene, one strand of the DNA double helix acts as a template for synthesis of an RNA molecule whose nucleotide sequence is determined, by the base-pairing rules, by the DNA sequence (Figure 2.8). The copying reaction is carried out by an enzyme called an **RNA polymerase**, which starts at the 3′ end of the gene and works step by step towards the 5′ end. Because base-paired polynucleotides must always be antiparallel, this means that the RNA molecule is synthesized in a 5′ to 3′ direction; in other words, the nucleotides are added one by one to the growing 3′ end of the molecule.

Although the large majority of the genes in any genome code for proteins, the RNA molecules transcribed from these genes, the **messenger RNAs** (**mRNA**) of the transcriptome, rarely make up more than 4% of the total RNA in a cell. The other 96% comprises **non-coding RNA**, the transcripts of a relatively small number of genes that do not code for proteins. The non-coding RNA molecules are not, however, unimportant: quite the contrary, they play vital roles in the cell. The two major types, **ribosomal RNA** (**rRNA**) and **transfer RNA** (**tRNA**), are central components of the machinery responsible for translating mRNA into proteins. Other non-coding RNAs are involved in the complex process by which expression of the protein-coding genes is regulated, so that only those genes whose protein products are needed by the cell at a particular time are transcribed.

In the second stage of genome expression, the mRNAs in the transcriptome are translated into protein molecules. In order to understand how proteins are synthesized, we must be familiar with their chemical structure, and we will therefore return to translation in Section 3.2. The important point to consider here is how the proteome can represent the end point in expression of the biological information needed to construct and maintain a living example of an organism. The key lies with the functional diversity of proteins (Table 2.3). Some proteins have structural roles (examples are collagen and osteocalcin in bone), others are **enzymes** and catalyze biochemical reactions that bring about the release and storage of energy and the synthesis of new compounds (such as RNA polymerase, which catalyzes the synthesis of RNA during transcription). Some proteins have transport roles (e.g. hemoglobin, which carries oxygen in the bloodstream), and others have storage, regulatory, or protective functions. Even complex activities such as the movement of an arm or leg can be explained purely in terms of protein activity, in this case the interactions between actin and myosin proteins in muscle cells. The development and functioning of a living organism therefore results from the coordinated activity of different protein molecules, and the biological information contained in the genes, the blueprint for life itself, is simply the instructions for synthesizing these proteins at the correct time and in the correct place.

2.4.2 How the genome specifies the biological characteristics of an organism

Although every member of a species possesses the same genome, each individual has their own specific characteristics, referred to as their **phenotype.** With humans, for example, the variable aspects of the phenotype include sex, height, eye color, ability to roll one's tongue, susceptibility to genetic diseases, and suchlike. Identical genomes can specify different phenotypes because most of the genes in a genome can exist in two or more forms, called **alleles**, the differences between the alleles of a gene being slight variations in nucleotide sequence, sufficient to alter the properties of the protein for which the gene codes. The alleles possessed by an individual make up its **genotype.**

An example of a human phenotype is the ability to digest lactose, the sugar present in milk. This phenotype results from the action of lactase, an enzyme secreted into the intestine. The lactase gene, which is expressed in the cells lining the intestine, has two allelic forms. In one form, the gene gradually becomes less active after weaning, so the amount of lactase that is synthesized declines as the child grows older. Adults who possess this allele are lactose intolerant, meaning that they cannot digest milk products. The second allele, in contrast, remains active after weaning, so adults display lactase persistence and are able to digest dairy foods. The two alleles of the lactase gene have slightly different nucleotide sequences, and by typing these variable sequences the lactase genotype of an individual can be determined. The gene is, of course, present in all cells, not just in the intestinal cells that secrete lactase, and so genotyping can be performed with cheek cells from a mouth swab, or with ancient DNA recovered from an archaeological specimen.

Lactase persistence/non-persistence is a relatively simple trait, one where the phenotype is conferred by a single gene. Even so, there are complications. Remember that humans are diploid, meaning that each of our cells has two copies of the genome and hence two copies of every gene. If both of the copies of the lactase gene are the same allele (the **homozygous** condition), then the phenotype is obvious. But what if one gene is the persistent allele and the other is the non-persistent version? With some genes, in this **heterozygous** state the phenotype is intermediate between the extremes specified by the two alleles. The alleles of the lactase gene, however, display **dominance** and **recessiveness**, meaning that when the individual is heterozygous it is the phenotype of the dominant allele that is expressed. In the case of the lactase gene, persistence is the dominant condition, because the allele that specifies continued synthesis of lactase into adulthood masks the presence of the second allele, whose lactase enzyme is only synthesized in early childhood. Although the diploid state complicates the link between genotype and phenotype, it does not affect the accuracy of DNA typing as a means of identifying the phenotype: it is simply necessary to ensure that the typing procedure will recognize both alleles if they are present.

Lactase persistence/non-persistence is one of approximately 6000 **monogenic** human characteristics, those whose alternative forms can be distinguished by typing the alleles at a single pair of genes. Unfortunately most of these monogenic characteristics are biochemical traits that are of little interest to the biomolecular archaeologist. Other characteristics, including most of those involved in the physical appearance of an individual, are **polygenic**, being specified by groups of genes, all of which contribute in different ways to the phenotype. Eye color is an example of a polygenic human trait, with at least two genes working together to determine where one's eyes fall in the continuum from dark brown to light blue. The phenotype for a polygenic trait can be identified only if the alleles at all the genes contributing to that trait are typed. The greater the number of genes, the greater the technical challenge, especially with ancient DNA, and the problem is exacerbated by the incomplete knowledge that geneticists have about many of these complex phenotypes. Clearly, if the exact number of genes controlling a phenotype is unknown, then it will not be possible to identify that phenotype accurately by typing alleles in ancient DNA. The same problems apply to other mammals and to plants, because several of the phenotypes associated with agricultural productivity (e.g. milk yield, seed size) are specified by polygenic systems. Additional complications arise from the interactions between genes and the environment, many phenotypes only being expressed in response to an environmental trigger of some kind. This is why individuals can possess alleles that give susceptibility to cancer, but never develop the disease, whereas others with the same alleles but unfortunate enough to be exposed to the environmental trigger, become ill.

2.4.3 How the genome provides a record of ancestry

The genes are not the only part of a genome that can exist in two or more forms. All regions display **polymorphisms**, positions where the nucleotide sequence is not the same in every member of the population. The polymorphic sites are so numerous that it is statistically highly unlikely that any two people alive today have exactly the same genome sequences, except for pairs of identical twins.

The basic principles of genetics state that children inherit part of their genome from their mother and part from their father. The polymorphisms in a child's genome are therefore a mix of the mother's and father's polymorphisms, which, in turn, are a mix of the grandmothers' and grandfathers' polymorphisms, and so on back in time. The polymorphic sites therefore provide a record of ancestry, and comparisons between individuals can enable family relationships to be inferred. If two individuals share a similar set of polymorphisms, then they are likely to be related; equally importantly, if their polymorphisms are different, then they are probably not related. This is called **kinship analysis**. It is relatively easy to perform with DNA from living people, and possible, but much more difficult, with ancient DNA (Chapter 11).

The more data that are included in a kinship analysis, the greater the degree of statistical confidence that can be assigned to the deductions that are made. For this reason, **single nucleotide polymorphisms (SNPs)**, variable positions in the genome sequence, are not the best choices for this type of study, because usually there are just two versions of each SNP (e.g. ..AGGTC.. and ..AGTTC). SNPs are therefore **biallelic**. Instead, kinship analysis is usually carried out by typing variations at short tandem repeats, as many of these loci are **multiallelic**, with 10 or more length variants found among the population as a whole. A genetic profile for an individual can therefore be built up by typing the alleles present at different

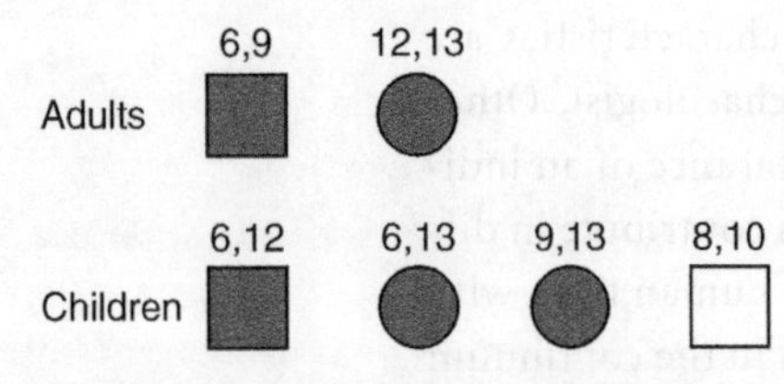

Figure 2.9 Using an STR to infer kinship between a set of skeletons. The numbers indicate the alleles for an STR that has been typed for two adults and four children. Squares represent males, and circles represent females. The allele combinations show that three of the children *could* be siblings and the two adults *could* be their parents. This is because each of these children possesses one allele present in the adult male and one in the adult female. The combinations also show that the fourth child is *not* the brother of the other children, nor is he the son of either adult. This child, shown as the open square, has alleles that are possessed by neither adult and by none of the other children.

STR loci within the genome, and comparisons between the profiles of different people, or different skeletons from an archaeological site, provide the data for the kinship analysis. A simple example with just a single STR is shown in Figure 2.9. Here, we see that three of the four children share alleles with a male and female adult. This observation in itself is not sufficient to deduce that these three children are siblings and the two adults are the parents, though the statistical chance would be quite high if the shared alleles were uncommon variants. To increase the degree of certainty, more STRs would have to be typed, but the analysis is by no means endless: forensic scientists currently use a set of 12 highly polymorphic STRs, which give a 1:1,000,000,000,000,000 chance that two individuals have exactly the same profiles. Note from Figure 2.9 that some deductions can be made with very few data: the alleles that the fourth child in the example possesses prove that neither the male nor female adult is a parent of this child.

STR analysis is ideal for studying close genetic relationships, such as exist in a family group, and can also be used to assign individuals to much broader populations. Broader population studies also make use of data acquired by typing SNPs, including ones on the Y chromosome to obtain information on the male lineage, or in the mitochondrial DNA (which is inherited only from the mother) to study the female lineage. With mitochondrial DNA, the usual strategy is to sequence the two **hypervariable regions** that lie adjacent to one another, 1150 bp in total, in the one part of the mitochondrial genome that does not contain any genes (Figure 2.10). Based on particular nucleotide sequence variations in the hypervariable regions, an individual or archaeological specimen can be assigned to one of the approximately 90 human mitochondrial **haplogroups** (Section 13.3).

What is true of humans is equally true of other animals and of plants, and indeed of all species that reproduce sexually. Pedigrees of domestic animals such as cattle can be determined by STR or SNP typing, and broader relationships, often relevant to understanding the development of prehistoric agriculture, can be established by mitochondrial DNA analysis. STRs and SNPs are also frequently used to assign crop plants to populations from which their origins can be deduced (Section 13.2).

2.5 How Ancient DNA is Studied

Although molecular biologists have devised a wide range of techniques for studying DNA, relatively few of these are used in biomolecular archaeology. Virtually all ancient DNA projects follow the same series of events: a DNA extract is prepared from the specimen being studied, the DNA in the extract is amplified by PCR, and then the new molecules produced by the PCR are cloned and sequenced. We will look at each of these techniques in turn.

2.5.1 Extraction and purification of ancient DNA from archaeological remains

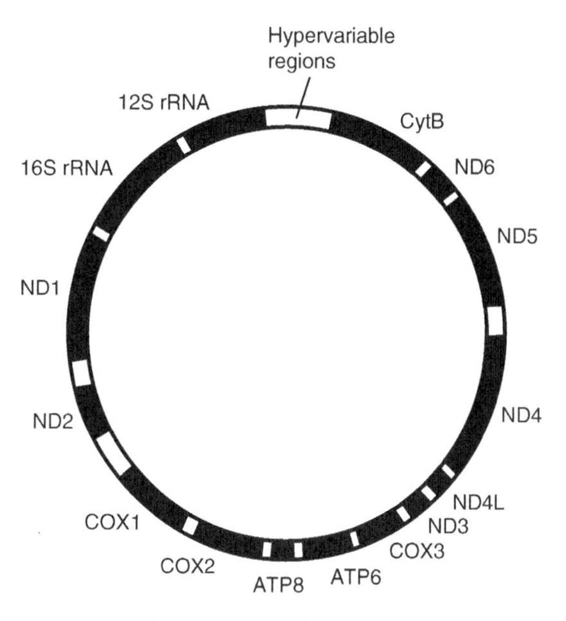

Figure 2.10 The human mitochondrial genome. The genes that are marked code for the two mitochondrial ribosomal RNA (rRNA) molecules and subunits of the cytochrome b (cytb), cytochrome oxidase (COX), NADH dehydrogenase (ND) and ATPase (ATP) proteins. The location of the two adjacent hypervariable regions is indicated. The other spaces left blank in the map contain transfer RNA genes.

The first step in an ancient DNA project is to prepare a sample of DNA from the specimen that is being studied. This can be looked on as a two-stage process, the first stage involving extraction of all the soluble molecules from the specimen and the second resulting in purification of the DNA molecules from this mixture.

DNA is readily soluble and can be extracted from most materials simply by resuspending a ground or pulverized sample in water or a weak buffer. After leaving the suspension for an hour or so to allow the soluble component to leach out, the preparation is briefly centrifuged so that the remains of the sample are discarded as a pellet at the bottom of the centrifuge tube, with the dissolved biomolecules retained in the supernatant. This is the approach that is used with modern biological samples and it appears to be effective with most archaeological specimens, but the assumption that ancient DNA is entirely soluble has never been proven and the possibility remains that a substantial proportion of this fraction remains in the sample, possibly because it has become tightly bound to the matrix during diagenesis, or possibly because it is present in structures that are not broken open by the grinding or pulverization processes. The latter appears to be a possibility with bone, recent research suggesting that a proportion of the ancient DNA is contained in crystalline deposits within the bone matrix, which are not accessed when the bone is ground into a powder. Soaking the bone powder in a chelating agent such as EDTA, which removes calcium ions and hence weakens the inorganic matrix, might therefore improve the yield of ancient DNA from at least some archaeological bones.

In extracts prepared from most types of modern tissue, the main constituents other than DNA are protein and RNA. The traditional way to remove these contaminants and hence purify the DNA is to add phenol or a 1 : 1 mixture of phenol and chloroform, as these organic solvents precipitate proteins and some types of RNA but leave the DNA in solution. Any remaining protein and RNA can be digested by treatment with protease and ribonuclease enzymes. This approach has been used with archaeological extracts, but ignores the fact that protein and RNA are unlikely to be major components of the aqueous extracts prepared from this type of material. Instead, the main contaminants will be products of diagenesis and compounds absorbed by the specimen from the surrounding environment. An important diagenetic process is the **Maillard reaction**, a complex interaction between proteins and carbohydrates that results in formation of a variety of organic substances. The nature of the environmental contaminants depends on the type of material being studied and its preservation conditions: for buried bones, the contaminants include ill-defined compounds such as

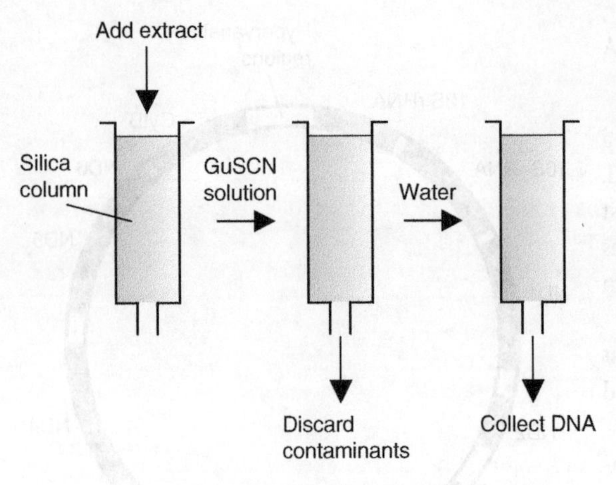

Figure 2.11 Purification of DNA by silica binding in the presence of guanidinium thiocyanate (GuSCN).

humic acids that result from plant decomposition and are present in soil. These diagenetic products and other contaminants are unlikely to be removed by treatments designed for purification of DNA from modern tissue extracts. Most procedures for ancient DNA purification therefore adopt a different strategy called **silica binding.** In the presence of guanidinium thiocyanate, DNA binds tightly to silica particles, providing an easy way of recovering the DNA from the aqueous extract. This is usually carried out using a chromatography column (Figure 2.11). The silica is placed in the column and the extract added. DNA binds to the silica and is retained in the column, whereas the contaminants pass straight through. After washing away the last contaminants with guanidinium thiocyanate solution, the DNA is recovered by adding water, which destabilizes the interactions between the DNA molecules and the silica.

Ideally, the ancient DNA extracted from an archaeological specimen will be contained in as small a volume as possible, not more than 100 μl. If the volume resulting from silica extraction is greater than this, then the ancient DNA can be concentrated by **ethanol precipitation.** In the presence of salt and at a temperature of −20°C or less, absolute ethanol efficiently precipitates polymeric nucleic acids. The precipitate can be collected by centrifugation, and then redissolved in the appropriate volume of water.

2.5.2 The polymerase chain reaction is the key to ancient DNA research

The amount of ancient DNA in an extract is so small that no information on its sequence or STR allele content can be obtained directly. The reason why ancient DNA research is so heavily dependent on PCR is because this procedure amplifies the small amounts of ancient DNA into quantities that can be analyzed by sequencing and STR typing.

PCR results in the repeated copying of a selected region of a DNA molecule. The two key components of a PCR are:

- ***Taq* DNA polymerase**, an enzyme that synthesizes new strands of DNA using an existing strand as the template. This particular type of DNA polymerase is obtained from *Thermus aquaticus*, a bacterium which, because it lives in hot springs, has thermostable enzymes, meaning that they are resistant to denaturation by heat treatment.
- A pair of short **oligonucleotides**, which bind to the target DNA molecule, one to each strand of the double helix. These oligonucleotides, which act as **primers** for the DNA synthesis reactions, delimit the region that will be amplified. They must therefore attach to the target DNA at either side of the segment that is to be copied: the sequences of these attachment sites must therefore be known so that primers of the appropriate sequences can be synthesized.

To carry out a PCR, the ancient DNA is mixed with *Taq* DNA polymerase, the two oligonucleotide primers, and a supply of nucleotides. The amount of ancient DNA can be very

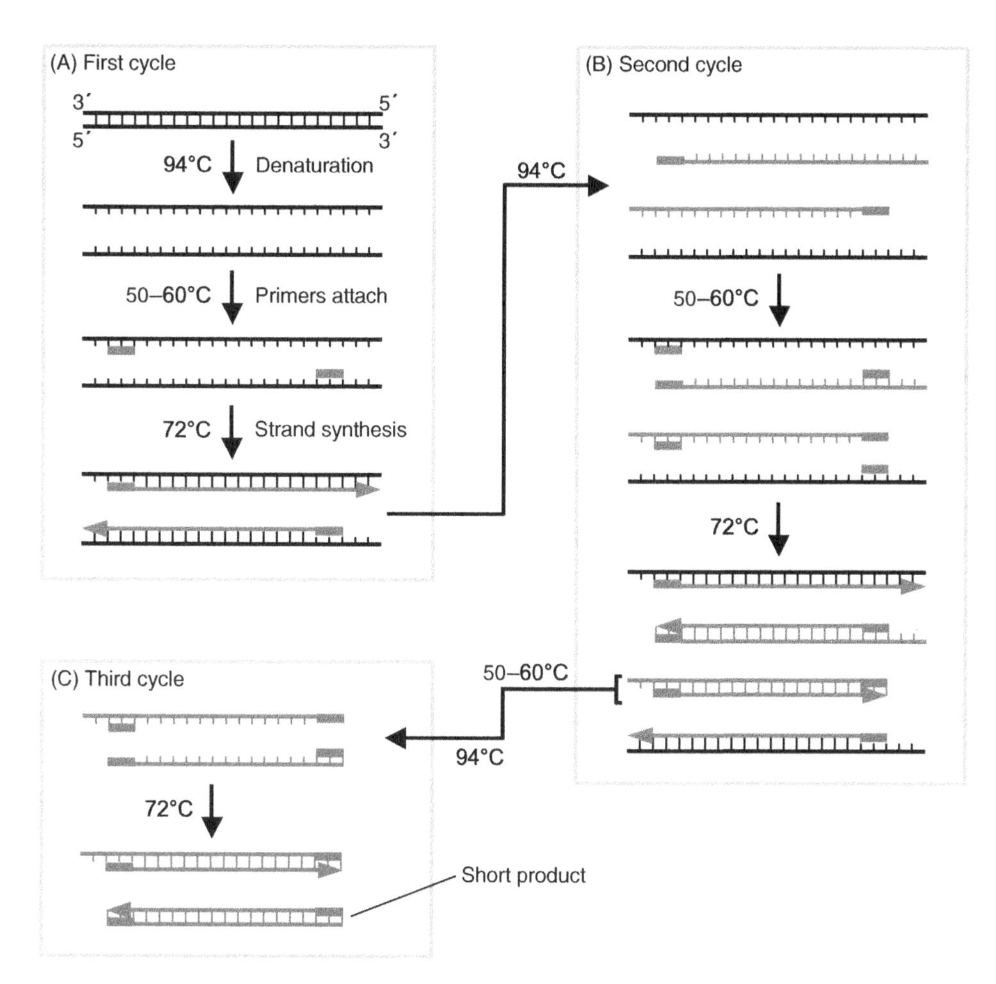

Figure 2.12 The first three cycles of a PCR. The original polynucleotides are shown in black and those synthesized during the PCR are in gray.

small because PCR is extremely sensitive and will work with just a single starting molecule. The mixture is then heated to 94°C, which causes the hydrogen bonds that hold together the two polynucleotides of the double helix to break, so the double-stranded DNA becomes separated into single-stranded polynucleotides (Figure 2.12A). The temperature is then lowered to 50–60°C, which results in some rejoining of the single strands of the target DNA, but also allows the primers to attach. Next, the temperature is raised to 72°C, the optimum for *Taq* polymerase, which now synthesizes new strands of DNA, their sequences determined by the base-pairing rules, using the primers as the starting point. The 5′ ends of these new polynucleotides are set by the primers, but their 3′ ends are at random positions, these being the positions where DNA synthesis terminates by chance.

The temperature cycle is now repeated. The first cycle products denature and the four resulting strands are copied during the DNA synthesis stage. This gives four double-stranded molecules, two of which are identical to the products from the first cycle and two of which

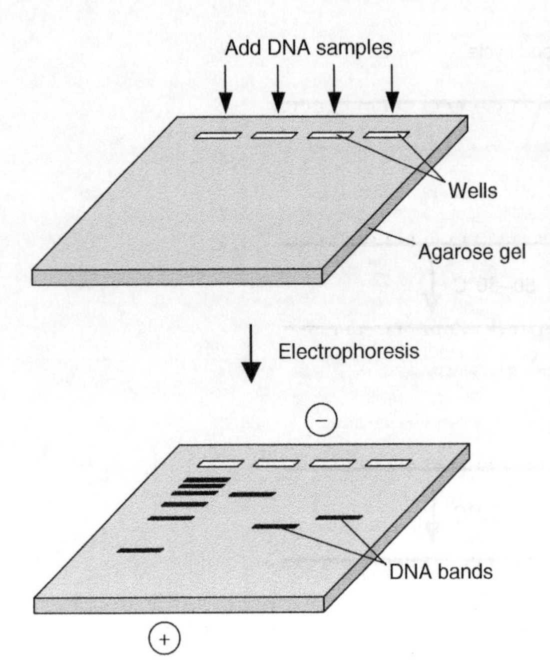

Figure 2.13 Agarose gel electrophoresis. The lane on the left contains size markers. The three other lanes contain PCR products of different sizes.

are made entirely of new DNA (Figure 2.12B). When the temperature cycle is repeated a third time, the latter give rise to "short products," the 5′ and 3′ ends of which are both set by the primer annealing positions (Figure 2.12C). In every subsequent cycle, the number of short products doubles until one of the components of the reaction becomes depleted. This means that after 30 cycles, there will be over 130 million short products derived from each starting molecule. In real terms, this equates to several micrograms of PCR product from a few picograms of target DNA.

The results of a PCR are usually analyzed by **agarose gel electrophoresis.** Electrophoresis is a method that uses differences in electrical charge to separate the molecules in a mixture. DNA molecules have negative charges, and so when placed in an electric field they migrate towards the positive pole. In solution, the rate of migration of a molecule depends on its shape and its charge-to-mass ratio, which means that all DNA molecules migrate at about the same speed, as they all are the same shape and all have very similar charge-to-mass ratios. If, however, electrophoresis is carried out in an agarose gel, then the length of the molecules also becomes a factor. The gel is made up of a network of pores, through which the DNA molecule must pass to reach the positive electrode. The smaller the DNA molecule, the faster it can migrate through the gel, so the molecules become fractionated according to their length (Figure 2.13). They can be visualized by soaking the gel in a stain such as ethidium bromide, which attaches to the DNA and fluoresces when the gel is exposed to ultraviolet radiation. Bands are seen, each band containing DNA molecules of the same length, the length being measurable by comparison with the positions of size markers. Agarose gel electrophoresis will therefore reveal a single band if the PCR has worked as expected and has amplified a single segment of the ancient DNA.

2.5.3 Careful design of the primers is crucial to success of a PCR

The primers are the key component of a PCR and the success or failure of an amplification depends largely on the design and use of the primers. It is simple enough to synthesize a pair of oligonucleotides whose sequences enable them to attach either side of the region to be amplified. It is much less straightforward to design a pair of primers that will give a specific and efficient amplification of this region of the template DNA.

The first important issue is the length of the primers. If a primer is too short, then there is a greater chance that it will attach to more than one site in the template DNA. Attachment sites for primers are expected to occur, on average, once every 4^n bp, where n is the length of the primer. So a sequence 8 nucleotides in length will occur, on average, once every $4^8 = 65{,}536$ bp, which means that primers of this length will have approximately 49,000 attachment sites, each one precisely complementary to the primer, in human DNA (3,200 Mb). There is therefore a high chance that a pair of 8-nucleotide primers would amplify more than one fragment from human DNA. In contrast, a 17-nucleotide primer sequence occurs only once every 4^{17} = 17,179,869,184 bp = 17,180 Mb and so would be expected to have a unique priming site in the human genome. In practice, primers of 15–20 nucleotides are suitable for most purposes. If they are greater than 30 nucleotides, then the hybridization rate, which increases with length, becomes a factor and complete attachment may not occur during the time allowed within the thermal cycle.

It is also important that the primers should not be able to attach to one another, as this will result in the primers copying one another during the PCR, leading to short "primer-dimers." Dimerization is promoted by the high concentrations of primers during the early cycles of the PCR, and can reduce synthesis of the amplification product by competing for the polymerase. Similar problems occur if the nucleotides at the 3′ end of a primer can attach to nucleotides within the *same* primer. The 3′ end is then extended by the polymerase, forming a hairpin structure, reducing the effective primer concentration and decreasing yield of the desired product.

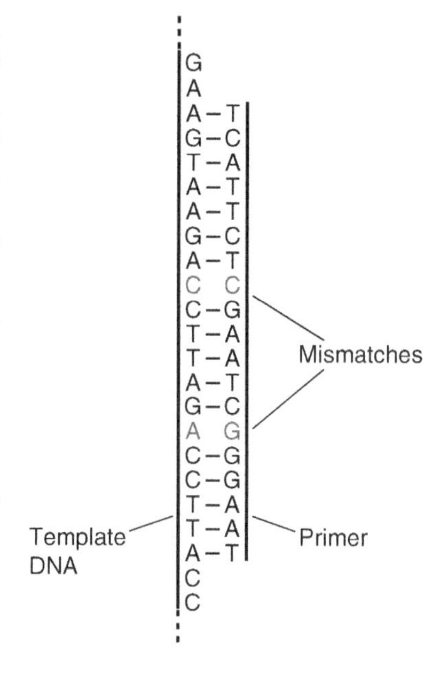

Figure 2.14 A mismatched hybrid. The sequence of the template DNA is similar but not identical to that of the primer. Mismatching can give rise to non-specific amplification if two primers anneal to incorrect positions that are located close to one another on the template DNA.

The final consideration is the exact temperature to use during the annealing stage of the thermal cycling, when the primers attach to the template DNA. DNA–DNA hybridization is a temperature-dependent phenomenon. If the temperature is too high, then no hybridization takes place, and instead the primers and templates remain dissociated. On the other hand, if the temperature is too low, mismatched hybrids – ones in which not all the correct base pairs have formed (Figure 2.14) – are stable, resulting in non-specific amplification. The ideal annealing temperature is determined from the **melting temperature** or T_m of the primer-template hybrid, this being the temperature at which the correctly base-paired hybrid dissociates ("melts"). A temperature 1–2°C below this is low enough to allow the correct primer-template hybrid to form, but too high for a hybrid with a single mismatch to be stable. The T_m of each primer can be estimated from its sequence, using the equation:

$$T_m = (4 \times [G + C]) + (2 \times [A + T])\,^{\circ}\mathrm{C}$$

in which $[G + C]$ is the number of G and C nucleotides in the primer sequence, and $[A + T]$ is the number of A and T nucleotides. The two primers should be designed so that they have identical melting temperatures, because if this is not the case, the appropriate annealing temperature for one primer may be too high or too low for the other member of the pair.

A variety of computer programs are available for scanning primer sequences for potential dimer and hairpin formation, some of these programs also taking account of other factors such as melting temperature in attempts to design the best primer pair for a

(A) The structure of a dideoxynucleotide

PHOSPHATES

BASE

SUGAR

(B) Incorporation of a dideoxynucleotide causes chain termination

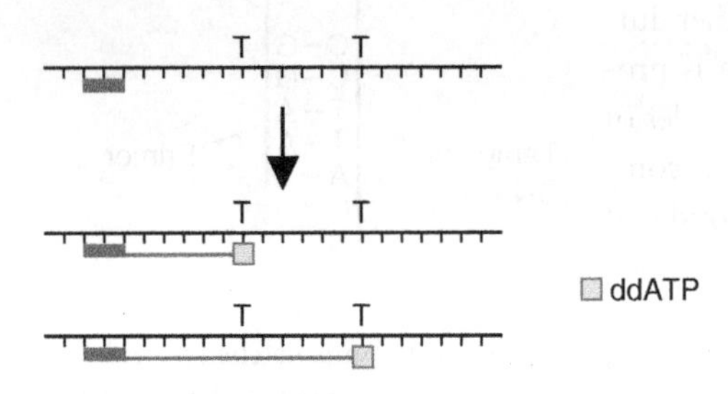

(C) Reading the sequence

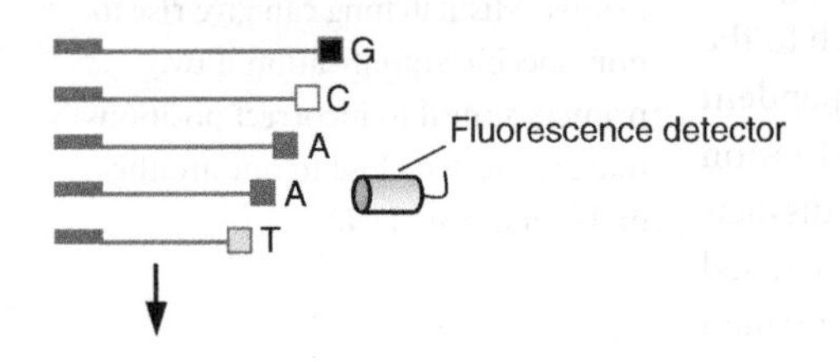

(D) The printout from an automated sequencer

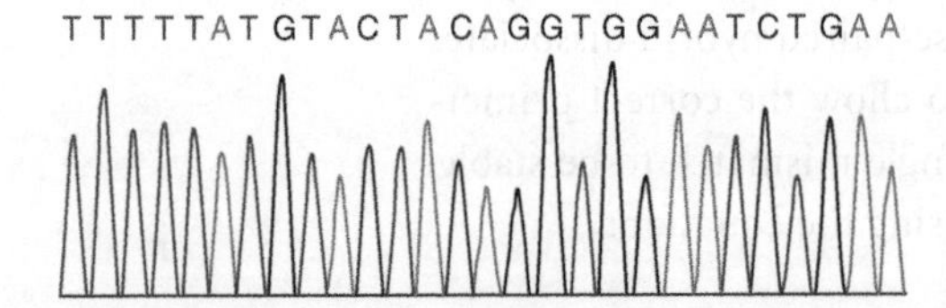

Figure 2.15 Chain termination DNA sequencing. In (A), the hydrogen atom that distinguishes the dideoxynucleotide from the normal version is boxed.

particular amplification. Once a primer has been designed, the uniqueness of its sequence within the genome under study can be checked by searching for it within the genome sequence. This can be done with the programme called **BLAST** (Basic Local Alignment Search Tool), which is used simply by logging on to the web site for one of the DNA databases and entering the primer sequence into the online search tool.

2.5.4 Obtaining the sequence of a DNA molecule

DNA sequencing is usually carried out by the **chain termination method**, which is based on the principle that single-stranded polynucleotides that differ in length by just a single nucleotide can be separated from one another by **polyacrylamide gel electrophoresis**. This means that it is possible to resolve a family of polynucleotides, representing all lengths from 10 to 1500 nucleotides, into a series of bands in a capillary gel. These polynucleotides are generated by carrying out a thermal cycling process, similar to PCR, but with two differences. The first is that only one primer is used, which means that only one of the strands of the starting molecule is copied, the product accumulating in a linear fashion, not exponentially as is the case in a real PCR.

The second difference between a sequencing reaction and PCR is that, as well as the standard nucleotides – dATP, dCTP, dGTP, and dTTP – the sequencing reaction contains four **dideoxynucleotides** – ddATP, ddCTP, ddGTP, and ddTTP. The difference between a normal nucleotide and its dideoxy version is that the latter has a hydrogen (–H) rather than hydroxyl (–OH) group attached to the 3′ carbon (Figure 2.15A). This alteration means that when a dideoxynucleotide is incorporated into a growing polynucleotide, it blocks further strand synthesis, because the 3′-OH group lacking in the dideoxynucleotide is the position to which the next nucleotide would normally be attached. Incorporation of a dideoxynucleotide therefore results in chain termination.

Each of the dideoxynucleotides is labelled with a different fluorescent marker. *Taq* polymerase does not discriminate between deoxy- and dideoxynucleotides, so either can be added into the polynucleotide that is being synthesized, but the deoxynucleotides are present in excess, so the strand synthesis does not always terminate close to the primer. In fact, several hundred nucleotides may be polymerized before a dideoxynucleotide is eventually incorporated. The result is a set of new molecules, all of different lengths, and each ending in a dideoxynucleotide whose fluorescent marker indicates the nucleotide – A, C, G, or T – that is present at the equivalent position in the template DNA (Figure 2.15B).

To read the DNA sequence, the mixture is loaded into a tube of a capillary gel system, and electrophoresis carried out to separate the molecules according to their lengths. After separation, the molecules are run past a fluorescent detector capable of discriminating the labels attached to the dideoxynucleotides (Figure 2.15C). The detector therefore identifies if each molecule ends in an A, C, G, or T, and in this way reads off the DNA sequence. The sequence can be printed out for examination by the operator (Figure 2.15D), or entered directly into a storage device for future analysis. Up to approximately 750 bp of sequence can be read in a single experiment. If a longer sequence is required (as is often the case when modern DNA is being studied) then a second sequencing experiment is carried out with a primer that attaches near the end of the sequence that has just been obtained.

2.5.5 PCR products obtained from ancient DNA should be cloned prior to sequencing

The standard thermal cycling sequencing reaction is ideal for obtaining the sequence of a PCR product, but this "direct" approach is not suitable for most ancient DNA applications. This is because of the danger that the starting mix for the PCR contained both ancient DNA and modern contaminating DNA. This would mean that a sequence obtained by the direct approach would be a mixture of the sequences of the ancient and modern DNA, and at every position where these sequences differ an ambiguity would appear in the readout. From the printout provided by the sequencing software it is usually possible to identify the two nucleotides present at each ambiguous position, but not to assign them to their correct sources – ancient or modern. If there is more than one type of contaminating DNA, then the problem becomes even more acute.

These considerations dictate that a different approach be taken to sequencing ancient PCR products, one that distinguishes the sequences of individual molecules in the product. This can be achieved by **cloning** the PCR product prior to sequencing. DNA cloning is a sophisticated technology, but we do not need to delve too deeply into it in order to understand how it is used in biomolecular archaeology. The central feature of a cloning experiment is the **vector**, a DNA molecule, often based on a naturally occurring **plasmid**, that is able to replicate inside a cell of the bacterium *Escherichia coli*. In its replicative state, the vector is a circular molecule, but at the start of the experiment it is obtained in a linear form. These linear vector molecules are mixed with the PCR product and a **DNA ligase** enzyme added, which joins double-stranded DNA molecules together end to end. Various combinations are produced, including some where a vector molecule has become linked to one of the PCR products and the two ends then joined together to produce a circular **recombinant** vector (Figure 2.16A).

In the next stage of the cloning procedure, the molecules resulting from ligation are mixed with *E. coli* cells, which have been chemically treated so that they are **competent**, a term used to describe cells that are able to take up DNA molecules from their environment (Figure 2.16B). Linear molecules, or circular molecules which lack the vector sequences, cannot be propagated inside a bacterium. Circular vectors, on the other hand, including the recombinant

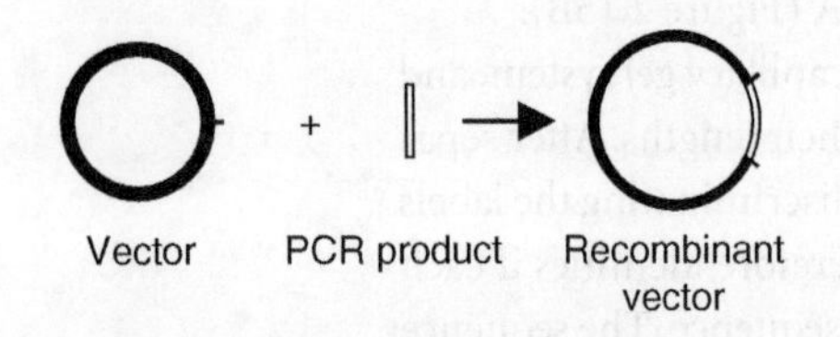

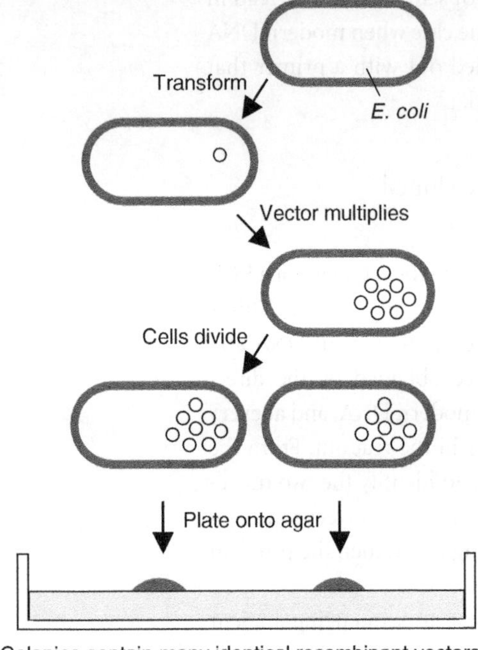

Figure 2.16 DNA cloning.

ones, are able to replicate, so that multiple copies are produced, exactly how many depending on the identity of the cloning vector, but usually several hundred per cell. Every time the host bacterium divides, copies of the vector are passed to the progeny cells, and further rounds of replication occur to maintain the copy number in the new cells. The bacteria are spread on to a solid agar medium so that individual cells give rise to colonies. Each colony will contain many copies of the vector, these molecules being the direct descendents of the original, single molecule taken up by the original, individual bacterium. If that original molecule was a recombinant vector, then the colony contains multiple copies of a single molecule from the PCR product, separate from all the other molecules in the PCR product. The sequences of individual molecules from the PCR product can now be obtained, without any confusion between them, from the DNA present in different recombinant colonies.

2.5.6 Examining the sequences of cloned PCR products

The first step in interpreting the set of sequences obtained from a cloned PCR product is to construct a **multiple alignment**. This simply involves typing out the sequences in a series so that equivalent positions are vertically aligned, possibly with gaps in one or more sequences to take account of insertions and deletions (Figure 2.17). This can easily be done by eye with a word processor, or alternatively a computer program such as BioEdit can be used.

From the multiple alignment it will immediately be obvious if all the sequences are identical or, as is expected with ancient DNA, each sequence has its own individual errors due to chemical modification of nucleotides during diagenesis (Section 8.2). If the ancient DNA was uncontaminated with modern DNA, then from the multiple alignment it should be possible to deduce a **consensus sequence**, by identifying which nucleotide occurs most frequently at each position. It is often assumed that the consensus sequence is the actual sequence of the ancient DNA, on the basis that errors resulting from damage are rare and so only a minority of clones display an error at any particular position in the sequence. Unfortunately, this assumption is not always valid, because any damaged molecule that is copied in the early cycles of the PCR will give rise to many **amplicons** containing the same error(s). If by chance the copies of a single damaged molecule predominate in the final PCR product, then the error(s) carried by this molecule will be seen in several sequences in the multiple alignment, and one or more of them could mistakenly be looked on as the "correct" nucleotide for that position in the consensus sequence.

```
Consensus sequence
AAGTACAGCAATCAACCCTCAACTATCACACATCAACTGCAACTCCAAAGCCACCCCT
Clone sequences
AAGTACAGCAATCAACCCTCAATTATCACACATCAACTGCAACTCCAAAGCCACCCCT
AAGTACAGCAATCAACCCTCAACTATCACACATCAACTGCAACTCCAAAGCCACCCCT
AAGTACAGTAATCAACCCTCAATTATCACACATCAACTGCAACTCCAAAGCCACCCCT
AAGTACAGCAATCAACCCTCAACTATCACACATCAACTGCAACTCCAAA-CCACCCCT
AAGTACAGCAATCAACCCTCAACTATCACACATTAACTGCAACTCCAAAGCCACCCCT
AAGTACAGCAATCAACCCTCAACTATCACACATCAACTGCAACTCCAAAGCCACCCCT
AAGTACAGCAATCAACCCTCAACTATCACACATCAACTGCAACTTTAAAGCCACCCCT
AAGTACAGCAATCAACCCTCAATTATCACACATCAACTGCAACTCCAAAGCCACCCCT
AAGTACAGCAATCAACCCTCAACTATCACACATCAACTGCAACTCCAAAGCCACCCCT
```

Figure 2.17 Multiple alignment of the sequences of 10 clones from a PCR directed at part of the human mitochondrial genome. Errors in the cloned sequences resulting from damage present in the ancient DNA templates are highlighted. The dash indicates a single nucleotide deletion in this particular sequence. The errors are sporadic, enabling a consensus sequence to be deduced. Note, however, that one position in the alignment has the same error in three of the cloned sequences, suggesting that a single template with this damaged nucleotide was copied early in the PCR and its amplicons are relatively frequent in the resulting product.

Careful examination of the multiple alignment might also reveal if there was a mixture of ancient and contaminating DNA at the start of the PCR. If the sequences of the ancient and contaminating DNAs are different (e.g. because they represent different mitochondrial DNA haplotypes), then two sets of sequences will be seen in the multiple alignment, each set with its own specific nucleotide variants. Two consensus sequences, one for the ancient DNA and one for the contaminant, can then be deduced. Problems such as how to work out which is the ancient consensus and which the contaminant, or whether both are contaminants, we will leave until Section 9.1.

Once one or more consensus sequences have been obtained, the haplotype or allele represented by each one can be identified simply by looking for the specific nucleotide variants that characterize that haplotype or allele. Sometimes, however, the consensus sequence will be unrecognizable because the PCR has inadvertently amplified the wrong sequence, either because the primers were poorly designed or because the annealing temperature used in the PCR was too low. Often the temptation is simply to change the primers or conditions and do the experiment again, but it is worth trying to identify the mystery sequence, as sometimes knowing what has been amplified can aid the redesign and help ensure the same mistake is not made again. To identify an unknown sequence a BLAST search can be made of one of the online DNA databases, which contain all the DNA sequences that have ever been obtained.

2.5.7 New methods for high throughput DNA sequencing

The chain termination method for DNA sequencing was invented in the mid-1970s and has not been greatly changed since then, other than by the introduction of automated versions of the technique. Over the last few years, however, entirely new approaches to DNA sequencing have been developed, ones that enable hundreds of thousands of short sequences to be obtained in a single experiment. These **next generation sequencing** methods are ideal for

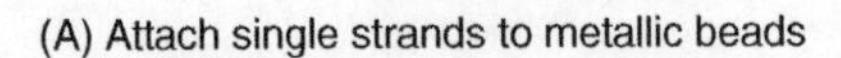

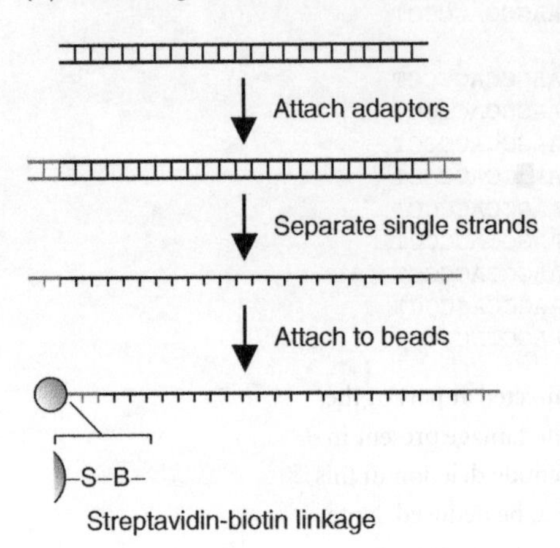

(B) PCRs in an oil emulsion

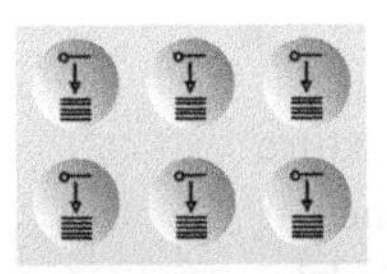

Figure 2.18 One approach to next generation sequencing.

studying entire genomes, because over 100 Mb of sequence can be obtained in a single run. The sequences are shorter than those provided by the chain termination method, between 10 and 100 bp, but computer techniques can be used to find overlaps between them and hence to generate **contigs** – long contiguous sequences that cover of all or part of a genome. As we will see in Section 16.4, this approach is being used to study the Neanderthal genome, as well as the genomes of extinct animals such as mammoths and cave bears.

The first stage of the high throughput sequencing approach is to prepare a library of the DNA fragments that will be sequenced. If modern DNA is being studied, then it must be broken into fragments of 300–500 bp, but this is not necessary with ancient DNA, as we expect these molecules already to be of short length. Two short **adaptors** are attached to each DNA fragment, one adaptor to either end (Figure 2.18A). These adaptors play two important roles. First, they enable the DNA fragments to be attached to small metallic beads. This is because one of the adaptors has a small protein called **biotin** attached to its 5′ end. Biotin has a strong affinity for a second protein, **streptavidin**, with which the beads are coated. DNA fragments therefore become attached to beads via biotin-streptavidin linkages. The ratio of DNA fragments to beads is set so that, on average, a single fragment becomes attached to each bead.

Each fragment will now be amplified by PCR so that enough copies are made for sequencing. The adaptors now play their second role as they provide the annealing sites for the primers for this PCR. The same pair of primers can therefore be used for all the fragments, even though the fragments themselves have many different sequences. If the PCR is carried out immediately, then all we will obtain is a mixture of all the amplicons, which will not enable us to obtain the individual sequences of each one. To solve this problem, PCR is carried out in an oil emulsion, each bead residing in its own aqueous droplet within the emulsion (Figure 2.18B). Each droplet contains all the reagents needed for PCR, and is physically separated from all the other droplets by the barrier provided by the oil component of the emulsion.

After PCR, the aqueous droplets are transferred into wells, so there is one droplet and hence one amplicon per well, and the sequencing part of the experiment is carried out. This is not done by the chain termination method but by a different technique, called **pyrosequencing**, which was developed several years ago but not used extensively until the introduction of these high throughput methods. The advantage of pyrosequencing is that it does not require electrophoresis or any other fragment separation procedure and so is simpler and more rapid than the chain termination method. In pyrosequencing, copies of the amplicon molecules are made using a DNA polymerase and a primer that recognizes one of the adaptors. As the new strands are being synthesized, the order in which the deoxynucleotides are incorporated is detected, so the sequence is read as the reaction proceeds. The addition of each deoxynucleotide to the end of the growing strand is detectable because it is accompanied by release of a molecule of pyrophosphate, which is converted by the enzyme sulfurylase into a flash of chemiluminescence

(Figure 2.19). If all four deoxynucleotides were present at the same time then continual flashes of light would be seen and no useful sequence information would be obtained. Each deoxynucleotide is therefore added separately, one after the other, with a nucleotidase enzyme also present in the reaction mixture. This means that if a deoxynucleotide is not incorporated into the polynucleotide, then it is rapidly degraded before the next one is added. During each cycle, a flash of chemiluminescence is seen only when the "correct" deoxynucleotide is added and incorporation occurs, making it possible to follow the order in which the deoxynucleotides are added onto the growing strand.

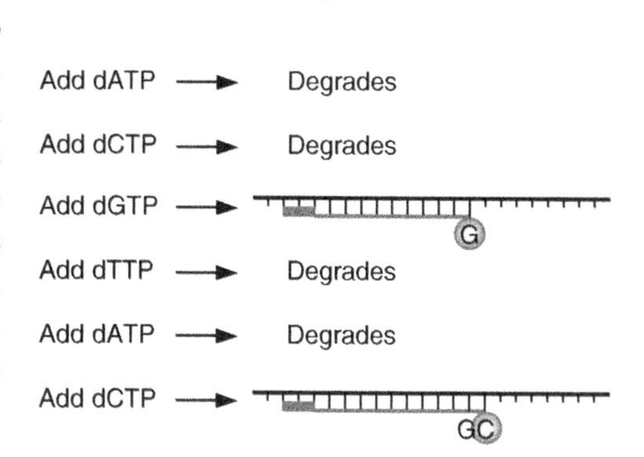

Figure 2.19 Pyrosequencing. The four nucleotides are repeatedly added in series. Incorporation results in a flash of chemiluminescence, which reveals which nucleotide has been added and enables the DNA sequence to be worked out – in this case, G followed by C.

2.5.8 Determining the evolutionary relationships between DNA sequences

Often the objective of an ancient DNA project is to understand the evolutionary relationship between an archaeological specimen and modern organisms, between prehistoric and modern human populations for example, or between prehistoric cattle and modern breeds.

Various methods can be used to deduce the evolutionary relationships between DNA sequences. When the sequences come from different species a **phylogenetic tree** is usually constructed. The tree comprises a set of external nodes, each representing one of the sequences that has been compared, linked by branches to internal nodes representing ancestral sequences (Figure 2.20A). The lengths of the branches indicate the degrees of difference between the sequences represented by the nodes. There are a number of different ways of constructing phylogenetic trees from sequence data, the main differences between them being the way in which the multiple alignment of the sequences is converted into numerical data that can be analyzed mathematically in order to produce the tree. The simplest method involves conversion of the sequence data into a **distance matrix**, which is a table showing the evolutionary distances between all pairs of sequences, these distances calculated from the number of nucleotide differences between each pair. The matrix is used by the tree-building software to establish the lengths of the branches connecting pairs of sequences in the tree that is drawn. Methods such as **neighbor-joining** and **maximum parsimony**, which construct trees from distance matrices, differ in the degree of rigor that they apply to the analysis, and whether they simply display the relative similarities between the sequences being studied, or if they genuinely attempt to infer an evolutionary history. The problem is that a completely rigorous analysis requires a great deal of data handling, so that even with datasets containing just ten sequences an immense amount of computing power is required. Most tree-building programs therefore employ various shortcuts to make the analysis feasible, but with the danger that the tree produced contains errors.

There are also occasions when conventional tree building will give errors because the underlying assumptions are inappropriate. This becomes important when the sequences being

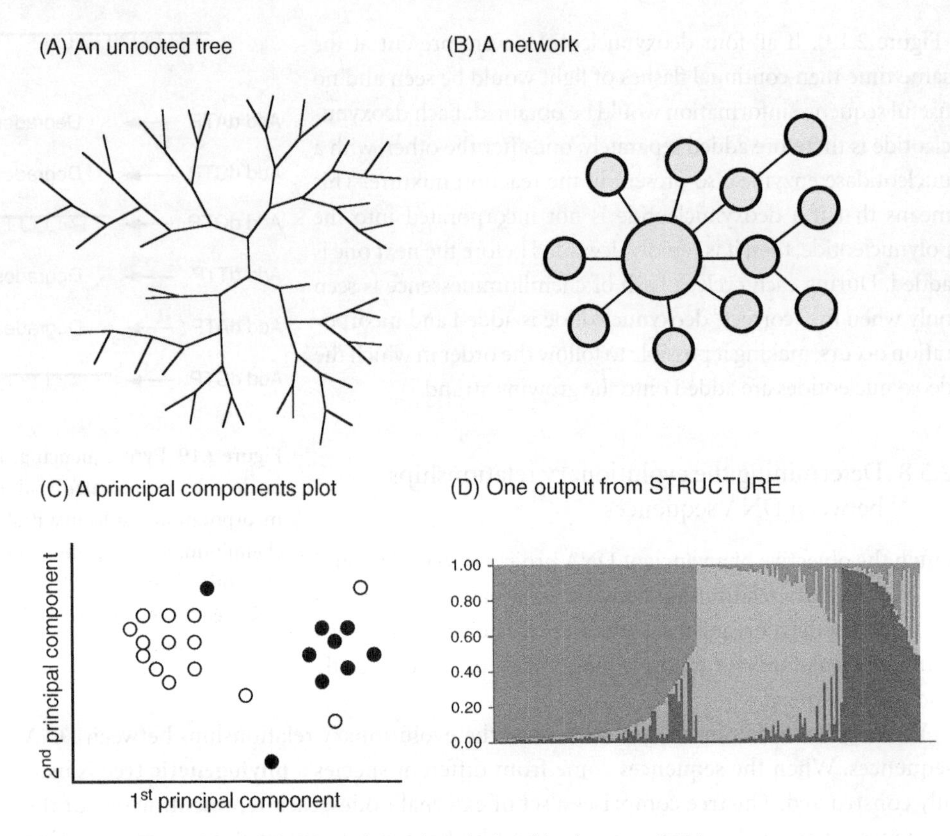

Figure 2.20 Various methods for studying the relationships between DNA sequences or genotypes. (A) An unrooted phylogenetic tree. (B) A network. The size of the larger circle indicates that this particular sequence is relatively common in the dataset. (C) A two-dimensional principal components graph. (D) An output from STRUCTURE, revealing three populations. Each individual is represented by a thin vertical line, its color(s) indicating the membership of the individual in each of the populations. The lines for those individuals that display admixture have segments of different color, the segments indicating the proportional membership of the individual in the different populations. The image in (D) is kindly provided by Claudia Grimaldo.

compared come not from different species but from members of the same species – as is usually the case in biomolecular archaeology projects. Now it is possible for sequences to recombine when individuals breed with one another, which might create a new sequence that is descended from two parent sequences that would occupy distant positions in a phylogenetic tree. With populations, it is also possible that ancestral sequences, represented by internal nodes in a tree, still exist. The solution to these problems is to illustrate the evolutionary relationships as a network rather than a tree, as now sequences that result from recombination can be depicted clearly, and the presence of ancestral sequences does not cause a problem (Figure 2.20B). We will encounter several examples of the use of networks to compare DNA sequences when we examine current research in biomolecular archaeology in Part III of this book.

There are also situations where we wish to know whether individuals – humans or domesticated animals and plants – fall into two or more population groups, the members of each group largely though not exclusively interbreeding with one another rather than with members of other groups. Although trees and networks show the relationships between individuals, they do not always enable those individuals to be assigned to populations. This is particularly true when short tandem repeat genotypes are being studied, because these data are multivariate (i.e. each genotype comprises multiple STRs, each of which has multiple length variants) and the algorithms used for constructing trees and networks are not designed for multivariate data. Other methods are therefore needed. One of these is **principal components analysis**, which searches for patterns within multivariate data in order to identify components which together account for all the variance in the dataset. Although there may be several of these components, in many STR datasets the two or three largest ones account for 75% or more of the data, and analysis just of these can give an accurate indication of the relationships between individuals. The results are depicted in the form of a two- or three-dimensional graph, each dimension representing the range of variability of one of the principal components, and points within the graph indicating each individual. Populations are indicated by clusters of points (Figure 2.20C).

Populations can also be identified using the program called **STRUCTURE**, which was written specifically to deal with STR genotype data. STRUCTURE identifies relationships between genotypes in order to assign individuals to *K* groups, where *K* can be any number, though in practice is usually constrained between one and nine. If *K* is known, then new individuals can be assigned to their correct population. If it is not known, then the operator can test different values in order to identify the number of populations that gives the best overall fit with the data. A particular advantage of STRUCTURE is that it indicates individuals whose genotypes have affinities with two or more groups (Figure 2.20D), such examples of **admixture** often being informative when contacts and other relationships between populations are considered.

Further Reading

Brown, T.A. (2007) *Genomes*, 3rd edn. Garland Science, Abingdon. [Details on DNA structure, genes and genomes.]

Brown, T.A. (2010) *Gene Cloning and DNA Analysis: An Introduction*, 6th edn. Wiley-Backwell, Chichester.

Hughes, S. & Moody, A. (eds.) (2007) *PCR*. Scion Publishing, Oxford.

Jolliffe, I.T. (2002) *Principal Component Analysis*, 2nd edn. Springer, New York.

Margulies, M., Egholm, M. & Altman, W.E., *et al.* (2005) Genome sequencing in microfabricated high-density picolitre reactors. *Nature*, 437, 376–80. [Next generation sequencing.]

Pritchard, J.K., Stephens, M. & Donnelly, P. (2000) Inference of population structure using multilocus genotype data. *Genetics*, 155, 945–59. [STRUCTURE.]

Prober, J.M., Trainor, G.L. & Dam, R.J., *et al.* (1987) A system for rapid DNA sequencing with fluorescent chain-terminating dideoxynucleotides. *Science*, 238, 336–41.

Salton, N. & Nei, M. (1987) The neighbor-joining method: a new method for reconstructing phylogenetic trees. *Molecular Biology and Evolution*, 4, 406–25.

3

Proteins

Biomolecular archaeology began with the study of ancient proteins, way back in the 1930s. It had been realized since the early years of the 20th century that human blood groups – the ABO system, for example – are specified by the identities of certain proteins present in the bloodstream, and by the 1920s methods for typing these blood groups in living people were well established. In 1933, William and Lyle Bond began to apply blood typing methods to Egyptian and Native American mummies. Almost immediately, the problem that would hamper ancient protein studies for most of the rest of the century raised its head. The Bonds found that the distribution of blood groups that they detected in their American specimens was quite different to that in modern Native Americans. The O blood type is commonest in Native Americans, with the A and B groups restricted to the northern parts of the New World, with B very rare indeed. But in the mummies that were tested, the Bonds found that although the B type was still rare, it was distributed throughout North and South America. The Bonds worried that this might indicate that the typing methods, developed for use with clinical samples, were giving inaccurate results with the partially degraded blood proteins in their mummies. This possible lack of specificity continued to raise question marks when attempts to identify ancient proteins were applied to other areas of biomolecular archaeology, and led to the "blood on stone tools" controversy. This is the debate, which has become quite heated at times, about whether proteins are preserved in what are thought to be blood residues on stone artifacts, and whether identification of these proteins can tell us what types of animal were being butchered at a particular site, or even if the stone implements were used on other humans (Section 9.2).

The specificity problems described above are an inherent issue with the **immunological methods** that were used in virtually all of the studies of ancient proteins carried out in the

 © 2011 Terry Brown & Keri Brown

20th century. We will understand the basis to this problem when we examine these methods later in this chapter. In the last few years, however, proteins have begun to play a much more central role in biomolecular archaeology, for two reasons. First, considerable effort has been put into understanding how proteins degrade in archaeological and paleontological material (Section 8.3). The information that has been obtained on the rates of decay and the types of chemical change that occur enables us to be much more confident about the accuracy – or otherwise – of protein detections made with different types of material. Second, molecular biologists have recently developed a whole new methodology for studying proteins, loosely called **proteomics**, which avoids immunological methods and is designed specifically for identification of small amounts of individual proteins in complex mixtures. This application of proteomics in biomolecular archaeology is promising to open up exciting new areas of research.

In this chapter, we will study the structure of proteins and the methods used, both immunological and proteomic, to study ancient proteins. Before doing this, we must understand exactly what ancient proteins can tell us about the past.

3.1 The Importance of Proteins in Biomolecular Archaeology

The examples of the early use of ancient proteins described above illustrate two of the applications of these compounds in biomolecular archaeology. First, ancient proteins can be used in species identification. This is because some proteins display small structural variations in different species, sufficient to enable the species from which the protein is derived to be identified. This is not possible with all proteins and in particular is difficult with collagen and osteocalcin, the two major protein components of bone, as these proteins have very similar three-dimensional structures in all vertebrates. Hemoglobin, on the other hand, does display species-specific variations, enabling the proteins produced by different animals to be distinguished by immunological tests, at least with modern specimens. The blood-on-stone-tools controversy arose from attempts to use these variations in hemoglobin structure to identify the species of origin of the blood residues thought to be present on some stone implements. The same approach has been used to identify the species responsible for coprolites – fossilized excrement – sometimes found in caves. With both stone tools and coprolites, if protein is preserved, then some traces of ancient DNA might also be present, so PCR and sequencing can be used with the same material to confirm the species identification.

Typing of blood groups in mummies is an example of the second of the applications of ancient proteins in biomolecular archaeology. In this type of work, proteins are used to distinguish variations between different people with the resulting data used in population studies. Back in the 1930s, when studies of blood groups were becoming popular, methods for typing DNA did not exist – in fact, it was not even realized at that time that DNA is the genetic material. Since the 1980s virtually all population studies on living organisms have been carried out using DNA data, simply because much more information can be obtained, more quickly, by typing DNA. The same is true in biomolecular archaeology and ancient proteins are now only infrequently used to identify variations between members of a single species.

So far we have examined applications where ancient proteins provide information that can also be obtained by studying DNA. This is not always the case, and research with proteins is increasingly taking biomolecular archaeology into realms that are inaccessible to ancient DNA. One example is the use of casein as a marker for milk products. Casein is found only in milk so its detection in cooking or storage vessels indicates that those vessels once contained milk products. The detection of casein is therefore important for understanding the development of dairying in prehistory (Section 12.3). There have also been attempts to detect disease pathogens in human remains by searching for protein markers that are specific for the pathogen. An example of this approach is the use of an immunological test for the histidine-rich protein II antigen of *Plasmodium falciparum* to assess if the person whose remains are being studied is likely to have died of malaria (Section 15.3).

All of the above applications of ancient proteins make use of the immunological methods that were developed, in their basic form, in the early part of the 20th century. The introduction during the last few years of new proteomics techniques provides an alternative way of carrying out this work, one that is less prone to inaccuracies and can give results with smaller amounts of more degraded material. **Paleoproteomics**, as it is called, might also open up entirely new avenues of research. With modern material, one of the aims of proteomics is to measure the relative quantities of all the proteins present in a particular tissue, so that changes in the protein profile that occur during diseases such as cancer can be charted. It will probably never be possible to replicate this kind of work with the small amounts of degraded proteins present in archaeological remains, however much we might like to use the approach to establish the prevalence of different cancers in the past. But it might be possible with some types of material to use the power of proteomics to obtain novel information relevant to biomolecular archaeology. The seeds of crops such as wheat and maize provide an intriguing example, as the nutritional quality of different varieties of these crops depends on the relative quantities of a few abundant proteins present in the grain. Using proteomics to type and quantify these proteins might therefore give new insights into past diet.

3.2 Protein Structure and Synthesis

A protein, like a DNA molecule, is a linear unbranched polymer. In proteins, the subunits are called **amino acids** and the resulting polymers, or **polypeptides**, can be up to 2000 units in length. The similarities end there: the chemical structure of an amino acid is completely different from that of a nucleotide, and there are 20 different amino acids in proteins, compared with just the 4 nucleotides in DNA.

3.2.1 Amino acids and peptide bonds

Twenty different amino acids are found in protein molecules (Table 3.1). Each has the general structure shown in Figure 3.1, comprising a central carbon atom, called the α-carbon, to which the following four chemical groups are attached:

Table 3.1 The 20 amino acids found in proteins.

Amino acid	*Three-letter abbreviation*	*One-letter abbreviation*
Alanine	Ala	A
Arginine	Arg	R
Asparagine	Asn	N
Aspartic acid	Asp	D
Cysteine	Cys	C
Glutamic acid	Glu	E
Glutamine	Gln	Q
Glycine	Gly	G
Histidine	His	H
Isoleucine	Ile	I
Leucine	Leu	L
Lysine	Lys	K
Methionine	Met	M
Phenylalanine	Phe	F
Proline	Pro	P
Serine	Ser	S
Threonine	Thr	T
Tryptophan	Trp	W
Tyrosine	Tyr	Y
Valine	Val	V

- A hydrogen atom.
- A carboxyl ($-COO^-$) group.
- An amino ($-NH_3^+$) group.
- The R group, which is different for each amino acid.

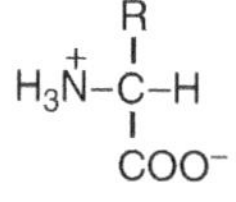

Figure 3.1 The general structure of an amino acid.

The R groups vary considerably in chemical structure and complexity (Figure 3.2). The simplest amino acid is glycine whose R group is just a hydrogen atom; the most complex are tyrosine, phenylalanine, and tryptophan, whose R groups contain aromatic ring structures. The differences between the R groups mean that although all amino acids are closely related, each has its own specific chemical properties. Some R groups are **polar** (e.g. serine and threonine) and prefer to be in an aqueous environment, others are **non-polar** (e.g. alanine, valine, and leucine) and lack an affinity for water. The majority of R groups are uncharged, though two amino acids have negatively charged R groups (aspartic acid and glutamic acid) and three have positively charged R groups (lysine, arginine, and histidine). Proteins can therefore display a vast range of chemical and biological properties, depending on the particular types of amino acid that they contain and the order in which the amino acids are joined together.

The polymeric structure of a polypeptide is built by linking a series of amino acids by **peptide bonds**, each one formed by a condensation reaction between the carboxyl group of one amino acid and the amino group of a second amino acid. The structure of a tripeptide,

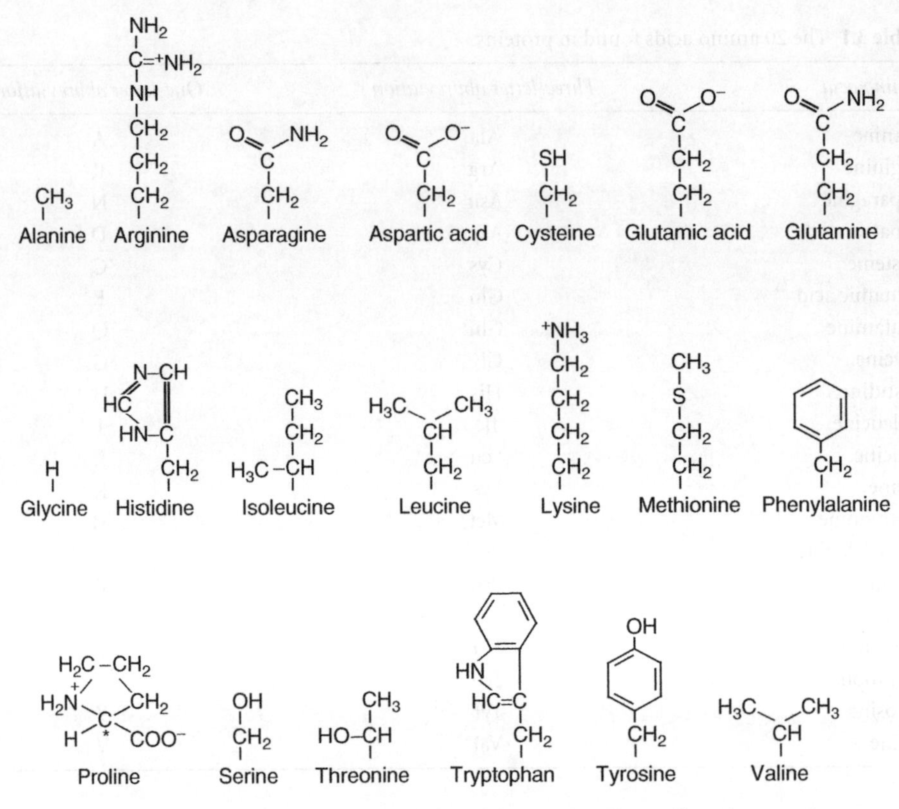

Figure 3.2 The R groups of the 20 amino acids found in proteins. For proline, the complete structure of the amino acid is shown, with the α-carbon indicated by the asterisk.

comprising three amino acids, is shown in Figure 3.3. Note that, as with a polynucleotide, the two ends of the polypeptide are chemically distinct: one has a free amino group and is called the **amino, NH_2–, or N terminus**; the other has a free carboxyl group and is called the **carboxyl, COOH–, or C terminus.** The direction of the polypeptide can therefore be expressed as either N→C (left to right in Figure 3.3) or C→N (right to left in Figure 3.3).

3.2.2 There are four levels of protein structure

Proteins are traditionally looked upon as having four distinct levels of structure. These levels are hierarchical, the protein being built up stage by stage, with each level of structure depending on the one below it:

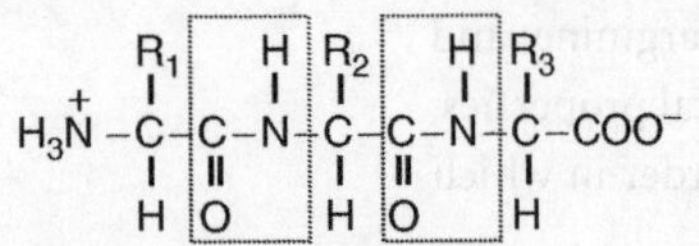

Figure 3.3 The structure of a tripeptide. The two peptide bonds are indicated by boxes.

- The **primary structure** of the protein is the linear sequence of amino acids.
- The **secondary structure** refers to the different conformations that can be taken up by the polypeptide. The two main types of secondary structure are the **α-helix** and **β-sheet.** These are stabilized mainly by hydrogen bonds that form between different amino acids in the polypeptide. Usually different regions of a polypeptide take up different secondary structures, so

that the whole is made up of a number of α-helices and β-sheets, together with less organized regions.

- The **tertiary structure** results from folding the secondary structural components of the polypeptide into a three-dimensional configuration (Figure 3.4). The tertiary structure is stabilized by various chemical forces, notably hydrogen bonding between individual amino acids, electrostatic interactions between the R groups of charged amino acids, and hydrophobic forces, which dictate that amino acids with non-polar side-groups must be shielded from water by embedding within the internal regions of the protein. There may also be covalent linkages called **disulfide bridges** between cysteine amino acids at various places in the polypeptide.

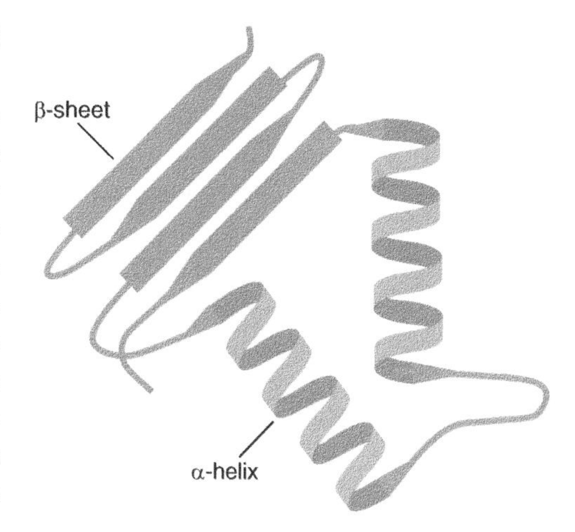

Figure 3.4 Schematic of the tertiary structure of a simple protein, comprising two α-helices and a small region of β-sheet.

- The **quaternary structure** involves the association of two or more polypeptides, each folded into its tertiary structure, into a multisubunit protein. This level of structure is not displayed by all proteins: many, especially those in bacteria, carry out their biological function as a single polypeptide. Some quaternary structures are held together by disulfide bridges between the different polypeptides, resulting in stable multisubunit proteins that cannot easily be broken down to the component parts. Other quaternary structures comprise looser associations of subunits, which means that these proteins can revert to their component polypeptides, or change their subunit composition, according to the functional requirements. The quaternary structure may involve several molecules of the same polypeptide or may comprise different polypeptides. An example of the latter is hemoglobin, which is made of two α-globin and two β-globin polypeptides. In some cases, the quaternary structure is built up from a very large number of polypeptide subunits, to give a complex array; the best examples are the protein coats of viruses, such as that of tobacco mosaic virus which is made up of 2130 identical protein subunits.

3.2.3 The amino acid sequence is the key to protein structure and function

Each of the higher levels of protein structure – secondary, tertiary, and quaternary – is specified by the primary structure, the amino acid sequence itself. This is most clearly understood at the secondary level, where it is recognized that certain amino acids, because of the chemical and physical properties of their R groups, stimulate the formation of an α-helix, whereas others promote formation of a β-sheet. Conversely certain amino acids more frequently occur outside regular structures and may act to determine the end-point of a helix or sheet. These factors are now so well understood that rules to predict the secondary structures taken up by amino acid sequences have been developed.

Although less well characterized, it is nonetheless clear that the tertiary and quaternary structures of a protein also depend on the amino acid sequence. The interactions between individual amino acids at these levels are so complex that predictive rules, although attempted, are still unreliable. However, it has been established for some years that if a protein is **denatured**, for instance by mild heat treatment or adding a chemical denaturant such as urea, so that it loses its higher levels of structure and takes up a non-organized conformation, it still retains the innate ability upon **renaturation** (e.g. by cooling down again) to reform spontaneously the correct tertiary structure. Once the tertiary structure has formed, subunit assembly into a multimeric protein again occurs spontaneously. This shows that the instructions for the tertiary and quaternary structures must reside in the amino acid sequence.

Through the secondary, tertiary, and quaternary structures the amino acid sequence also specifies the protein's function. This is because a protein, in order to perform its function, must interact with other molecules, and the precise nature of these interactions is set by the overall shape of the protein and the distribution of chemical groups on its surface. Understanding the interactions of proteins with other molecules (including other proteins) is a complex area of modern biochemistry, but the basic principle that amino acid sequence determines function is easily illustrated. Let us consider the **helix-turn-helix** proteins, which make up one family of the diverse group of proteins that must attach themselves to a DNA molecule in order to perform their function. The name "helix-turn-helix" indicates that within each of these proteins there is motif made up of two α-helices separated by a turn. The latter is not a random conformation but a specific structure made up of four amino acids, the second of which is usually glycine. This turn, in conjunction with the first α-helix, positions the second α-helix on the surface of the protein in an orientation that enables it to fit inside one of the grooves of a DNA molecule. This second α-helix is therefore called a recognition helix because it makes the vital contacts with the DNA.

The active form of most helix-turn helix proteins is a dimer. The two recognition helices (one from each polypeptide in the dimer) are exactly 3.4 nm apart and therefore fit into two adjacent sections of a DNA molecule. Hence the shape of the protein is critical to its function: if either recognition helix were absent, or if they were orientated incorrectly on the surface of the protein, then the protein would not be able to bind to DNA. But this is only part of the story. Most helix-turn-helix proteins have regulatory roles in gene expression and so attach to specific nucleotide sequences adjacent to the genes whose activity they control. This requires that precise contacts be made between chemical groups in the recognition helices and the bases in the DNA sequence to which the protein binds. If the helices have the wrong chemical groups, or if the groups are positioned incorrectly, then the binding sequence will not be recognized.

The function of a helix-turn-helix protein therefore depends on its amino acid sequence in three ways. First, the recognition structure is present because the amino acid sequence of that particular part of the polypeptide promotes formation of an α-helix. Second, the amino acid sequence of the protein as a whole specifies a quaternary structure in which the two recognition helices of the dimeric protein are orientated in precisely the correct way. Third, the amino acid sequence of each recognition helix provides the particular combination of R groups that enables the protein to attach to its specific binding sequence. Hence there is a precise link between the amino acid sequence of the polypeptide and the function of the dimeric helix-turn-helix protein. An equivalent description of the link between sequence and function can be given for many other types of protein.

3.2.4 The amino acid sequence of a protein is specified by the genetic code

There is an equally direct link between the nucleotide sequence of a gene and the amino acid sequence of the protein for which it codes. This is the key to genome expression, because it means that the nucleotide sequence of the gene is, in essence, a set of instructions for constructing a protein with a particular function. The language in which those instructions are written is called the **genetic code** (Figure 3.5).

The genetic code must provide a means of converting the 4-letter language of the gene into the 20-letter language of proteins. When molecular biologists first began to decipher the genetic code, back in the 1950s, they realized immediately that a triplet code – one in which each codeword or **codon** comprises three nucleotides – is required to account for all 20 amino acids found in proteins. A two-letter code would have only $4^2 = 16$ codons, which is not enough to account for all 20 amino acids, whereas a three-letter code would give $4^3 = 64$ codons. When the genetic code was fully worked out, it was realized that the 64 codons fall into groups, the members of each group coding for the same amino acid (Figure 3.5). Only tryptophan and methionine have just a single codon each: all others are coded by two, three, four, or six codons. This feature of the code is called **degeneracy**. The code also has an **initiation codon**, usually 5′–AUG–3, which indicates the start of the gene, and three **termination codons** – 5′–UAG–3′, 5′–UAA–3′, and 5′–UGA–3′ – one of which comes at the end of a gene. The termination codons do not themselves code for amino acids, but the initiation codon specifies methionine, which means that most newly synthesized polypeptides start with this amino acid.

Codon	Amino acid	Codon	Amino acid	Codon	Amino acid	Codon	Amino acid
TTT	phe	TCT	ser	TAT	tyr	TGT	cys
TTC		TCC		TAC		TGC	
TTA	leu	TCA		TAA	stop	TGA	stop
TTG		TCG		TAG		TGG	trp
CTT	leu	CCT	pro	CAT	his	CGT	arg
CTC		CCC		CAC		CGC	
CTA		CCA		CAA	gln	CGA	
CTG		CCG		CAG		CGG	
ATT	ile	ACT	thr	AAT	asn	AGT	ser
ATC		ACC		AAC		AGC	
ATA		ACA		AAA	lys	AGA	arg
ATG	met	ACG		AAG		AGG	
GTT	val	GCT	ala	GAT	asp	GGT	gly
GTC		GCC		GAC		GGC	
GTA		GCA		GAA	glu	GGA	
GTG		GCG		GAG		GGG	

Figure 3.5 The genetic code.

Recall that during the first stage of genome expression the process called transcription results in synthesis of RNA copies of genes. The mRNA copies of the protein-coding genes are then translated into protein during the second stage of genome expression. Translation is more complex than transcription because the laws of chemistry make it impossible for the amino acid sequence of a polypeptide to be built up directly on its mRNA: there is simply no way that the necessary chemical contacts can be made. Instead, an adaptor molecule is needed to form a bridge between the nucleotide sequence of the mRNA and the amino acid sequence of the growing polypeptide (Figure 3.6). This adaptor molecule is transfer RNA. Each organism synthesizes a number of different types of tRNA, at least one for each of the 20 amino acids. Each tRNA forms a covalent attachment with its particular amino acid, and a base-pairing attachment with one or more of the codons that specifies that amino acid. The process takes place within

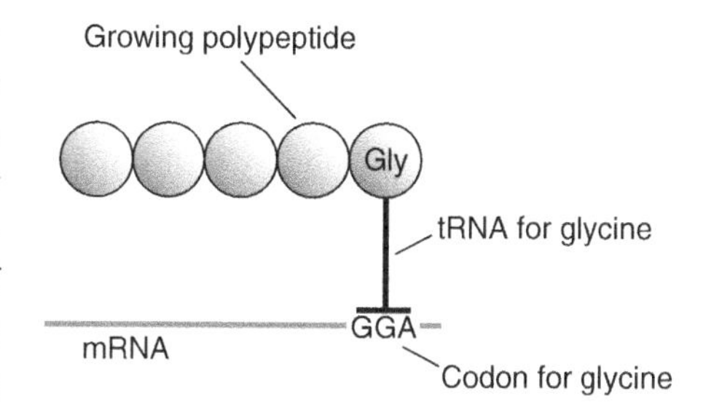

Figure 3.6 The role of tRNA as an adaptor molecule in protein synthesis. The mRNA is being read from left to right. A codon for glycine has been reached, which is recognized by a tRNA that is attached to a glycine amino acid. The glycine is therefore placed at the end of the polypeptide that is being synthesized.

one of the structures called **ribosomes**, with the mRNA read in the 5′→3′ direction and the polypeptide synthesized unit by unit in the N→C direction.

3.2.5 Post-translational modifications increase the chemical complexity of some proteins

The standard genetic code specifies 20 different amino acids. This repertoire is increased dramatically by post-translational chemical modification, which results in a vast array of different amino acid types. This means that the amino acid sequence of a mature, functioning protein might be different to that of the polypeptide coded by the gene.

The simplest types of chemical modification occur in all organisms and involve addition of a small chemical group (e.g. an acetyl, methyl, or phosphate group) to an amino acid side chain, or to the amino or carboxyl groups of the terminal amino acids in a polypeptide. Over 150 different modified amino acids have been documented in different proteins. Some proteins undergo an array of different modifications, an example being the chromosomal protein histone H3, which can be modified by acetylation, methylation, and phosphorylation at a number of positions along its polypeptide chain. Chemical modification often affects the activity of a protein, possibly changing its function in a subtle but important way, or possibly activating a protein whose function is only needed at a particular time.

A more complex type of modification, found predominantly in eukaryotes, is **glycosylation**, the attachment of large carbohydrate side chains to polypeptides. There are two general types of glycosylation, **O-linked glycosylation** in which the sugar is attached to the hydroxyl group of a serine or threonine amino acid, and **N-linked glycosylation** in which the attachment is to the amino group on the side chain of asparagine (Figure 3.7). Glycosylation can result in attachment to the protein of large structures comprising branched networks of 10–20 sugar units of various types. These side chains help to target proteins to particular sites in cells and determine the stability of proteins circulating in the bloodstream. Another type of large-scale modification involves attachment of long-chain lipids, often to serine or cysteine amino acids. This process is called **acylation** and occurs with many proteins that become associated with membranes.

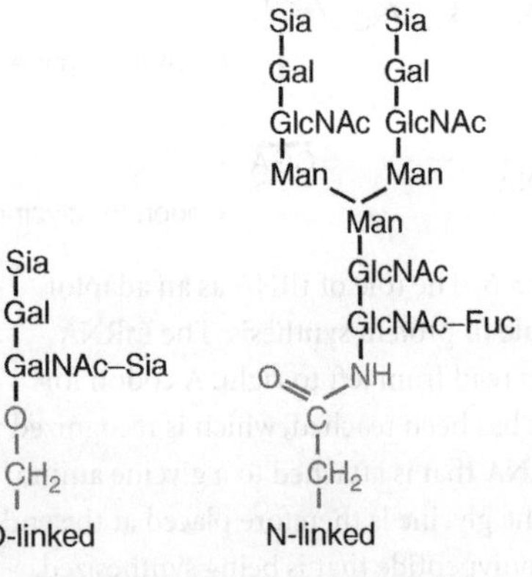

Figure 3.7 O- and N-linked glycosylation. The O-linked structure is attached to the R group of a serine amino acid, and the N-linked structure is attached to an asparagine. Abbreviations: Fuc, fucose; Gal, galactose; GalNAc, *N*-acetylgalactosamine; GlcNAc, *N*-acetylglucosamine; Man, mannose; Sia, sialic acid.

3.3 Studying Proteins by Immunological Methods

Immunological methods for protein detection have been used in clinical settings since 1900, when Karl Landsteiner devised the first tests for distinguishing the A, B, and O blood groups. Very similar procedures were used by the Bonds in the 1930s to identify the blood types of Egyptian and Native American mummies. The sophisticated immunoassays of the present day are direct descendents of those early blood tests. All of these methods are based on the specific reaction between **antibodies** and **antigens** that occurs during the natural immune response of mammals and other animals.

3.3.1 Immunological methods depend on the reaction between antibody and antigen

If a purified sample of a protein is injected into the bloodstream of a rabbit, the immune system of the animal responds by synthesizing antibodies that bind to and help degrade the foreign molecule. This is a version of the natural defence mechanism that the animal uses to deal with antigens, which include not only proteins and other compounds but also invading bacteria, viruses, and other infective agents. Once a rabbit is challenged with a protein, the levels of antibody present in its bloodstream remain high enough over the next few days for substantial quantities to be purified. After purification, the antibody retains its ability to bind to the protein with which the animal was originally challenged.

Antibodies, which themselves are proteins called **immunoglobulins**, are synthesized by the B lymphocytes in the bloodstream and secreted into the blood serum. Each B cell makes a slightly different type of immunoglobulin protein, providing a huge diversity of antibodies able to respond to the equally diverse range of antigens that an animal is likely to encounter during its lifetime. If the antigen is relatively large, as is the case with most proteins, then it will react with several different immunoglobulins in the bloodstream, each one recognizing a different surface feature or **epitope**. This collection of reacting immunoglobulins is called a **polyclonal antibody**. It is important to appreciate that to be absolutely specific for the protein antigen, an individual immunoglobulin must recognize a surface epitope that is itself unique to that protein. All polyclonal antibodies will contain at least some immunoglobulins with this level of specificity, but there will others that recognize epitopes that are common surface features shared between different proteins. These immunoglobulins will also cross-react with those other proteins, which means that polyclonal antibodies are rarely entirely specific for the protein against which they are raised. **Monoclonal antibodies**, on the other hand, which are prepared by cloning individual B cells and therefore contain just a single type of immunoglobulin, can be totally specific for their target antigen if the immunoglobulin recognizes a unique surface feature of the antigen. However, monoclonals are expensive to prepare and those that are available commercially are not designed against protein antigens likely to be of interest in biomolecular archaeology.

A number of different types of immunological method have been developed over the years. These differ in the way in which the reaction between antibody and antigen is detected, some being purely qualitative – giving a simple yes/no answer – and others enabling the amount of the antigen to be quantified with differing degrees of precision. The quantitative tests are called **immunoassays**. We will now turn our attention to the details of the most important of these methods.

3.3.2 Methods based on precipitation of the antibody–antigen binding complex

With most polyclonal antibodies, reaction with the antigen forms an insoluble antibody–antigen binding complex, which precipitates out of the solution. This is called the **precipitin reaction**, and most of the early immunological methods made use of this property to detect the presence of the antigen. In the simplest method, the test is carried out in a solution and precipitation detected by eye, either from the increased cloudiness, or by centrifuging the

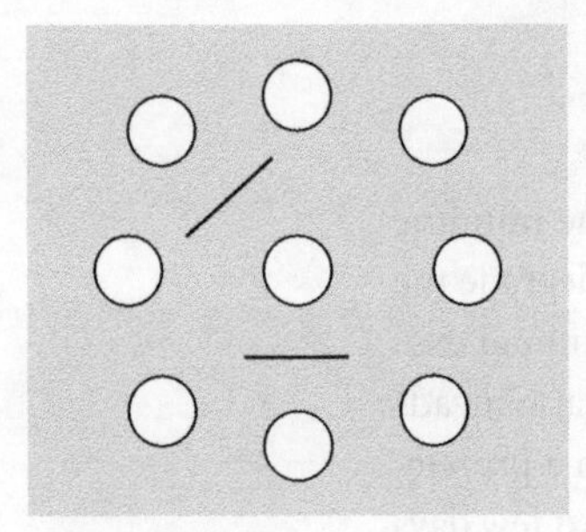

Figure 3.8 One version of the Ouchterlony technique. The antibody is placed in the central well and test solutions in the outer circle of wells. Two precipitin reactions can be seen, indicating the two outer wells that contain antigens that cross-react with the antibody.

solution and checking for the presence of an insoluble pellet. If this approach is used, then serial dilutions of the antigen solution must be assayed because the amount of precipitation is dependent on the relative concentrations of antibody and antigen. The yield of precipitate does not simply increase with increasing amounts of antigen, but instead reaches a maximum within the "zone of equivalence" and then tails off. So simply adding as much antigen as possible is not advisable, and indeed can lead to non-specific reactions, as the incorrect antigen can give a precipitate if added at an inappropriately high concentration.

Some of these problems can be alleviated by carrying out the precipitin reaction in a thin slab of agar or agarose gel rather than in solution. In the **Ouchterlony technique**, which at one time was the most frequently used immunological method, the antibody and antigen solutions are placed in wells that are cut out of the agarose (Figure 3.8). The relative concentrations of the antibody and antigen are not so critical as in a solution test, because both form a concentration gradient as they diffuse through the gel. The wells are placed close enough together for the two diffusion zones to overlap, and a precipitate forms at the position within the overlap where the antibody and antigen concentrations are at equivalence. The precipitate can sometimes be seen in the gel without any further treatment, but for greater sensitivity the gel can be stained with Coomassie Blue or some other protein-specific dye, which will clearly reveal the line of precipitation.

Because it is dependent on the natural process of diffusion, the Ouchterlony technique is quite slow. In the original method, it took days for the concentration gradients to become set up, and the wells had to be continually replenished with antibody and antigen in order to maintain the process. Even when reduced to a microscale it still takes hours. The various types of **immunoelectrophoresis** technique are designed to speed up the meeting of antibody and antigen in a gel. Most proteins have a net negative charge at pH 8.0 and migrate towards the positive electrode during electrophoresis. As with DNA, when carried out in a gel, electrophoresis results in separation of proteins according to their sizes. This applies to most protein antigens but, unusually, not to antibodies. Immunoglobulin G molecules migrate in the opposite direction, towards the positive electrode, not by electrophoresis but by a type of electrically stimulated diffusion called **electroendosmosis**. Therefore, if antibody and antigen are placed in wells with a pair of electrodes to either side, then when the electric current is applied the two compounds move towards one another. The antigen migrates as a single sharp band but the antibody, being subject to electroendosmosis rather than electrophoresis, forms a concentration gradient. At some position within the overlap the zone of concentration equivalence is reached, and the antibody–antigen binding complex forms a line of precipitation (Figure 3.9). This is called the **crossover immunoelectrophoresis** (**CIP** or **CIEP**) technique. It is much quicker than the Ouchterlony method, and more sensitive because the antigen moves as a single band rather than becoming diluted out within a concentration gradient.

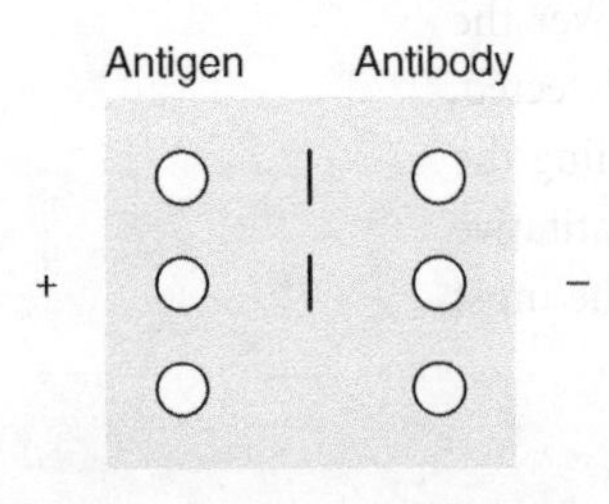

Figure 3.9 Crossover immunoelectrophoresis. Two precipitin reactions are seen, indicating the two wells on the left that contain antigens that cross-react with the antibody.

3.3.3 Enzyme immunoassays enable more sensitive antigen detection

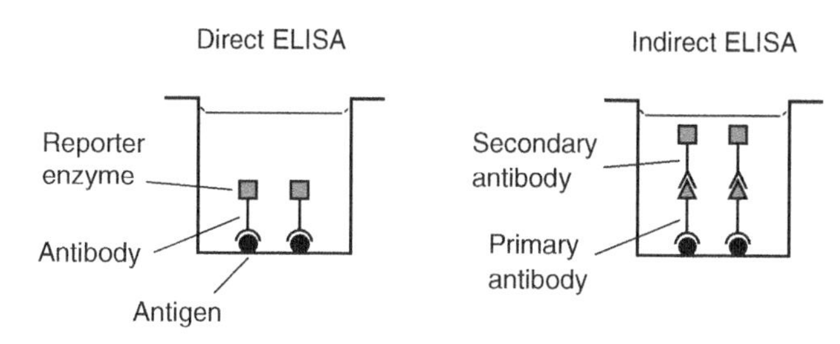

Figure 3.10 Direct and indirect ELISA.

Immunological methods based on the precipitin reaction are primarily qualitative. They indicate if the specific antigen is present or absent but provide only approximate information on the amount of antigen in a sample. Qualitative information is of only limited use in a clinical setting, so most of the more recently developed methods are genuine immunoassays able to yield quantitative data. These newer methods are also more sensitive than the precipitation based ones, and hence are more applicable to biomolecular archaeology, where we are almost always working with small amounts of material.

The enzyme assay called **ELISA** (enzyme-linked immunosorbent assay) is the most popular of these methods, largely because it is quick and relatively easy to carry out, and does not use harmful chemicals. In ELISA, antibody–antigen binding is detected by measuring the activity of a **reporter enzyme** that is conjugated to the antibody. Horseradish peroxidase is a popular reporter enzyme because its activity can be monitored by adding any one of various chromogenic or chemiluminescent substrates, such as tetramethylbenzidine (TMB), which is converted by the enzyme into a blue product, or luminal, which is oxidized with the release of chemiluminescence. The amount of the reporter enzyme that is present can therefore be measured by assaying for the color change or intensity of emitted light.

ELISA can be carried out in a direct or indirect format. In both methods, the antigen is first absorbed to the walls of a well in a microtiter plate. For direct detection, the antibody is conjugated to the reporter enzyme and added to the well (Figure 3.10). After washing away any unbound antibody, the amount of the antigen-antibody complex is assayed by measuring the activity of the reporter enzyme. The indirect method begins in the same way, but the initial antibody that is added, now called the **primary antibody**, is not conjugated to the enzyme. The amount of antigen-antibody complex that is formed is measured by adding a **secondary antibody**, conjugated to the enzyme, that recognizes not the antigen but the primary antibody. This might seem odd, but an antibody is itself a protein and so if injected into an animal will give rise to an immunological reaction, providing that the animal is of a different species from the one from which the primary antibody was obtained. So if the primary antibody was prepared in a rabbit, for example, then the secondary antibody could be obtained by injecting a sample of the primary antibody into a goat.

The indirect method is clearly more work, so what advantages does it have? One advantage is that the researcher need only prepare the primary antibody in the usual way, by challenging a rabbit with the antigen. The conjugated secondary antibody can then be purchased from a commercial supplier. This secondary antibody is simply "anti-rabbit immunoglobulin," so can be used with a whole range of different primary antibodies, so long as all of these are obtained from rabbits. Also, if the secondary antibody is polyclonal, then it will recognize different epitopes on the surface of the rabbit immunoglobulin protein, so that more than one molecule of the secondary antibody attaches to each molecule of the primary antibody.

As the secondary antibody carries the reporter enzyme, this multiple binding results in a higher amount of signal and hence a greater sensitivity.

The direct and indirect detection strategies illustrated by ELISA can also be carried out with antibodies that are labelled with a radioactive marker. The resulting **radioimmunoassay (RIA)** can, under some circumstances, be more sensitive than ELISA, but suffers from the health and disposal problems of all radioactive procedures.

3.3.4 Potential and problems of immunological methods in biomolecular archaeology

From the above discussion of immunological methods, you will appreciate that these techniques have great potential in biomolecular archaeology. The ability of an antibody to distinguish one protein from another means, for example, that immunological methods could be used to detect protein markers such as casein, indicative of the presence of milk products, in residues from potsherds collected from archaeological sites. The ability of antibodies to distinguish closely related proteins also underlies the attempts to use immunological tests to identify the species of origin of the supposed blood residues sometimes found on the surfaces of stone tools.

By now you should also be aware that the specificity of immunological methods depends on the perfect match between the antibody and the epitope that it recognizes. The epitope is a structural feature on the surface of the antigen, which for a protein antigen will be determined in part by the tertiary structure and in part by the distribution of chemical groups on the protein surface. If these surface features of the protein change during diagenesis, then it is quite possible that the antibody will become less effective at recognizing it. If a polyclonal antibody is used, then the multiplicity of epitopes that are recognized will often be sufficient to retain at least some cross-reactivity with a partially degraded protein, but more relevant is the possibility that the protein now displays a non-specific cross-reaction with other antibodies. If this occurs, then protein antigens might be misidentified.

The possible misidentification of ancient proteins when immunological tests are used is not a purely hypothetical problem. Several studies have shown that there is poor agreement between the results of different immunological methods applied to the same protein residue, with techniques such as CIEP, dependent on the precipitin reaction, often giving different results compared with immunoassays such as RIA. Even more worrying are the results of experiments where artifacts have been coated with blood of a known origin and then tested by CIEP. In one such study, only 20 of 54 samples were correctly identified, with many of the misidentifications not being negative results but positive identification of blood protein from a different species. These studies provide a clear warning that the outcomes of immunological tests of ancient proteins must always be treated with caution.

3.4 Studying Proteins by Proteomic Methods

In recent years, molecular biologists have devised a number of new methods for studying proteomes, the complex collection of proteins present in a living cell. **Proteomics** refers to all of these techniques, and the term **paleoproteomics**, used to describe the application of these techniques in biomolecular archaeology, is a misnomer because in reality only one of the collection of proteomics techniques is actually used. This technique is called **protein profiling**.

The approach employed by protein profiling is quite different to that followed when an immunological method is used. Rather than attempting to identify a single chosen protein, in profiling an entire protein mixture is first separated into its components and then individual proteins selected and identified on the basis of their amino acid contents.

3.4.1 Various methods are used to separate proteins prior to profiling

The first step in protein profiling is to separate all the individual proteins in the preparation that is being studied, or at least as many as is feasible. In the most challenging type of protein profiling, when the entire proteome of a human cell is being examined, the starting extract may contain as many as 20,000 different proteins. The most suitable separation process for such a complex mixture is **two-dimensional gel electrophoresis**. Electrophoresis in a polyacrylamide gel is one of the standard methods for separating proteins. Depending on the composition of the gel and the conditions under which the electrophoresis is carried out, different chemical and physical properties of proteins can be used as the basis for this separation. The most frequently used technique makes use of the detergent called sodium dodecyl sulfate (SDS), which denatures proteins and confers a negative charge that is roughly equivalent to the length of the unfolded polypeptide. Under these conditions, the proteins separate according to their molecular masses, the smallest proteins migrating more quickly towards the positive electrode. Alternatively, proteins can be separated by **isoelectric focusing** in a gel that contains chemicals that establish a pH gradient when the electrical charge is applied. In this type of gel, a protein migrates to its **isoelectric point**, the position in the gradient where its net charge is zero. These methods are combined in two-dimensional gel electrophoresis (Figure 3.11). In the first dimension, the proteins are separated by isoelectric focusing. The gel is then soaked in SDS, rotated by 90° and a second electrophoresis, separating the proteins according to their sizes, carried out at right angles to the first. This approach can separate several thousand proteins in a single gel. After electrophoresis, staining the gel with a protein dye reveals a complex pattern of spots, each one containing a different protein.

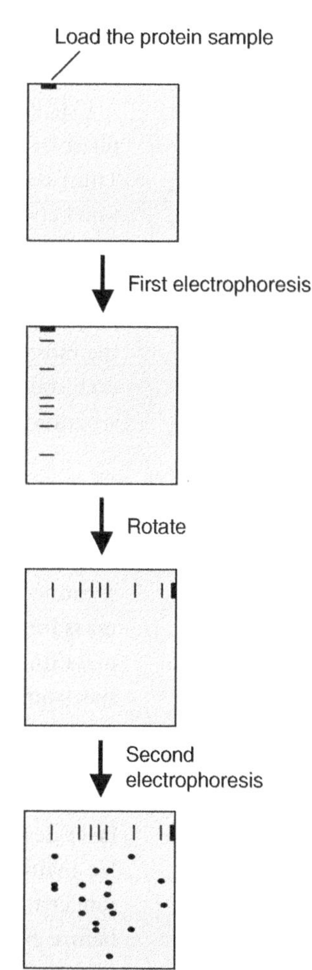

Figure 3.11 Two-dimensional gel electrophoresis.

For less complex protein mixtures, including most of those likely to be obtained from archaeological contexts, a technique as discriminatory as two-dimensional electrophoresis might not be needed to separate all the proteins that are present. One-dimensional electrophoresis, using either an SDS or isoelectric focusing gel, might be sufficient, but **column chromatography** is more commonly used. In this procedure, the protein solution is passed through a column packed with a solid compound of some kind, the identity of which is determined by the type of separation that is being carried out. There are several different methods, the most important of which are:

- **Gel filtration chromatography** uses a matrix made up of small porous beads, often of silica. As the proteins pass through the column, the smaller ones are delayed as they enter and exit the pores of the matrix. The proteins therefore become separated out according to their size, the largest ones eluting from the column first, and the smallest ones last.
- **Ion-exchange chromatography** separates proteins according to their net electric charges. The matrix consists of positively or negatively charged polystyrene beads, and the pH or ionic strength of the liquid phase – the solution passed through the column to elute the proteins – is gradually increased or decreased to achieve separation of proteins with different charges.
- **Reverse phase chromatography** uses a matrix whose surface is covered with non-polar chemical groups such as hydrocarbons. Unlike the other two methods, the liquid phase is not entirely aqueous. Instead, a mixture of water and an organic solvent such as methanol or acetonitrile is used, possibly as a gradient. Separation occurs because most proteins have hydrophobic areas on their surfaces, which bind to the non-polar matrix. The stability of this attachment decreases as the organic content of the liquid phase increases, so proteins are separated according to their net surface hydrophobicities.

Usually **high performance liquid chromatography** (HPLC) is used, with the liquid phase being pumped at high pressure into a column with an internal diameter of less than 1 mm, contained within a reinforced steel sleeve, a format designed to achieve a high resolution between individual proteins. Resolution and speed of separation can also be increased by using a **monolith** column, which contains a single solid block of polymer instead of the tiny beads used in the traditional methods. Despite these embellishments, groups of proteins might still not be separated, in which case they will elute from the column together. In the most sophisticated systems, these mini-mixtures are further separated by transferring each consecutive fraction into an array of secondary columns, in which a further round of separation using a different procedure is carried out.

3.4.2 Identifying the individual proteins after separation

The second stage of protein profiling is to identify the individual proteins that have been separated from the starting mixture. This used to be a difficult proposition but **peptide mass fingerprinting** has provided a rapid and accurate identification procedure. Peptide mass fingerprinting was itself made possible by advances in **mass spectrometry**. Mass spectrometry was originally designed as a means of identifying a compound from the mass-to-charge ratios of the ionized forms that are produced when molecules of the compound are exposed to a high energy field. The standard technique could not be used with proteins because they are too large to be ionized effectively, but **matrix-assisted laser desorption ionization time-of-flight** (MALDI-TOF) gets around this problem, at least with peptides of up to 50 amino acids in length. Of course, most proteins are much longer than 50 amino acids, and it is therefore necessary to break them into fragments before examining them by MALDI-TOF. The standard approach is to digest the protein with a sequence-specific protease, such as trypsin, which cleaves proteins immediately after arginine or lysine residues. With most proteins, this results in a series of peptides 5–75 amino acids in length.

Once ionized, the mass-to-charge ratio of a peptide is determined from its "time of flight" within the mass spectrometer as it passes from the ionization source to the detector. The mass-to-charge ratio enables the molecular mass of the peptide to be worked out, which in turn allows its amino acid composition to be deduced. This compositional information is then compared with a database to identify the protein from which the peptide is derived. This database contains the amino acid sequences of all proteins that are known, including many that have never been directly studied but whose amino acid sequences can be predicted from the nucleotide sequences of their genes. With very good fortune, a unique match can be made between just one peptide and a single protein in the database, but to make a confirmed identification it is usually necessary to have compositional information for two or more peptides from the same protein. Accurate identification is also aided if there is some *a priori* knowledge about the range of proteins in the original mixture. If necessary, the entire Swissprot or Genbank protein databases can be searched, but these are so huge that a unique match is unlikely, even if several peptides have been analyzed. When used with modern extracts, the organism from which the protein extract is obtained is almost always known, so just the proteins of that species need be present in the database.

When used in biomolecular archaeology, protein profiling is hampered by the degraded nature of ancient proteins, which can result in changes in their peptide mass fingerprints due, for example, to loss of chemical groups from individual amino acids or attachment of additional groups during the diagenesis process. Paleoproteomics is still very much in its infancy, and the extent to which this problem can be solved is as yet uncertain. As we will see in later chapters, in certain applications where the range of possible proteins in the starting mixture is quite limited, peptide mass fingerprints prepared from archaeological samples have already yielded information of great value.

Further Reading

Bailey, C.S. (1994) Radioimmunoassay of peptides and proteins. *Methods in Molecular Biology*, 32, 449–60.

Brown, T.A. (2007) *Genomes*, 3rd edn. Garland Science, Abingdon. [Details on the genetic code and protein synthesis.]

Crowther, J.R. (2001) *The ELISA Guidebook*. Humana Press, Totowa, N.J.

Görg, A., Weiss, W. & Dunn, M.J. (2004) Current two-dimensional electrophoresis technology for proteomics. *Proteomics*, 4, 3665–85.

Hay, F.C. & Westwood, O.M.R. (2002) *Practical Immunology*, 4th edn. Wiley-Blackwell, Oxford.

Leach, J.D. & Mauldin, R.P. (1995) Additional comments on blood residue analysis in archaeology. *Antiquity*, 69, 1020–2. [Describes the problems in attempted identification of blood residues.]

Lesk, A. (2004) *Introduction to Protein Science: Architecture, Function, Genomics*. Oxford University Press, Oxford.

Twyman, R.M. (2004) *Principles of Proteomics*. Garland Science, Abingdon.

4

Lipids

Lipids are a broad group of compounds that include the fats, oils, waxes, steroids, and various resins. They have diverse structures but all are hydrophobic and most are lipophilic, meaning that they are insoluble in water but soluble in organic solvents such as acetone and toluene. Their functions in nature are equally diverse and go far beyond the familiar role of fats and oils in storage of energy. Phospholipids and sterols are major structural components of the membranes of virtually all organisms, and other lipids have specialist roles in the cell walls of bacteria. Waxes are secreted onto the surfaces of the leaves and fruits of plants, where they protect against dehydration and attack by small predators such as insects; and some animals and birds secrete waxes and other lipids, which have similar protective functions on fur and feathers. Steroids and prostaglandins include many important hormones, and vitamins A, D, E, and K are lipids. One single group of lipids, the **terpenes**, is the largest class of natural products and includes approximately 25,000 different compounds, synthesized mainly by plants and with a variety of functions, including disease resistance, signalling, and protection against predator attack, as well as involvement in important physiological processes such as photosynthesis.

Lipids are relatively resistant to degradation by chemical and microbial processes. The ability of some lipids to survive with little structural change for long periods was first recognized by organic geochemists working with geological specimens, and in the 1970s the first attempts were made to study ancient lipids in archaeological contexts. This work suffered initially from the application of techniques that were not entirely suitable for analysis of the small amounts of partially degraded material that were available, but the field began to mature in the early 1990s when high resolution separation and identification methods based around gas chromatography and mass spectrometry were first used. Since then, ancient lipid research has moved forward in dramatic fashion, and is now, arguably, the most productive and informative area of biomolecular archaeology.

 © 2011 Terry Brown & Keri Brown

The importance of lipids in biomolecular archaeology arises largely because of their species specificity. Some lipids are made by just a single or small number of species, so they can be used as biomarkers, detection of the lipid indicating the past presence of the species of origin. So, for example, the detection of particular lipids absorbed within the matrix of potsherds from archaeological cooking vessels often enables the food that was processed in that vessel to be identified, both vegetables such as brassicas as well as animal products (Section 12.2). Identification of residues derived from oils and other lipids can be used in a similar way to determine the past contents of storage vessels. The presence of organic matter in soils, possibly indicative of land management practices such as manuring, can be assessed by searching for lipids associated with mammalian waste products (Section 14.1). Prehistoric technology can be explored, for example by studying the composition of adhesives used to attach arrow heads to their shafts (Section 14.1). We will study many of these uses of ancient lipids in later chapters.

4.1 The Structures of Lipids

It is not possible to make general statements about lipid structure. There is an almost bewildering variety of different types of compound, so many that lipid structure is rarely dealt with comprehensively even in textbooks of biochemistry. We will examine the structures of the most important types of lipid, focussing of course on those of greatest importance in biomolecular archaeology.

4.1.1 Many lipids are fatty acids or fatty acid derivatives

Fatty acids are both important lipids in their own right and precursors of other lipids such as fats, oils, soaps, and waxes. Fatty acids are a type of **carboxylic acid**. This is a compound made up of a central carbon atom attached to an oxygen radical (=O), a hydroxyl group (–OH), and an R group that is different in each carboxylic acid (Figure 4.1A). The general formula is therefore R–COOH. The simplest carboxylic acids are familiar natural products: formic acid (H–COOH), which is present in ant bites and bee stings, and the acetic acid of vinegar (CH_3–COOH), but of course these water soluble compounds are not lipids. In a fatty acid, the R group of the carboxylic acid is a much more complex **hydrocarbon** chain made up of 4–36 carbons with their attached hydrogen atoms. Because of this highly hydrophobic R group, fatty acids are almost insoluble in water, but readily soluble in many organic solvents.

Fatty acids are divided into two classes depending on the structure of the hydrocarbon chain. If all the links between adjacent carbons are single bonds, which will mean that every carbon carries two hydrogen atoms, then the fatty acid is said to be **saturated**. If, on the other hand, there are one or more pairs of carbons linked by double bonds, then the fatty acid is **unsaturated** (Figure 4.1B). The nomenclature used to describe the particular structure of an individual fatty acid is

$$M:N(\Delta^{a,b,\ldots})$$

In this formula, M is the number of carbons in the chain and N is the number of double bonds. If double bonds are present then the ($\Delta^{a,b,\ldots}$) component is included, with a, b, ... ,

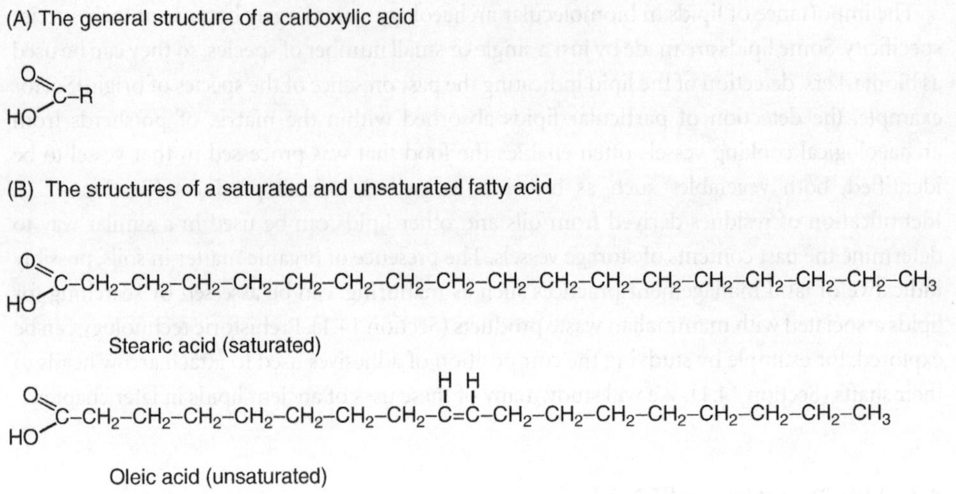

(A) The general structure of a carboxylic acid

$HO-C(=O)-R$

(B) The structures of a saturated and unsaturated fatty acid

$HO-C(=O)-CH_2-CH_2-CH_2-CH_2-CH_2-CH_2-CH_2-CH_2-CH_2-CH_2-CH_2-CH_2-CH_2-CH_2-CH_2-CH_2-CH_3$

Stearic acid (saturated)

$HO-C(=O)-CH_2-CH_2-CH_2-CH_2-CH_2-CH_2-CH_2-CH=CH-CH_2-CH_2-CH_2-CH_2-CH_2-CH_2-CH_2-CH_3$

Oleic acid (unsaturated)

Figure 4.1 The general structure of a carboxylic acid (A), and the structures of two fatty acids (B).

Table 4.1 Fatty acids.

Structural designation	*Name*
Saturated	
12:0	Lauric acid, dodecanoic acid
14:0	Myristic acid, tetradecanoic acid
16:0	Palmitic acid, hexadecanoic acid
18:0	Stearic acid, octadecanoic acid
20:0	Arachidic acid, eicosanoic acid
22:0	Behenic acid, docosanoic acid
24:0	Lignoceric acid, tetracosanoic acid
Monounsaturated	
$16:1(\Delta^{9})$	Palmitoleic acid
$18:1(\Delta^{9})$	Oleic acid
Polyunsaturated	
$18:2(\Delta^{9,12})$	Linoleic acid
$18:3(\Delta^{9,12,15})$	Linolenic acid
$20:4(\Delta^{5,8,11,14})$	Arachidonic acid

indicating the number(s) of the carbons immediately preceding the double bond(s). According to this nomenclature a fatty acid with 18 carbons and no double bonds is designated 18:0 or $C_{18:0}$; this is stearic acid (also called octadecanoic acid), a common component of the fats and oils produced by a variety of animals and plants. The designation $18:1(\Delta^{9})$ is oleic acid, the major component of olive oil, with an 18-carbon chain and one double bond immediately after carbon number 9. Arachidonic acid, found in peanut oil, is designated $20:4(\Delta^{5,8,11,14})$. Other common fatty acids are listed in Table 4.1. Although fatty

Table 4.2 Triglycerides.

Fatty acid composition	*Name*
Simple triglycerides	
12:0, 12:0, 12:0	Trilaurin
16:0, 16:0, 16:0	Tripalmitin
18:0, 18:0, 18:0	Tristearin
18:1(Δ^9), 18:1(Δ^9), 18:1(Δ^9)	Triolein
Complex triglyceride	
18:1(Δ^9), 18:1(Δ^9), 16:0	Component of olive oil

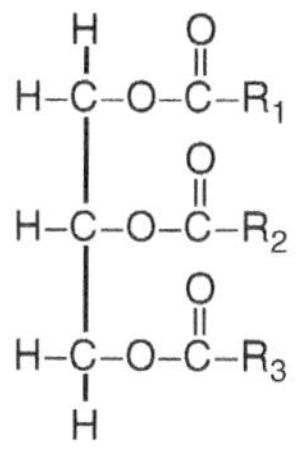

H–C–O–C(=O)–R_1
H–C–O–C(=O)–R_2
H–C–O–C(=O)–R_3

Figure 4.2 The general structure of a triglyceride. Each R group is a fatty acid chain.

acids have the potential for vast diversity, not all possible structures are found in the natural world. Most fatty acids have an even number of carbons, reflecting their biochemical mode of synthesis, which involves linking together two-carbon units. Rarely are there more than four double bonds in the hydrocarbon chain, and there are preferred positions for these bonds, most frequently Δ^9, Δ^{12}, and/or Δ^{15}.

The absence of double bonds means that the hydrocarbon chain of a saturated fatty acid has a linear structure. These linear molecules are able to pack together closely, which means that most of the saturated types have melting points above 40°C and hence are fatty solids at room temperature. The presence of a double bond introduces a kink into the hydrocarbon chain, preventing the molecules of an unsaturated fatty acid from forming such closely associated arrays. Unsaturated fatty acids therefore have lower melting points and most are oily liquids at room temperature.

4.1.2 Fats, oils, soaps, and waxes are derivatives of fatty acids

Most natural fats and oils contain both fatty acids and derivatives of these compounds called **triacylglycerols** or **triglycerides**. A triglyceride consists of three fatty acids that have each formed an ester linkage (–O–) with a single molecule of glycerol (Figure 4.2). In some triglycerides, the three fatty acid chains are identical, examples being tripalmitin which has three 16:0 chains, and triolein containing three 18:1(Δ^9) chains (Table 4.2). These are called **simple triglycerides**. In **complex triglycerides**, the chains are mixed. As with the free fatty acids, fully saturated triglycerides have relatively high melting points and some are fats at room temperature. Those with one or more unsaturated chains are usually oils. Note, however, that fatty acids and triglycerides rarely occur in nature as pure compounds. Most naturally occurring fats and oils are mixtures of different compounds, usually both saturated and unsaturated types, their complexity and ubiquity in our diet providing the bread and butter for the health food industry.

Soaps are derivatives of triglycerides, formed by heating with an alkali such as potassium hydroxide, this process called **saponification**. The treatment breaks the ester linkages and converts the triglyceride back to its component fatty acids, which form salts with the cation of the alkali: potassium salts if potassium hydroxide is used (Figure 4.3A). The presence of the cation increases the hydrophilic properties of the carboxyl end of the fatty

(A) Formation of a soap

Triglyceride ($H_2C(OCOR_1)$–$HC(OCOR_2)$–$H_2C(OCOR_3)$) → Glycerol ($H_2C(OH)$–$HC(OH)$–$H_2C(OH)$) + Soaps ($K^+\ ^-O$–$C(=O)$–R_1, $K^+\ ^-O$–$C(=O)$–R_2, $K^+\ ^-O$–$C(=O)$–R_3)

Triglyceride Glycerol Soaps

(B) A micelle

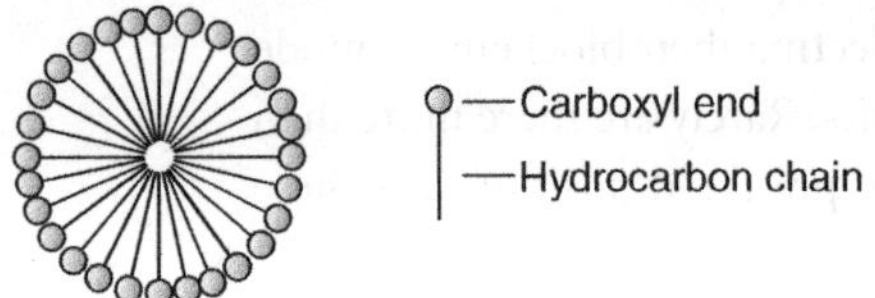

Figure 4.3 Soaps. (A) Formation of a soap by saponification of a triglyceride with potassium hydroxide. (B) A micelle formed by a soap.

CH_3–$(CH_2)_{28}$–CH_2–OH + HO–C(=O)–$(CH_2)_{14}$–CH_3

Triacontanol Palmitic acid

↓

CH_3–$(CH_2)_{28}$–CH_2–O–C(=O)–$(CH_2)_{14}$–CH_3

Triacontanylpalmitate

Figure 4.4 Reaction of palmitic acid with triacontanol to form triacontanylpalmitate, the major component of beeswax. The ester linkage (–C–O–C–) is boxed.

acid, meaning that this end of the structure has an affinity for water, whereas the other end does not. Soaps can therefore form aggregates called **micelles,** spheres with the carboxyl groups on the surface and the hydrocarbon chains embedded within the structure, away from the surrounding water (Figure 4.3B). The cleansing properties of soap are due to their ability to take the water-insoluble compounds that constitute "dirt" out of solution by trapping them within the micelle. Soaps have been made for over 4000 years by mixing various animal fats with carbonates obtained from burnt wood. In our modern society, natural soaps have been supplemented with synthetic detergents in which the fatty acid chain is linked to any one of several types of hydrophilic group, giving a more effective micelle forming activity.

Fatty acids also form products when reacted with long chain alcohol compounds. An alcohol is any compound with the general structure R–CH_2–OH. The simplest alcohol is methanol (H–CH_2–OH) and next most complex is the ethanol of fermented and distilled products (CH_3–CH_2–OH). The alcohols that we are interested in have much longer R groups, such as triacontanol, whose formula is CH_3–$(CH_2)_{28}$–CH_2–OH. Alcohols can form an ester linkage with the carboxyl group of a fatty acid, the ester between triacontanol and palmitic acid (the 16:0 fatty acid, see Table 4.1), being better known as beeswax (Figure 4.4). Waxes generally have higher melting points than fatty acids or triglycerides, most in the range 60–100°C.

4.1.3 Fatty acid derivatives are important components of biological membranes

Certain types of lipid are important integral components of biological membranes, making up some 5–10% of the dry weight of the average cell. The most important of these are the **glycerophospholipids,** also called phosphoglycerides or simply phospholipids, although the last is not a specific name and could be applied to other types of lipids that contain a phosphate group. A glycerophospholipid resembles a triglyceride, but one of the fatty acids is replaced by a hydrophilic group attached to the glycerol moiety by a phosphodiester bond (Figure 4.5). This hydrophilic group is referred to as the "head group" because it is located at the head of the molecule, from which the two fatty acid chains emanate. The simplest glycerophospholipid is phosphatidic acid, in which the head group is a hydrogen atom, but others are more complex, such as phosphatidylcholine, with a choline head group, and phosphatidylserine, in which the head group is a serine amino acid. Quite elaborate structures are known: the glycerophospholipid with a glycerol head group is called phosphatidylglycerol, which can be further modified to give the highly complex glycerophospholipid called cardiolipin, a major component of the inner mitochondrial membrane.

(A) The general structure of a glycerophospholipid

```
   H   O
   |   ||
H–C–O–C–R1
   |   O
   |   ||
H–C–O–C–R2
   |   O
   |   ||
H–C–O–P–O–X
   |   |
   H   O–
```

(B) The structures of important head groups

Head group	Glycerophospholipid
–H	Phosphatidic acid
$–CH_2–CH_2–\overset{+}{N}H_3$	Phosphatidylethanolamine
$–CH_2–CH(COO^-)–\overset{+}{N}H_3$	Phosphatidylcholine
$–CH_2–CH(OH)–CH_2–OH$	Phosphatidylglycerol

Figure 4.5 Glycerophospholipids. (A) The general structure of a glycerophospholipid, with the phosphodiester bond shown in gray. (B) The head groups of four important glycerophospholipids.

The presence of the highly hydrophilic head group attached to the highly hydrophobic fatty acids creates an **amphipathic** molecule with the special properties that underlie the architecture of membranes. Glycerophospholipids attempt to protect their hydrophobic regions from water by aggregating into bilayers, with the fatty acid tails inside the bilayer and the head groups on the upper and lower surfaces (Figure 4.6). This is the basic membrane structure, able to exist within an aqueous environment but forming a barrier across which hydrophilic compounds cannot readily pass.

A second important group of membrane lipids, the **sphingolipids**, have similar shapes to glycerophospholipids and hence form bilayers on their own or when mixed with glycerophospholipids, but they have a different chemical structure. The basic unit of a sphingolipid is sphingosine, a long chain hydrocarbon with an internal hydroxyl group (Figure 4.7). A hydrophilic head group is attached to the last carbon of the chain, and a fatty acid to the second last. The molecule therefore has a hydrophilic head group and two hydrophobic tails, the tails being the fatty acid and the bulk of the sphingosine component. The head group is either a phosphate containing compound such as phosphocholine, a simple sugar such as glucose, or a more complex sugar structure. The ones with the complex sugars are called **gangliosides**, which are present in membranes found in mammalian brains.

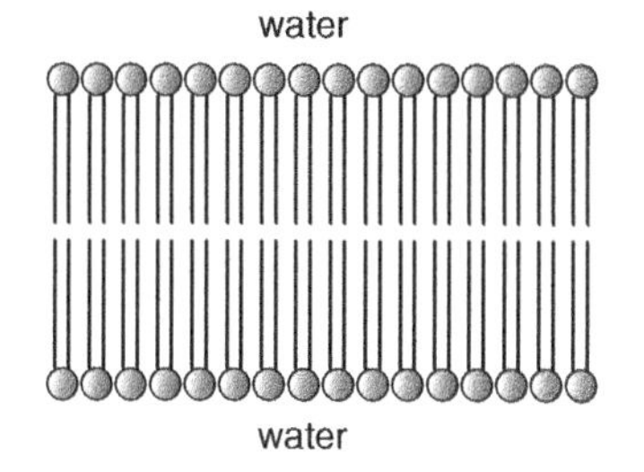

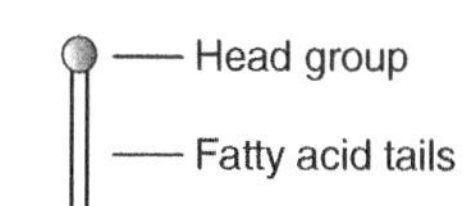

Figure 4.6 The bilayer structure formed by glycerophospholipids.

4.1.4 Terpenes are widespread in the natural world

Now we move away from the fatty acid derivatives to other types of lipid. This leads us to the terpenes, the most variable of all types of natural product with over 25,000 different compounds known. Most of these are made by plants, and many are specific for a single or small groups of species. The resins secreted by trees and other plants are largely composed of terpenes, and these compounds are equally important components of derivatives of resins such as adhesives, varnishes, and some types if incense and perfume. Terpenes are therefore excellent biomarkers that can be used both to identify a species of origin and to detect a product obtained from resin. Many are of proven or potential importance in biomolecular archaeology.

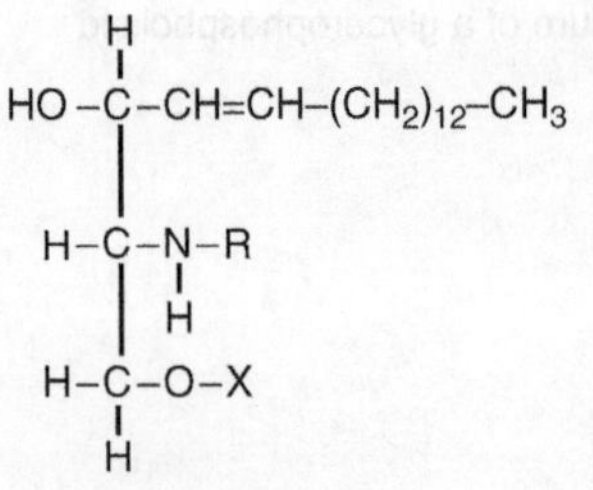

Figure 4.7 The general structure of a sphingolipid. This structure shows sphingosine with an "R" indicating a fatty acid and an "X" a head group.

Terpenes are hugely variable compounds, but are all based on the small hydrocarbon called **isoprene** (Figure 4.8A). Different terpenes are distinguished by the number of isoprene units that they contain, which can be anything from one in the hemiterpenes to hundreds in the polyterpenes. The latter include thick resinous substances such as rubber and gutta-percha. The different chain lengths account for one part of the great diversity of these compounds, but immense additional variability arises from the vast range of structural derivatives that exist for each class of terpene. These modified terpenes are called **terpenoids** and their variabilities are so great that it can be difficult to realize that they are members of the same class of compounds. Consider, for example, a few common monoterpenoids, each consisting of two isoprene units (Figure 4.8B). With myrcene and geraniol, fragrant chemicals obtained from the oils of bay and rose plants, respectively, the structures are relatively simple and the underlying isoprene units are easily identified. This becomes less easy when the derivatization has given rise to a terpenoid with a hydrocarbon ring component, as is the case with carvone from caraway, and terpineol from pine oil. Finally, with compounds such as camphor, from laurel wood, the terpene basis to the structure becomes quite elusive.

Terpenoids are so variable that it is tempting to continue examining more and more complex and elaborate structures *ad infinitum*. Instead, we will focus on just four additional compounds that we will encounter in later chapters because of their importance in biomolecular archaeology. These are the resin terpenoids of pine, spruce, and birch, which were exploited in the past as pitches, tars, and adhesives (Section 14.1). The coniferous resins of pine and spruce are largely composed of diterpenoids, in particular abietic acid and pimaric acid (Figure 4.9). The two compounds are closely related, both comprising three six-membered hydrocarbon rings derived from a four-unit terpene backbone. The resin of birch bark contains betulin and lupeol, triterpenoids containing five-ring structures. As well as its uses in prehistoric technology, betulin and related compounds have clinical applications as anti-inflammatory agents, and may have been used for this purpose by some prehistoric groups.

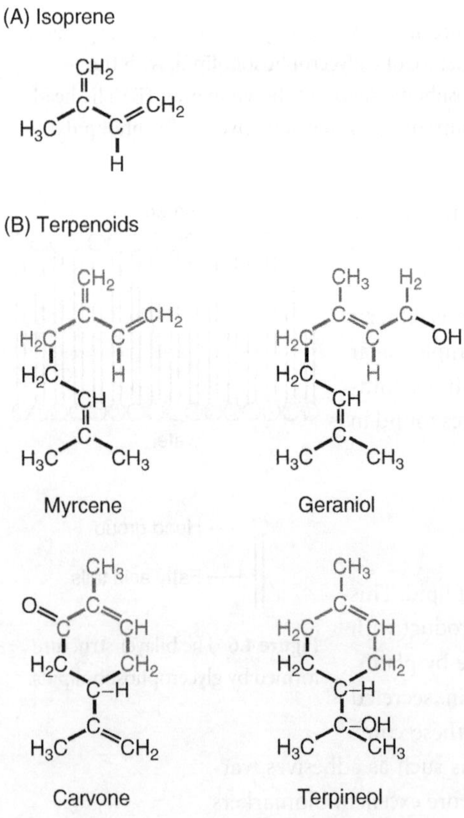

(A) Isoprene

(B) Terpenoids

Myrcene

Geraniol

Carvone

Terpineol

Figure 4.8 Isoprene and terpenoids. In (B), one of the isoprene units is shown in gray.

4.1.5 Sterols are derivatives of terpenes

The complex ring derivatives of terpenes that we have just examined lead us towards the next type of lipid that we must consider. The **sterols** are formed by cyclization of squalene, which is a triterpene comprising six isoprene units – note that squalene is not a *terpenoid*, as it has the basic triterpene structure with no additional chemical modification. The core sterol structure produced by squalene cyclization

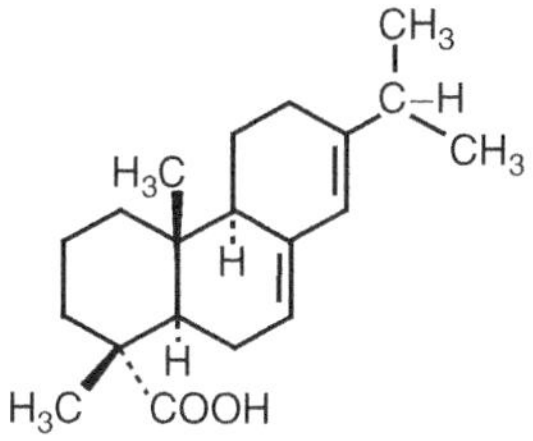

Abietic acid

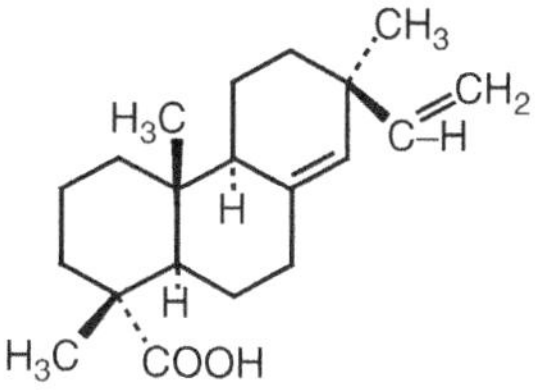

Pimaric acid

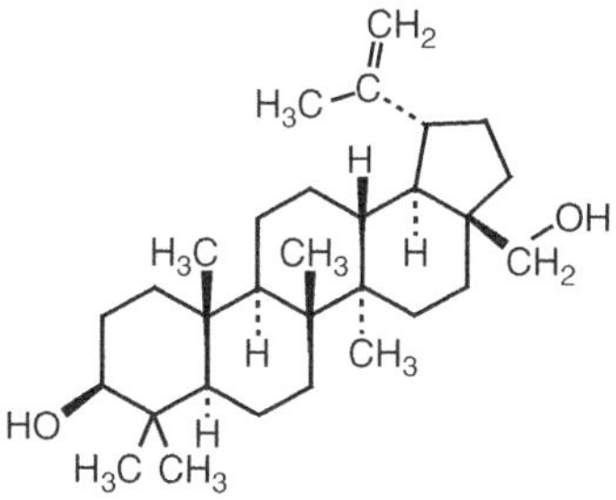

Betulin

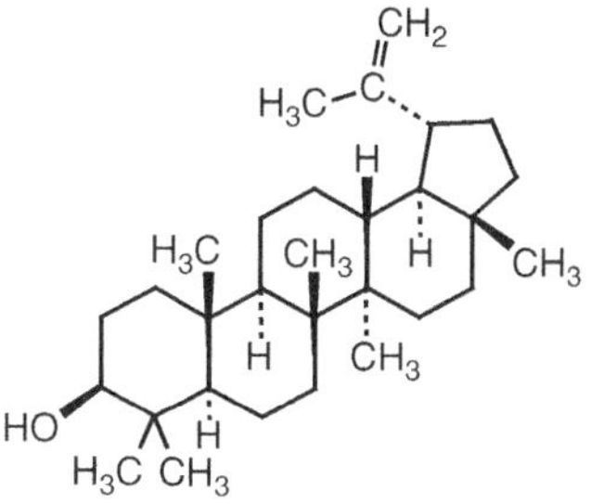

Lupeol

Figure 4.9 Important resin terpenoids. In these drawings, triangular bonds are pointing out of the plane of the paper, and dotted bonds are pointing below the plane.

has four hydrocarbon rings, three of which have six carbons each and one possessing just five carbons.

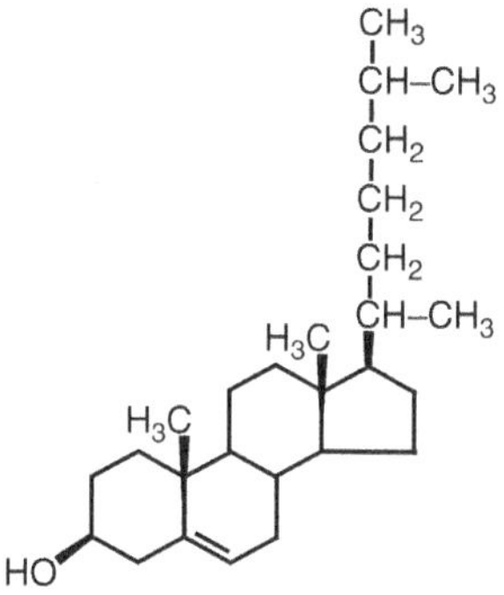

Sterols are another major lipid constituent of cell membranes. Like other membrane components, sterols are amphipathic, having a hydrophilic head group provided by a hydroxyl group (–OH), and, in most cases, a hydrophobic hydrocarbon chain as an R group at the other end of the molecule. They are therefore able to participate with glycerophospholipids and sphingolipids in formation of bilayers. Cholesterol, the best known animal sterol, is a typical example, with an eight-member hydrocarbon R group comprising six carbons in a chain with two attached as short internal branches (Figure 4.10). The equivalent compound in plants is stigmasterol, whose R group is similar in size to that of cholesterol but has a slightly different hydrocarbon configuration. As well as these membrane constituents, some sterols have hydrophilic R groups and are readily soluble in water. These include the **bile acids**, which have side chains that terminate in a carboxylate group (–COOH), the simplest example being cholic acid. Bile acids are synthesized in the liver and secreted into the large intestine, where they help to emulsify fats in the diet and hence aid their breakdown by lipase enzymes. Bile acids are therefore excreted in faeces and can act as biomarkers for the use of manure in land management (Section 8.4).

Figure 4.10 Cholesterol.

The **steroids**, which themselves are another large class of lipids, are sterol derivatives, the basic steroid unit being identical to that of sterols but with another chemical group replacing the hydroxyl. Because this group is variable in steroids, the sterols are, strictly speaking, a subclass of steroids, and the two names are occasionally confused in the literature. The R group possessed by a steroid is usually hydrophilic, and these molecules are water soluble. They include a number of important hormones in humans and other mammals, including the male and female sex hormones, cortisol and aldosterone. Anabolic steroids, notorious in our modern world but not as yet important in biomolecular archaeology, include testosterone and other natural hormones which have roles in the regulation of bone and muscle synthesis.

4.1.6 Tying up the loose ends

Having covered most of the important classes of natural lipids, we will now briefly consider the two final types, both of which have important biological functions but are produced in such small amounts that their recovery from archaeological settings is unlikely.

First, the **eicosanoids** are compounds derived from the $20:4(\Delta^{5,8,11,14})$ fatty acid arachidonic acid. Eicosanoids are synthesized from arachidonic acid molecules released from membrane glycerophospholipids in response to hormone stimulation, and they themselves have hormone-like activity, controlling a number of biological processes, including reproduction and the pain response – the common painkillers aspirin and ibuprofen act by preventing formation of certain types of eicosanoid. They are not true hormones, as they stay within the tissues in which they are synthesized, rather than circulating to distant parts of the body in the bloodstream. Examples are prostaglandin and the thromboxanes.

Vitamins A, D, E, and K are lipids. Vitamins A, E, and K are related to terpenoids, and vitamin D has a steroid structure but with one of the hydrocarbon rings broken open. All four are essential nutritional requirements for vertebrates, vitamin A being a vision pigment, D a hormone precursor, E involved in prevention of oxidative damage within cells, and K a part of the blood clotting system.

4.2 Methods for Studying Ancient Lipids

Over the years, a number of methods have been used to study ancient lipids, but modern research is based around a single approach based on separation of the lipids in a mixture by **gas chromatography** (GC) followed by structural characterization of the purified compounds by **mass spectrometry** (MS). There are variations in the way in which the two parts of the procedure are carried out, and for some applications **liquid chromatography** has proved to be a more suitable alternative for the separation phase. But for virtually all archaeological projects, and indeed studies of modern lipids, this basic approach is effective and accurate.

A number of factors underlie the success of GC-MS in biomolecular archaeology. These include technical issues such as the sensitivity of the procedure, as little as 1 pg of material detectable under the best operating conditions, meaning that the small amounts of lipids obtainable from ceramics and other archaeological materials can be analyzed. Most important though has been the effort put into understanding the degradation products for important archaeological lipids. None of these molecules is fully stable, so the detection of an ancient lipid often depends not on the ability to identify the original compound, but on demonstrating the presence of its breakdown products in an archaeological sample. Without a knowledge of how the lipid degrades, very little effective use could be made of this type of compound in biomolecular archaeology. An example will illustrate the importance of understanding the breakdown products. Mention was made earlier of the coniferous resins abietic acid and pimaric acid, used in the past as pitches and tars. Preparation of these compounds requires heating the wood, which converts them into pyrolysis products even before diagenesis begins to have an effect. A knowledge of the structures of these pyrolysis products enables the original presence of the parent compounds to be inferred, and the identity of the products gives information on the temperatures reached during the heat treatment (Section 14.1). In this case, therefore, the background research on the breakdown pathways has been both essential for identification of the original compounds and valuable as a source of specific information about the past technology that made use of them.

4.2.1 Separating ancient lipids by gas chromatography

Gas chromatography separates lipids according to their differential partitioning between a carrier gas, the **mobile phase**, usually hydrogen or helium, and a liquid **stationary phase** contained within a chromatography column with an internal diameter of 0.1–0.7 mm and overall length of 10–100 m. The liquid phase is chosen for the particular application but can, for example, be polysiloxane if the compounds to be separated are fatty acids. The liquid phase is stationary because it is bonded to a silica coating on the inner surface of the column.

The lipid sample must be volatile in order to be carried by the gas through the column, so extracts are prepared in an organic solvent such as a chloroform and methanol mixture. It is also usually necessary to carry out **derivatization** of the sample prior to analysis. This is a chemical process that modifies hydrophilic and reactive groups in the lipid molecules so that the sample is more readily volatile. Derivatization also reduces the possibility of molecules in the sample reacting with one another, or themselves, during separation. A number of different chemicals can be used for derivatization, a popular one being *N*, *O*-bis(trimethylsilyl)-trifluoroacetamide (BSTFA) which replaces the reactive groups with trimethylsilyl units (Figure 4.11). This treatment is immaterial as far as the eventual identification of the lipids is concerned, because the structures of the BSTFA derivatives of all lipids are known or can be predicted, so the analysis can easily search for these derivatives rather than the original unmodified compounds.

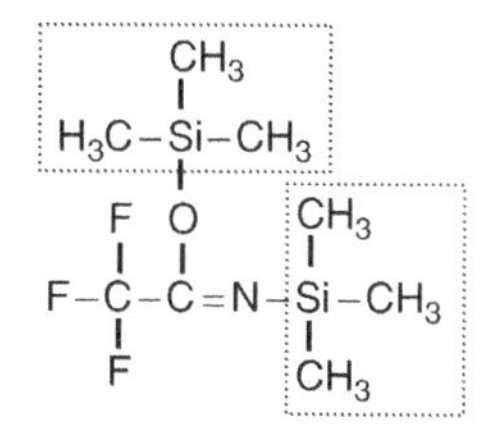

Figure 4.11 BSTFA. The two trimethylsilyl units, which can be transferred to reactive groups on the target molecule during derivatization, are boxed.

The rate at which a lipid passes through the column depends on its **partition coefficient**. This term refers to the relative solubility/volatility of the lipid in the liquid and gas phases of the chromatography system and is loosely related to the boiling point of the compound. Any compound that displays zero solubility in the liquid phase will pass through the column without hindrance, but if the liquid phase has been chosen carefully, then at least those lipids that one is interested in will interact with it to some extent. This interaction takes the form of a continual absorption into the liquid phase, release into the gas phase, and reabsorption back into the liquid phase. For each lipid, the equilibrium between absorption and release is dictated by the compound's partition coefficient. The more the equilibrium is set towards absorption, the slower the

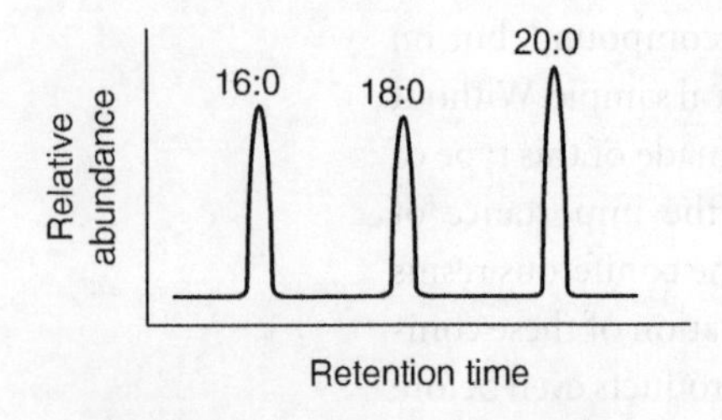

Figure 4.12 An illustration of a gas chromatogram, showing the separation of three fatty acids according to their retention times in the chromatography column.

compound will move through the column. The lipids in the sample are therefore separated according to their partition coefficients and emerge from the end of the column as purified fractions, except of course for any compounds whose partition coefficients are too close for them to be resolved (Figure 4.12).

The separation process must be carried out at an elevated temperature in order for the sample to move through the column at a reasonable rate. The temperature has to be chosen carefully as the degree to which individual compounds are separated depends on elution rate, so too high a temperature might speed the process up to the extent where individual compounds will be poorly resolved. An isothermal separation is carried out by maintaining a set temperature throughout the run, but usually the temperature is programmed to increase from, for example, 40° to 300°C at a rate of 5°C per minute. This means that at the start of the run, when the temperature is relatively low, those compounds that pass most quickly through the column are retained for sufficient time for them to be resolved, and during the later stages, at the higher temperatures, the more slowly moving compounds that otherwise might never be eluted from the column can also be obtained. A greater range of the lipids in the sample can therefore be resolved in a single run.

The peak height ratios in the chromatogram indicate the relative proportions of the individual compounds in the starting mixture, so the amount of each one can be estimated by comparison with internal standards of known concentration. In those situations where there is some understanding of the likely content of the starting material, such as is often the case when a food residue is being examined, the characteristic retention time of different lipids enables individual compounds to be identified in the chromatogram. Alternatively if there is insufficient information to identify individual compounds solely from the GC results, the eluates can be analyzed further by mass spectrometry.

4.2.2 Identifying ancient lipids by mass spectrometry

Mass spectrometry is a means of identifying a compound from the **mass-to-charge ratio** (designated *m/z*) of the ionized form that is produced when molecules of the compound are exposed to a high-energy beam of electrons. Most lipids are naturally uncharged, so exposure to the energy beam results in each molecule losing one electron, to give the molecular ion M^+. As this molecule has a charge of +1, its mass-to-charge ratio is directly proportional to its molecular weight. Exposure to the energy beam also causes some molecules to break up, in predictable ways, to give a series of fragment ions. The *m/z* values of the molecular ion and its daughter fragment ions provide a fingerprint that enables the compound to be identified. A chemical ionization process can also be used, which gives a slightly different molecular ion, $[M+H]^+$, with much less fragmentation. Chemical ionization is used as a complement to electron ionization with those compounds that become completely fragmented during the latter, harsher, treatment. The absence of the molecular ion in these completely fragmented fingerprints can make it difficult to identify the compound, so the $[M+H]^+$ ion is generated by the chemical method to give the missing piece of data.

There are several types of mass spectrometer, differing in the configuration of the mass analyzer, the part of the instrument that separates the molecules according to their *m/z* values (Figure 4.13). In a **magnetic sector mass spectrometer**, the mass analyzer is a single or series of magnets through which the ionized molecules are passed. The magnets are arranged so that each ion must follow a curved trajectory in order to avoid hitting the walls of the analyzer. The degree of deflection of an ion depends on its *m/z* value, so most ions hit the walls and only a few pass through the magnet to the detector at the other end. The magnetic field can therefore be set up specifically to allow ions of a particular *m/z* to reach the detector, or the field can be gradually changed so that molecules of differing *m/z* values can be collected. A **quadrupole** mass analyzer uses a slightly different strategy. This type of instrument has four magnetic rods placed parallel to one another, surrounding a central channel through which the ions must pass. Oscillating electrical fields are applied to the rods, deflecting the ions in a complex way so their trajectories "wiggle" as they pass through the quadrupole. Again, the field can be set so that only ions of the selected *m/z* value can emerge and be detected, or alternatively the field can be gradually changed so all ions are detected.

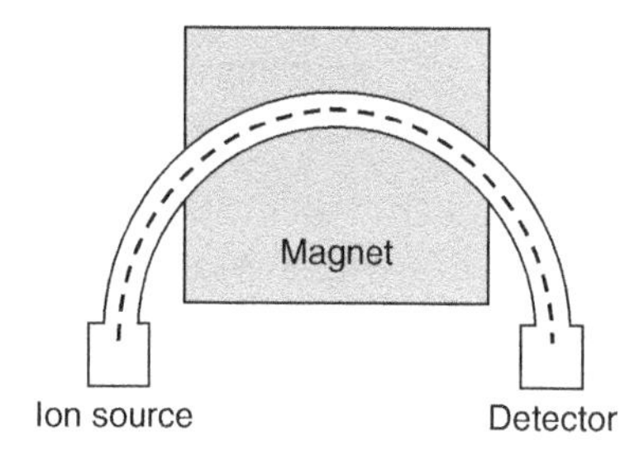

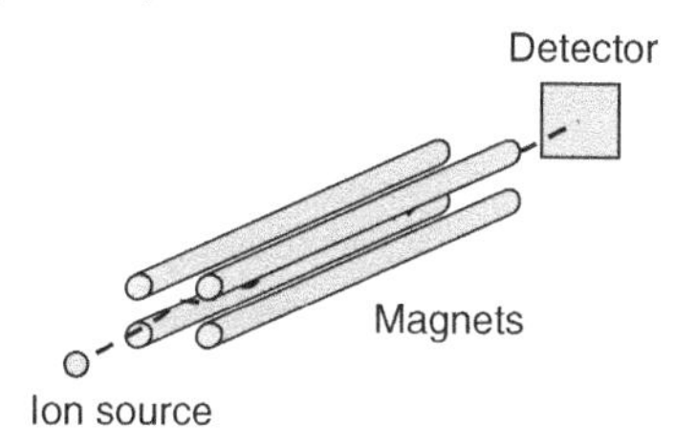

Figure 4.13 Configurations of the magnetic sector and quadrupole mass analyzers.

When coupled to a GC, the mass spectrometer analyzes the individual peaks emerging from the chromatography column, possibly performing several runs per second. Each analysis produces a **mass spectrum** showing the *m/z* values for the molecular and fragment ions derived from an individual peak, along with their relative abundances, this information usually being sufficient to deduce the structure of the starting compound. In this format, the mass spectrometer is being used purely and simply as a tool for compound identification, and we will study several examples of this type of application of GC-MS in later chapters. However, when mixtures of lipids are being studied, the MS component of GC-MS can be used to provide additional information about the individual compounds that are present. Two important applications are:

• The **total ion content** (TIC) is a measure of all the ions (molecular and fragment) produced from a particular GC fraction. The mass spectrometer sums the amounts of the individual ions and plots not a spectrum but a chromatogram showing TIC versus retention time in the GC column. In this mode, the mass spectrometer is acting primarily as a detector, the advantage being that the resulting chromatogram has a much greater degree of sensitivity than that produced simply by GC. Minor components of the starting mixture, which might be missed by GC alone, are therefore detected. In TIC mode, the mass spectrometer still compiles mass spectra for each individual peak, enabling ones of interest to be examined in more detail.

• In **single ion monitoring** (SIM), the mass spectrometer constructs a chromatogram identifying those peaks from the GC that contain ions of a particular *m/z* value. Although called "single" ion monitoring, more than one *m/z* value can be monitored at the same time, so the distribution in the chromatogram of related molecular and fragment ions can be determined. SIM can therefore be used when the objective is to find out if a particular compound is present in the starting mixture, its advantage over TIC being that as just a few ions are being monitored the chromatogram has less background "noise," enabling a more accurate and sensitive identification of the desired peaks.

4.2.3 Modifications to the basic GC-MS methodology

GC-MS methodology is extremely flexible, and a number of variations have been made to the basic formats described above. We end the chapter by looking at the most of important of these.

The first important modification that we will consider concerns the process used to generate the mobile-phase lipid mixture that is entered into the GC. In the conventional methodology, a volatile lipid sample is produced by mixing the extract with an organic solvent. The disadvantage with this approach is that some of the lipids in the starting material might be excluded from the analysis because they are too involatile to enter the mobile phase when treated in this way. This is often the case when natural samples such as soil are being examined. **Pyrolysis GC-MS** (PyGC-MS) helps to solve this problem by adopting a different approach to injection of the sample into the GC. The sample is heated in an oxygen-free atmosphere at a high temperature, typically 610°C for 10 seconds, which causes involatile high-molecular-weight compounds to break down into smaller fragments. These fragments usually have a greater volatility and so enter the mobile phase and pass down the GC in the normal way. The peaks in the resulting chromatogram will, of course, represent the breakdown products of the starting molecules, not those molecules themselves, but identification of the breakdown products is usually sufficient to enable the composition of the original sample to be deduced.

Moving to the other end of the GC-MS procedure, in **tandem MS** (MS/MS), the mass spectrometer has two or more mass analyzers linked in series, so several rounds of MS can be carried out in tandem. Usually each new round of MS involves additional fragmentation of the ions, so further information on the structure of the starting molecule can be obtained. If different types of fragmentation are used (e.g. electron ionization in one stage and chemical ionization in a second stage), then different types of structural information can be gleaned, enabling complex molecules, or members of a closely related family of compounds, to be identified. In many machines, the linked mass analyzers are of different format (e.g. quadrupole followed by sector), giving even greater flexibility to the types of analysis that can be carried out.

The most drastic modification of GC-MS is to replace the GC part with a different separation format. The most common of these is LC-MS, in which separation is achieved by high performance liquid chromatography (HPLC). The underlying basis to HPLC is the same as GC except that the mobile phase is a liquid and the stationary phase a solid, such as silica, which is packed into the chromatography column. HPLC is a better choice than GC if the compounds to be analyzed are liable to break down at the high temperatures used during GC, or if they are very hydrophilic and so likely to pass quickly through a GC column with little resolution. If the compounds are unstable, then the method used to generate the ions immediately prior to the MS stage must be relatively gentle to avoid their complete breakdown. **Electrospray ionization** is a mild ionization procedure that results initially in charged droplets of the solution emerging from the HPLC, these droplets evaporating and transferring their charges to the molecules dissolved within them.

The final modification to GC-MS that is important in biomolecular archaeology is **gas chromatography combustion isotope ratio mass spectrometry** (GC-C-IRMS). This technique is used specifically to measure the stable isotope ratios within a variety of ancient biomolecules, not just lipids. We will therefore postpone our consideration of GC-C-IRMS until we examine the methods used to study stable isotopes in Chapter 6.

Further Reading

Fenn, J.B., Mann, M., Meng, C.K., Wong, S.F. & Whitehouse, C.M. (1989) Electrospray ionization for mass spectrometry of large biomolecules. *Science*, 246, 64–71.

McLafferty, F.W. (1981) Tandem mass spectrometry. *Science*, 214, 280–7.

McNair, H.M. & Miller, J.M. (2009) *Basic Gas Chromatography*, 2nd edn. Wiley, Hoboken, N.J.

Pond, C.M. (1998) *The Fats of Life*. Cambridge University Press, Cambridge.

Regert, M., Garnier, N., Descavallas, O., Cren-Olivé, C. & Rolando, C. (2003) Structural characterization of lipid constituents from natural substances preserved in archaeological environments. *Measurement Science and Technology*, 14, 1620–30.

Sparkman, O.D. & Penton, Z. (2010) *Gas Chromatography and Mass Spectrometry: A Practical Guide*. Academic Press, London.

5

Carbohydrates

Carbohydrates are the most difficult of the biomolecules studied by biomolecular archaeologists. Some carbohydrates are extremely resistant to degradation, the modified polysaccharide called lignocellulose being one of the most stable of all natural products, but those that do survive in the archaeological record have relatively little variability, so the information that can be obtained from them is limited. The methods used for their study are also less well developed than those for DNA, proteins, and lipids. Ancient carbohydrates do, however, have one very important application in biomolecular archaeology. **Starch grains** are synthesized in the cells of many types of plants and can be recovered both from sediments and from the surfaces of implements used for processing plants. The morphologies and other features of these grains can sometimes be used to identify the species of origin, though account must be taken of the changes that grains might undergo both during processing of the plant tissue and subsequently during diagenesis. If a secure identification can be made, then starch grains can act as biomarkers for plant species, both for inferring that a species was present at an archaeological site and also in understanding which plants were processed during food production. Frequently, starch grains are studied alongside **phytoliths**, microscopic deposits of silica or calcium oxalate found in many plant cells and, like starch grains, sometimes recovered from archaeological sites. Fossil pollen might also be present. The combined information obtained from these three types of **plant microfossil** can often enable a species to be identified more accurately than is possible if just one type of material is studied.

In this chapter, we will examine the chemical structure of carbohydrates, with the emphasis very much on starch, and then examine the methods used to study starch grains.

 © 2011 Terry Brown & Keri Brown

Table 5.1 Examples of monosaccharides.

Name	*Number of carbons*	*Description*
Erythrose	4	Precursor for synthesis of some amino acids
Ribose	5	Component of RNA
Ribulose	5	Precursor of glucose in photosynthesis
Glucose	6	Main product of photosynthesis
Galactose	6	Found in dairy products and sugar beet
Fructose	6	Common in fruits and vegetables

5.1 The Structure of Carbohydrates

Starch is a **polysaccharide**, a polymeric carbohydrate made up of **monosaccharide** units. Therefore, to understand the structure of starch and other polysaccharides we must first study the structures of monosaccharides.

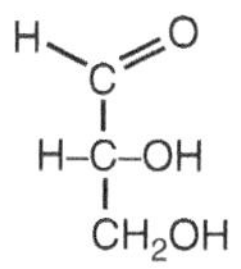

H–C=O
H–C–OH
CH_2OH

Figure 5.1 The flat structure of glyceraldehyde.

5.1.1 There are left and right handed versions of each monosaccharide

Monosaccharides include a number of familiar compounds, such as ribose and 2′-deoxyribose, which we met in Chapter 2 as components of RNA and DNA nucleotides, and glucose, which is the main product of photosynthesis and the primary energy source for most living organisms (Table 5.1). One of the simplest monosaccharides is glyceraldehyde, which is synthesized in many cells as an intermediate in the metabolism of other carbohydrates. Glyceraldehyde comprises a central carbon atom attached to –H, –OH, –CHO and $–CH_2OH$ groups (Figure 5.1). The –CHO is called a carbonyl group, the characteristic feature of those compounds called **aldehydes**. Glyceraldehyde is therefore an aldehyde sugar or **aldose**, or more specifically an **aldotriose**, as it has three carbon atoms.

Although shown as a flat structure in Figure 5.1, in reality glyceraldehyde has a three-dimensional configuration. To understand this configuration we must first consider the way in which chemical bonds are orientated around a carbon atom. Carbon has a valency of four, meaning that it can form four single bonds. These bonds have a tetrahedral arrangement, as shown in Figure 5.2A. This means that there are two versions of glyceraldehyde, differing in the positioning of the four groups around the carbon (Figure 5.2B). It is not possible to go from one to the other simply by rotating the molecule – they are mirror images and so genuinely different. The two versions are called dextro- or D-glyceraldehyde and laevo- or L-glyceraldehyde, meaning, respectively, right and left handed glyceraldehyde, the names deriving from the ability of solutions of these compounds to rotate the plane of plane- polarized light in different

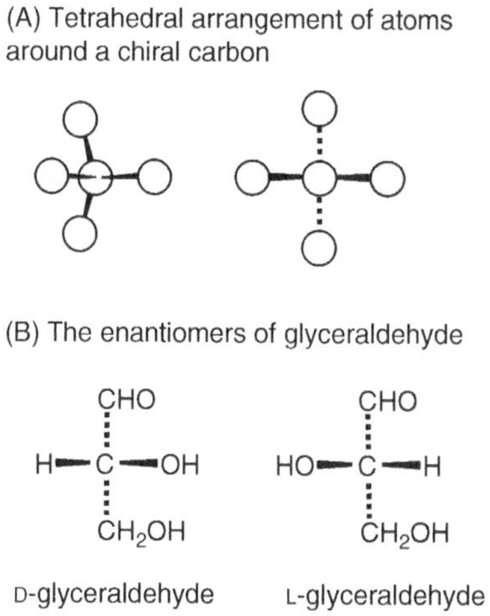

(A) Tetrahedral arrangement of atoms around a chiral carbon

(B) The enantiomers of glyceraldehyde

CHO
H–C–OH
CH_2OH
D-glyceraldehyde

CHO
HO–C–H
CH_2OH
L-glyceraldehyde

Figure 5.2 (A) The arrangement of atoms around a central carbon. The drawing on the right shows the convention for indicating the orientations of the bonds in a tetrahedral structure of this kind. (B) The two enantiomers of glyceraldehyde.

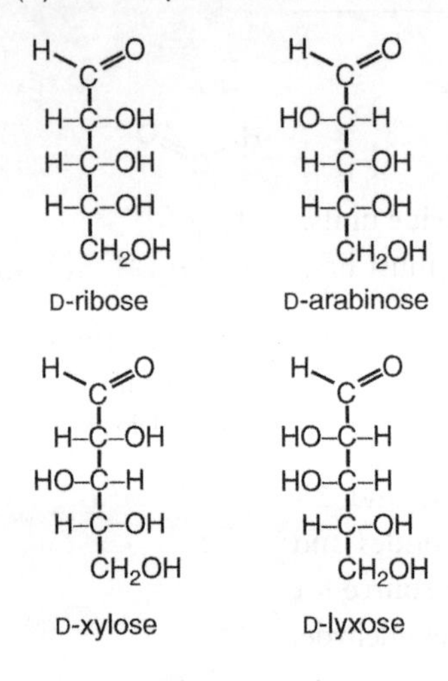

Figure 5.3 The D-enantiomers of the aldotetroses and aldopentoses.

directions. Because of this feature, a carbon atom that is attached to four different groups, as in glyceraldehyde, is said to be **chiral**, from the Greek for "hand." The D- and L- forms of glyceraldehyde differ only in their effect on plane-polarized light: they have identical chemical properties and are, in essence, the same compound. We refer to them as **enantiomers.**

Now we will move up a level and consider the **aldotetroses**, the aldehyde sugars that contain four carbon atoms. These have two chiral carbons, which means that there are four possible configurations, comprising two pairs of enantiomers. The compounds are called D- and L-erythrose and D- and L-threose (Figure 5.3A). Note that the structures of erythrose and threose are not themselves mirror images, as the relative positioning of the hydroxyl groups is different. Erythrose and threose are therefore different compounds with distinct chemical properties. Technically, they are **diastereomers.**

The five-carbon **aldopentoses** (Figure 5.3B) each have three chiral carbons, giving four pairs of enantiomers. These include three sugars that are common in nature – ribose, present in nucleotides, arabinose, and xylose. The eight **aldohexose** enantiomer pairs include glucose, mannose, and galactose (Figure 5.4). There is also a different series of monosaccharides, in which the oxygen atom is attached to carbon number 2 rather than carbon number 1. The resulting C=O group is characteristic of a **ketone**, so these are the ketone sugars or **ketoses.** The simplest of these is the four-carbon ketose called erythrulose, which has one chiral carbon and so exists as D- and L-forms. One of the diastereomers of the six-carbon ketose family is fructose (Figure 5.5), which, like glucose, is an important dietary sugar obtained from many fruits and vegetables.

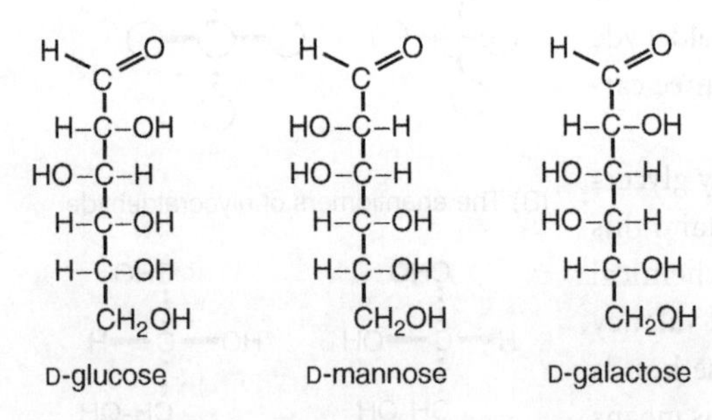

Figure 5.4 Three naturally occurring aldohexoses.

5.1.2 Some monosaccharides also exist as ring structures

As well as the linear structures that we have considered so far, those monosaccharides with five or more carbons can also form ring or cyclic molecules. We are already aware of this because the form of ribose present in nucleotides is a ring molecule (Figure 2.1), not the linear chain shown in Figure 5.3B. The cyclic version of ribose is formed by reaction between the carbonyl group at carbon number 1 with the hydroxyl at carbon 4 (Figure 5.6). A similar reaction can occur for the aldohexoses such as glucose. The cyclic form of an aldohexose is called a **pyranose** because it has a structural similarity with an unrelated organic compound called pyran. The specific chemical name for the ring form of glucose is therefore glucopyranose.

Formation of the ring structure provides further opportunities for creating variations in the structure of a monosaccharide. The cyclic form of D-glucose, for example, has α and β versions which differ in the positioning of the hydroxyl group attached to carbon 1, the carbon originally present in the carbonyl group of the linear form and which participated in the chemical reaction that led to cyclization (Figure 5.7). The two structures, α-D-glucopyranose and β-D-glucopyranose, have slightly different optical properties, but otherwise are chemically identical. They are called **anomers**. The α form can readily convert into β, and vice versa, so solutions of D-glucose contain a mixture of the two types, usually with a small amount of the linear form also present.

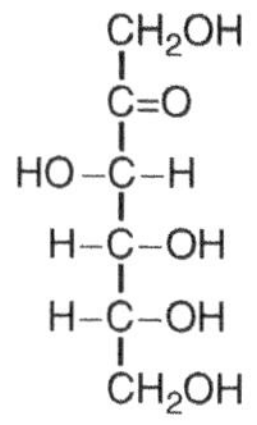

The five- and six-carbon ketose monosaccharides can also form ring structures. Reaction is between the ketone group at carbon 2 and the hydroxyl at carbon 5 to give a five-membered **furanose** ring. The cyclic derivative of fructose is therefore called fructofuranose. As with glucopyranose, there are α and β anomeric forms.

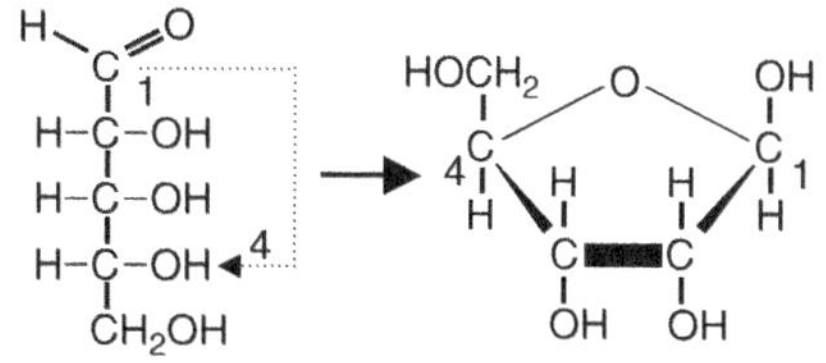

D-fructose

Figure 5.5 Fructose.

Figure 5.6 Conversion of the linear form of ribose into its ring structure.

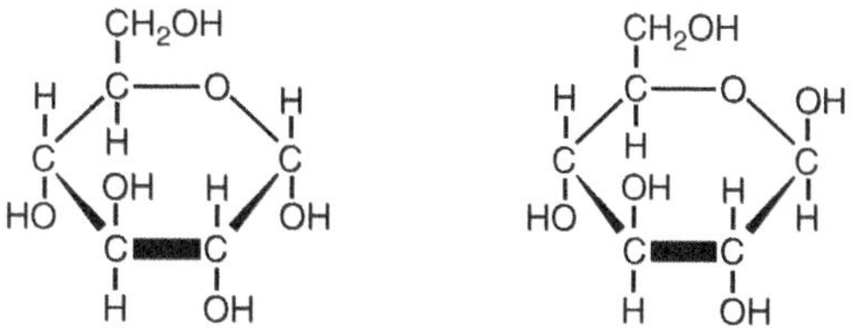

α-D-glucopyranose

β-D-glucopyranose

Figure 5.7 The anomers of D-glucose.

5.1.3 Disaccharides are made by linking together pairs of monosaccharides

Now we move on to the way in which individual monosaccharide units are linked together to form longer chain carbohydrates. The simplest of these are the **disaccharides**, which comprise two joined monosaccharide units. Some disaccharides are very common in nature (Table 5.2), including sucrose, which is the type of sugar obtained from sugar cane or beet, which we put in our coffee, and which is made up of glucose and fructose units. Lactose, from milk, has glucose and galactose units, and maltose from malt barley has two glucose units.

The link between the two monosaccharide units in a disaccharide is called an ***O*-glycosidic bond**. This type of bond is formed between pairs of hydroxyl groups, one on each monosaccharide, so a great deal of variability is possibile. In maltose, which contains two glucose units, the link is between the hydroxyls attached to carbon 1 of one glucopyranose and carbon 4 of the second unit (Figure 5.8). The link is therefore denoted as "1→4." Carbon 1 is the anomeric carbon, so we must also distinguish if the bond involves the α or β version. In maltose, it is α, so the correct chemical name of this disaccharide is α-D-glucopyranosyl-(1→4)-D-glucopyranose.

Table 5.2 Examples of disaccharides.

Name	*Component sugars*	*Description*
Sucrose	Glucose + fructose	From sugar cane and sugar beet
Lactose	Glucose + galactose	Milk sugar
Maltose	Glucose + glucose	Malt sugar, from germinating cereals
Trehalose	Glucose + glucose	Made by plants and fungi
Cellobiose	Glucose + glucose	Breakdown product of cellulose

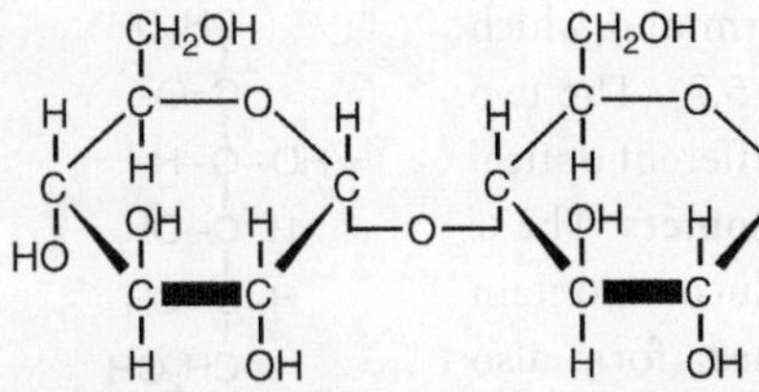

Figure 5.8 Maltose.

In maltose, the second glucopyranose is free to interconvert between its α and β anomers, as its carbon number 1 is not involved in the glycosidic bond. This is not always the case. Trehalose, for example, is also made up of two glucose units, but with this disaccharide the bond is between the pair of number 1 carbons. Trehalose is therefore α-D-glucopyranosyl-(1→1)-α-D-glucopyranose. Of course, the two monosaccharide units do not always have to be glucose. Sucrose is α-D-glucopyranosyl-(1→2)-β-D-fructofuranose, the link being between the α version of the anomeric carbon 1 of glucose and β version of carbon 2 (the anomeric carbon) of fructose. Lactose is β-D-galactopyranosyl-(1→4)-D-glucopyranose.

5.1.4 Polysaccharides are long chain carbohydrates

Polysaccharides are made up of many cyclic monosaccharide units linked by glycosidic bonds. The chains can be linear or branched, and the monosaccharide units can be identical or mixed. If all the same, then the compound is a **homopolysaccharide**; and if mixed, then it is a **heteropolysaccharide**.

Starch is a homopolysaccharide made entirely of D-glucose units. There are two types of starch molecule, called amylose and amylopectin. The difference between the two is that amylose is a linear polymer of D-glucose units linked by (α1→4) glycosidic bonds, whereas amylopectin has a branched structure made up of (α1→4) chains and (α1→6) branch points, the branches occurring every 24–30 units along each linear chain (Figure 5.9). All plants synthesize both amylose and amylopectin, with the latter usually being the predominant form. Unlike the other biomolecules we have studied, starch molecules are variable in size, amylose chains containing anything from a few hundred to 2500 glucose units, and amylopectin spanning a similar range but with a higher upper limit, possibly 6000 units in the largest molecules. The factors that determine the sizes of individual molecules are not well understood. The polysaccharide chains can take up various conformations, the most stable of which gives

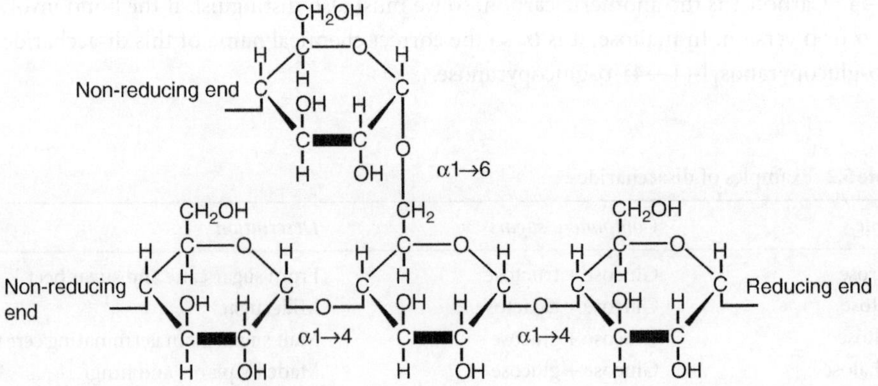

Figure 5.9 The structure of the branch point in a starch molecule.

the (α1$\rightarrow$4) chain a fairly tight curvature. Amylose and amylopectin therefore have compact coiled structures that enable the molecules to pack tightly within starch grains.

Starch is a storage polysaccharide, the monosaccharide units being utilized for energy generation by cleaving them from the ends of the amylose and amylopectin molecules. Only the "non-reducing" ends are cut back, these being the ends that terminate in the non-anomeric carbon number 4 (Figure 5.9). Amylose, being a linear chain, has one reducing and one non-reducing end, but in amylopectin the branches give rise to a molecule with many non-reducing ends that can be utilized concurrently. The non-reducing ends are attacked by amylase and other enzymes that release individual glucose molecules, possibly along with maltose disaccharides and maltotriose – the trisaccharide comprising three glucose units linked by (α1$\rightarrow$4) bonds.

Starch is by no means the only polysaccharide made by plants. Cellulose is also a linear D-glucose homopolysaccharide, but with (β1$\rightarrow$4) linkages. This subtle difference compared with amylose results in a molecule with completely different properties. The most stable conformation of cellulose is a straight chain, rather than the coli of the amylose molecule, with individual cellulose chains being able to line up side by side and attach to one another by hydrogen bonding. This gives rise to the rigid networks that play an important role in plant cell walls. Animals also produce a number of important polysaccharides, including the storage carbohydrate glycogen, which has a similar structure to amylopectin but with more frequent branch points.

5.2 Studying Starch Grains

Glycobiologists – those biochemists who specialize in the study of carbohydrates – have devised sophisticated methods for analysis of polysaccharides, in particular for determining the sequences of the monosaccharide units in complex chains such as those present in glycoproteins (see Section 3.2). We need not concern ourselves with any of these high technology methods because, at present, the only carbohydrate that is relevant in biomolecular archaeology is starch, and we are lucky in being able to use relatively simple methods to characterize the various types of starch deposits found in archaeological contexts. These deposits are in the form of starch grains that are synthesized within plant tissues and which can, under some circumstances, be used as biomarkers for the plants that make them. We will look first at the process by which these grains are synthesized and then at how they are studied.

5.2.1 Starch grains are stores of energy produced by photosynthesis

Photosynthesis is the process by which plant cells utilize the energy contained in sunlight to convert carbon dioxide and water into carbohydrates. The energy now contained in the carbohydrate molecules can subsequently be released by breakdown of the carbohydrates via the linked pathways of glycolysis, the citric acid cycle and oxidative phosphorylation, which together constitute the process of aerobic respiration. The released energy can then be used to drive other cellular processes, either within the plant itself or within the cells of herbivores that eat the plants.

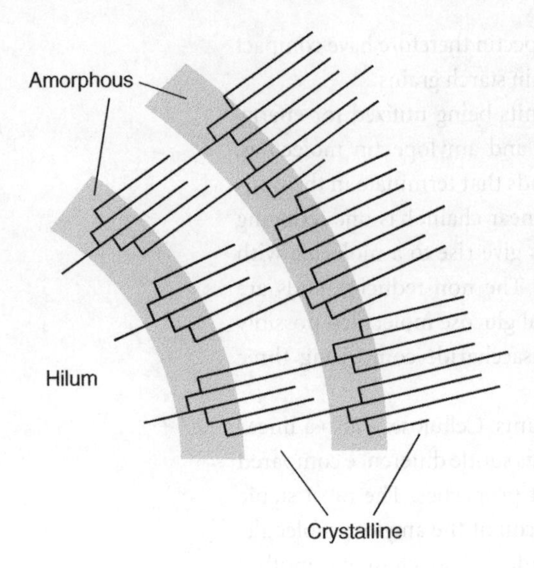

Figure 5.10 The amorphous and crystalline layers within a starch grain.

The primary product of photosynthesis is glucose, some of which is used immediately to provide the energy needs of the leaves and other metabolically active parts of the plant. The remainder is converted into stored energy in the form of starch, which is synthesized by polymerizing the glucose molecules into amylose and amylopectin. Two distinct types of starch formation take place in plants:

- **Transient** or **transitory starch synthesis** occurs in the chloroplasts, the organelles that carry out photosynthesis. These reserves are short lived, most being used up to supply energy for the plant during the following period of darkness.
- **Stored starch synthesis** gives rise to the long-lived grains that are sometimes recovered from archaeological contexts. Stored starch is not made in the chloroplasts but in related structures called **amyloplasts** which are present in roots, developing seeds, and storage organs such as fruits and tubers. The monosaccharides used to produce stored starch are therefore transported to these tissues from the sites of photosynthesis within the leaves and other aerial parts of the plant.

Within an amyloplast, starch synthesis initiates at a formation centre called the **hilum** and radiates out as layer is placed upon layer. Both amylose and amylopectin are made, but the latter predominates in most species. Each amylopectin molecule is placed in the developing grain with its single reducing end pointing in towards the hilum, and its multiple non-reducing ends pointing outwards. As the grain grows, a series of concentric layers is formed, made up of alternating amorphous and crystalline regions, the amorphous layers largely containing the branch points of adjacent amylopectin molecules, and the crystalline component built up of parallel unbranched chains (Figure 5.10). The density of packing of the amylopectin molecules in the individual layers depends on the water content of the grain, which in turn is influenced by factors such as the degree to which the amylopectin molecules are modified by addition of chemical groups such as phosphates.

Within individual amylopectin molecules, pairs of unbranched glucose chains wind around one another to form a double helical structure with a pitch of one turn for every six glucose units. The ordered association between adjacent helices gives rise to the crystalline layers. At least two different types of helix packing occur in different species: in potato, for example, the helices take up a relatively open arrangement, but in many cereals they are more densely packed. In legumes such as peas and beans, the arrangement appears to be intermediate between these two extremes.

5.2.2 Starch grains can be used as biomarkers to distinguish between different groups of plants

The variations in starch grain structure noted above lead us to the most important aspect of these grains as far as biomolecular archaeology is concerned. The starch grains synthesized by different species have characteristic features that are usually sufficient to enable the

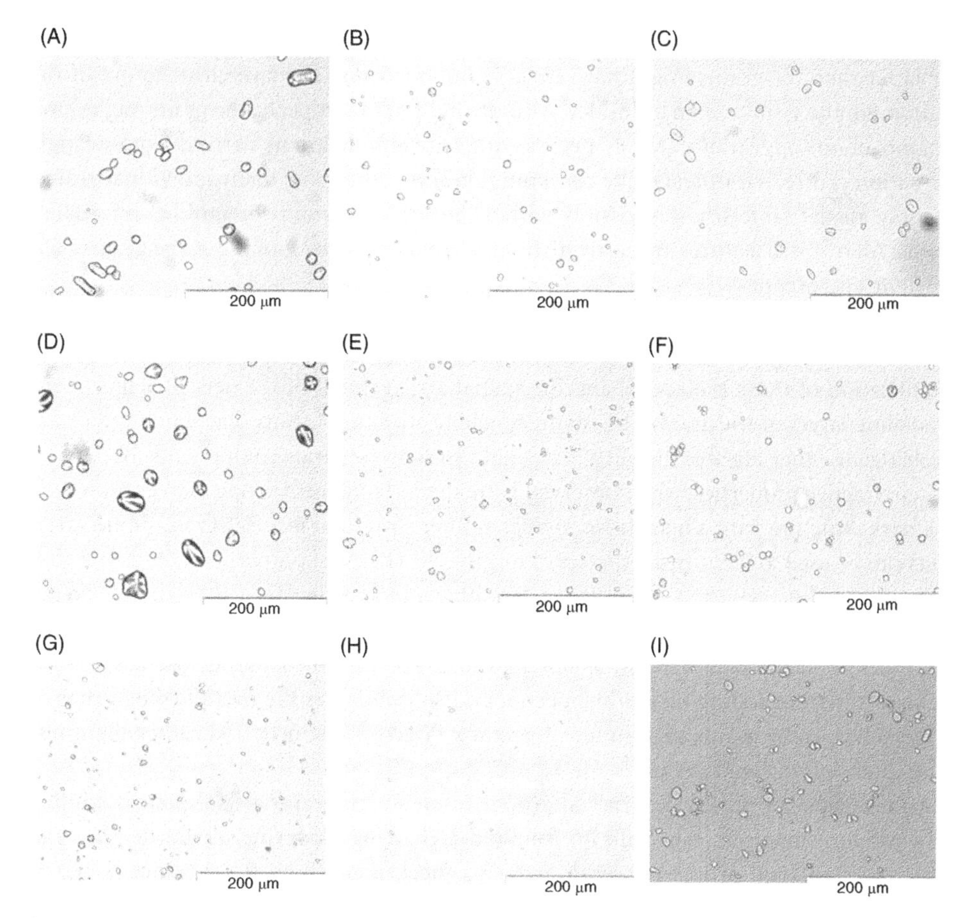

Figure 5.11 Typical starch grains. (A) Lotus root, (B) maize, (C) mung bean, (D) potato, (E) water chestnut, (F) cassava, (G) sweet potato, (H) oats and (I) plantain. Reprinted from Julie Wilson, Karen Hardy, Richard Allen, Les Copeland, Richard Wrangham, and Matthew Collins, "Automated classification of starch granules using supervised pattern recognition of morphological properties," *Journal of Archaeological Science*, 37, 594–604, copyright 2010, with permission from Elsevier and the authors.

genus of origin to be identified and may, in some cases, be sufficiently specific to allow an individual species to be pinpointed. Starch grains can therefore be used as biomarkers for the plants that make them.

Botanists in the 19th century were aware of the variations between starch grains from different types of plant, and over the years a number of schemes have been devised for the classification of these grains. In modern botany, these taxonomic schemes make use of the size and shape of the grains, and their optical and chemical properties. Sizes vary over tenfold from less than 10 to over 150 μm in diameter. In cultivated maize, the range is approximately 5–25 μm, in wheat and barley it is 30–45 μm, and in potato 30–50 μm and larger. Shapes are equally variable (Figure 5.11), from smooth and round through ovoid and pear-shaped to more complex angular structures. In some species, the grains are flat discs and in others small spheres. The central hilum might be elongated, or absent, and if the latter then the grains might have sizeable hollow cores. The layered structure of the grain might be visible when examined with the light microscope, or the grain might have an amorphous appearance. Most species synthesize simple grains derived from a single formation centre, but others make compound grains comprising many individual grains, possibly all the same size, or possibly with a central large grain surrounded by smaller grains. All of these features are easily distinguishable simply by looking at the grains with the light microscope.

The optical and chemical properties of different types of grain reflect variations in the density of packing of the amylopectin molecules within the grain, the degree of chemical modification of those molecules, and the spatial arrangement of the helical chains in the crystalline layers of the grain. The semi-crystalline structure means that starch grains are birefringent – they are able to split a ray of light into two separate components. Because of these refractive properties, when observed with a polarizing microscope each grain is seen as a dark structure with a light cross, giving another structural feature whose subtle variations can be used to identify the species of origin. The chemical properties include the ability to bind iodine, which stains amylose dark blue and amylopectin light red. The overall appearance of a stained starch grain therefore depends on the relative proportions of these two types of starch, although not as much as might be imagined, as amylose has a much higher affinity for iodine than amylopectin, and so dominates the overall color reaction. Nonetheless, some grains high in amylopectin will give a distinctly red rather than blue color after staining with iodine.

Although the optical and chemical properties of starch grains are frequently studied with modern material, especially to follow the changes occurring during baking and other types of food processing, with archaeological material greater reliance is placed on size and shape as a means of distinguishing grains from different species. In order to use size and shape to classify archaeological grains, it is first necessary to establish the amount of variation present in equivalent modern material, to provide reference data against which the archaeological specimens can be compared. In some cases, the archaeological questions demand that closely related species be distinguished, one notable example being in studies of the origin of maize cultivation, in which it is necessary to distinguish the remains of cultivated maize from those of its wild progenitor teosinte. Starch grains could only be used for this purpose if their sizes and/or morphologies include distinctive features that separate maize from teosinte. Studies of modern plants have shown that starch grains in maize are, on average, larger than those in teosinte, reflecting the higher starch content of cultivated maize, the mean lengths being 8.8–9.5 μm for most types of teosinte and 11.1–15.8 μm for maize. This suggests a clear distinction between the two, but problems emerge when we consider the range of sizes in the two species, which is 4–18 μm for teosinte and 4–26 μm for maize. Similar complications arise when the shape and internal and surface features of the grains are examined. For the majority of these features there are no significant differences between teosinte and maize grains, and those features that display differences between the two are not diagnostic. Maize grains more frequently have an "irregular" shape compared with the rounder grains of teosinte, and maize grains have a greater proportion of deeper and more clearly defined compression facets, but these distinctions are not absolute. The conclusion is that it is possible to use starch grains to distinguish teosinte and maize, but only if sufficient numbers of grains are available for an accurate estimate of the size range and morphological features of the assemblage. If only a few grains are available, as might often be the case with archaeological material, the analysis will be much less precise.

5.2.3 A brief word on phytoliths and fossil pollen

One of the strengths of plant microfossil research is that we are not dependent solely on study of starch grains in order to make a tentative identification of the presence of a particular plant in an archaeological setting. Starch grains are frequently studied alongside phytoliths and fossil pollen in order to obtain complementary lines of evidence relating to plant identification. Although these two types of microfossil are not, strictly speaking, ancient biomolecules, we should, before concluding this chapter, briefly understand how they are used in archaeological science.

Phytoliths are microscopic inorganic deposits, usually of silica, which are formed in plant cells if the groundwater being taken up by the roots contains dissolved silicon compounds. The silica contained in the water is deposited around the periphery of the cells, forming structures with characteristic shapes and sizes. It is debatable if phytoliths have any function within the plant, though it possible that they provide cells with some degree of structural rigidity. When the plant decomposes, the phytoliths are deposited in the soil and some have sufficient stability to survive in the archaeological record. As with starch grains, they have been recovered not just from sediments but also from the surfaces of cooking and food processing implements. The shapes and sizes of phytoliths have some degree of species specificity and so can be used as biomarkers for different types of plant. The same considerations as those that we have discussed with starch grains apply also to phytoliths: the size and morphological features of phytoliths from related species might overlap, and accurate identification of fossil phytoliths might not be possible unless substantial numbers are available.

Pollen is the plant's male reproductive cell and is produced by all seed plants. Even those plants that are self-fertilizing release a certain amount of pollen, and out-breeders might release huge quantities as windborne particles, or particles designed to adhere to pollinating insects. Pollen grains are highly resistant to decay and survive in paleontological as well as archaeological settings. The surfaces of pollen grains are exquisitely sculpted, so that individual types can be recognized, although among some types of plant, notably grasses (which include the cereals important in agriculture) the extent of variation between species is relatively low. When pollen from teosinte and maize are compared, for example, both size and morphological features overlap to such an extent that the two species cannot be distinguished. Pollen studies, have, however, been valuable in tracing the spread of maize cultivation into areas of the New World where teosinte is not found, and has been used to follow forest clearance and other changes in vegetation patterns during prehistory. Recently, techniques for extracting ancient DNA from fossil pollen have been developed, opening up a new avenue of plant microfossil research.

Further Reading

BeMiller, J. & Whistler, R. (eds.) (2009) *Starch Chemistry and Technology*, 3rd edn. Elsevier, Burlington.

Buléon, A., Colonna, P., Planchot, V. & Ball, S. (1998) Starch granules: structure and biosynthesis. *International Journal of Biological Macromolecules*, 23, 85–112.

Davis, B.G. & Fairbanks, A.J. (2002) *Carbohydrate Chemistry*. Oxford University Press, Oxford.

Holst, I., Moreno, J.E. & Piperno, D.R. (2007) Identification of teosinte, maize, and *Tripsacum* in Mesoamerica by using pollen, starch grains, and phytoliths. *Proceedings of the National Academy of Sciences USA*, 104, 17608–13. [Describes the morphologies and sizes of maize and teosinte starch grains.]

Lindhorst, T.K. (2007) *Essentials of Carbohydrate Chemistry and Biochemistry*. Wiley VCH, Weinheim.

Piperno, D. (2006) *Phytoliths: A Comprehensive Guide for Archaeologists and Paleoecologists*. AltaMira Press, Lanham.

Smith, A.M., Denyer, K. & Martin, C. (1997) The synthesis of the starch granule. *Annual Review of Plant Physiology and Plant Molecular Biology*, 48, 67–87.

6

Stable Isotopes

The analysis of stable isotopes of carbon, nitrogen and, to a lesser extent, oxygen, forms an important component of biomolecular archaeology. These isotopes are not themselves biomolecules, but are contained within biomolecules, and their analysis is conducted with bulk protein or lipid preparations from archaeological materials, and also with individual compounds purified from archaeological extracts. The relative amounts of the different isotopes of a particular element can be used to study certain aspects of the diet of past peoples. A diet rich in marine resources can be distinguished from a diet largely made up of terrestrial animal protein, and the utilization of maize as a foodstuff can be distinguished because this plant has a different carbon isotope ratio to many other cereals and vegetables. The potential of stable isotopes in archaeology was first established back in 1977, by studies of carbon isotope ratios in bone collagen from native American skeletons, which showed that maize was not a significant component of the diet of the inhabitants of the northeast United States until after 100 BC.

Several examples of the use of stable isotopes in archaeology will be discussed later in this book, especially in Chapter 12 when we will look specifically at how biomolecular archaeology is used to study past diet. To prepare the ground for those discussions we must, in this chapter, examine the chemical and biological basis to the variations in stable isotope ratios that underlie their utility in biomolecular archaeology, and investigate the methods used to measure those variations in archaeological materials.

6.1 Isotopes and Isotopic Fractionation

The existence of alternative forms of the same element was first appreciated by the radiochemist Frederick Soddy, the name **isotopes** being coined by the Scottish doctor, feminist, and romantic author Margaret Todd in 1913. The fact that not all isotopes are radioactive, some being **stable isotopes**, was discovered in the same year by J.J. Thomson.

 © 2011 Terry Brown & Keri Brown

Table 6.1 Naturally occurring isotopes of elements relevant to biomolecular archaeology.

Element	*Isotope mass numbers*	*Proportion in nature*	*Stability*
Carbon	12	98.93%	Stable
	13	1.07%	Stable
	14	one part per trillion	Half-life of 5730 ± 40 years
Nitrogen	14	99.64%	Stable
	15	0.36%	Stable
Oxygen	16	99.76%	Stable
	17	0.04%	Stable
	18	0.20%	Stable
Strontium	84	0.56%	Stable
	86	9.86%	Stable
	87	7.00%	Stable
	88	82.58%	Stable

6.1.1 Isotopes are different versions of a single element

An atom consists of a nucleus containing positively charged protons and neutral neutrons, surrounded by a cloud of negatively charged electrons. The chemical identity of the element is determined by its **atomic number**, which is the same as the number of protons. This number is invariant for all atoms of that particular element. Every hydrogen atom has just a single proton and an atomic number of 1, every carbon atom has six protons and an atomic number of 6, and so on.

Although the number of protons is invariant, different atoms of the same element can have different numbers of neutrons. These different versions of an element are called isotopes. Carbon, for example, has three naturally occurring isotopes, each containing six protons but with six, seven and eight neutrons, respectively (Table 6.1). The number of protons and neutrons in a nucleus is called the **mass number**, so the three isotopes of carbon have mass numbers of 12, 13, and 14, and are described as carbon-12, carbon-13, and carbon-14, or ^{12}C, ^{13}C, and ^{14}C. These are the isotopes of carbon that are found in nature: there are also 12 isotopes from ^{8}C to ^{22}C that do not exist in measurable amounts in the environment but that can be created under laboratory conditions. The majority of elements, but not all, have naturally occurring isotopes, the largest numbers being nine for xenon and ten for tin.

Some isotopes are radioactive and decay with emission of subatomic particles at a characteristic rate described as the **half-life.** Carbon-14 is an example of a radioisotope, decaying with a half-life of 5730 ± 40 years. Carbon-14 is of immense importance in archaeological science as it is the basis to radiocarbon dating, but this type of dating is not a part of biomolecular archaeology and we do not need to concern ourselves with the details of radioactive decay. The isotopes that interest us are those that are stable and do not undergo decay, such as ^{12}C and ^{13}C. Stable isotopes coexist in nature, but not in equal amounts. Usually the lightest isotope is highly predominant and the other(s) present in trace amounts. This is the case with the stable carbon isotopes, ^{12}C comprising

98.93% of all the carbon atoms in existence, and ^{13}C making up just 1.07%. Carbon-14 is present at only one part per trillion.

6.1.2 Isotope fractionation can change the relative amounts of the stable isotopes of a particular element

Although all of the isotopes of an element have identical chemical properties, their different masses influence the precise ways in which they behave during physical and chemical processes. For the lighter elements such as carbon, these mass differences are not trivial – ^{13}C is 8.3% heavier than ^{12}C – and can have a significant effect on the behavior of different isotopes. These effects can lead to **isotope fractionation**, which results in a change in the relative proportions of the isotopes in the products of a reaction compared to the proportions in the initial substrates.

There are two major types of isotope fractionation. Equilibrium fractionation results from the differential exchange of isotopes between two physical phases that are in equilibrium with one another. This process is less important in biomolecular archaeology but is responsible for establishing the temperature-dependent oxygen isotope ratios in ice, which form the basis to paleoclimate studies aimed at following temperature fluctuations in the past. The second process, the one that is important in biomolecular archaeology, is **kinetic fractionation**, which occurs during a unidirectional physical or chemical reaction, and usually involves preferential reaction of lighter isotopes compared with heavier ones. The reaction products therefore become enriched for the lighter isotopes. Examples of kinetic fractionation that we will meet later in the chapter include the evaporation of water to form clouds, and many enzymatic reactions in living cells.

The difference between the stable isotope contents of environmental and biological fractions is usually a few parts per thousand. The notation used to describe and to calculate this difference is as follows:

$$\delta^{13}C = \left(\frac{(^{13}C/^{12}C)_{sample}}{(^{13}C/^{12}C)_{standard}} - 1 \right) \times 1000‰$$

A positive $\delta^{13}C$ value therefore means that the ^{13}C content of the sample is enriched compared with the standard or, conversely, the ^{12}C is depleted, and a negative $\delta^{13}C$ indicates a depletion in ^{13}C and enrichment in ^{12}C. For this notation to be of general utility there clearly needs to be universal agreement on the identity of the standard against which sample results will be compared. For carbon the standard is a marine limestone called Peedee belemnite (PDB), which has now been used up in its original form, and replaced by other standards whose $\delta^{13}C$ values relative to PDB are known, such as a special graphite preparation supplied by the US National Bureau of Standards. Oxygen ratios can also be measured against PDB, or alternatively with "standard mean ocean water" (SMOW). Atmospheric air is usually used as the standard for nitrogen isotope measurements.

6.2 Carbon and Nitrogen Isotope Fractionations Enable Past Human Diets to be Studied

For biomolecular archaeologists, the most important isotope fractionations occurring in nature are those that enable the types of food eaten by humans in the past to be studied. This work currently depends almost exclusively on analysis of the stable carbon isotopes ^{12}C and ^{13}C, and those of nitrogen, ^{14}N and ^{15}N.

6.2.1 Carbon fractionation in the biosphere enables the presence of maize in the diet to be identified

The principles behind the use of stable isotopes to study diet are illustrated by the use of $\delta^{13}C$ measurements taken from human skeletons to determine when maize was first used as a foodstuff in different parts of the New World. This application was mentioned above as forming the very first demonstration of the potential of stable isotope research in biomolecular archaeology, and will be discussed again in Section 12.3 when we look at the way biomolecular archaeology has subsequently been used to study diet in the Americas.

Identification of a maize diet is possible because of the carbon isotope fractionations that occur during photosynthesis, leading to enrichment in the ^{12}C isotope. This occurs in either one or two stages, depending on the type of photosynthetic pathway operating in a particular plant species (Figure 6.1). In all plants, there is an initial enrichment during the entry of atmospheric carbon dioxide into the plant, as the lighter isotope diffuses more rapidly through the microscopic pores on the outer surfaces of the leaves. In those plants that use the C_3 photosynthetic pathway, there is a subsequent enrichment during the photosynthesis process itself, as carbon dioxide molecules containing ^{12}C are converted more readily into glucose by the reaction catalyzed by ribulose-1,5-bisphosphate carboxylase (Rubisco). Because of these two ^{12}C enrichments, the $\delta^{13}C$ values of the tissues of C_3 plants are lower than the equivalent value for atmospheric carbon dioxide: $\delta^{13}C$ for atmospheric carbon dioxide is approximately −8‰ but within C_3 plant tissues the mean $\delta^{13}C$ is −26.5‰ with a range of −24 to −36‰, the precise value depending on environmental conditions such as temperature, humidity, day length, and intensity of sunlight. C_3 plants are found naturally in temperate regions of the world such as Europe, Asia, and North America, and include wheat, barley, rice, potato, and many other important crops.

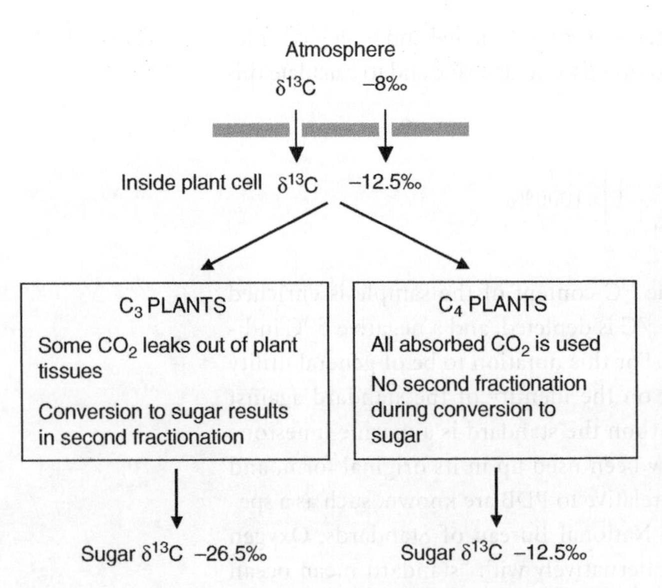

Figure 6.1 The differences between the carbon isotope fractionations occurring during photosynthesis in C_3 and C_4 plants.

Maize, on the other hand, uses the C_4 photosynthetic pathway, which is confined to a relatively small number of tropical species (sorghum and millet are other examples) and which plant biologists believe enable these plants to fix carbon dioxide more efficiently in the arid environments in which they grow. One feature of the C_4 pathway is that the carbon dioxide absorbed by the plant becomes concentrated in special cells prior to its use by Rubisco. This means that most if not all of the absorbed carbon dioxide is eventually converted into sugar, so although the molecules containing ^{12}C still pass more readily through the Rubisco reaction, this does not result in enrichment of the lighter isotope because, in the end, the enzyme uses all the carbon dioxide that is available. In a C_3 plant, a part of the carbon dioxide that is absorbed by the leaves is lost again by leakage before being converted into sugar, so the more rapid utilization of ^{12}C by Rubisco leads to an actual isotope fractionation. The second fractionation step is therefore absent in C_4 plants. Their $\delta^{13}C$ values are still lower than atmospheric carbon dioxide – because of the fractionation occurring during absorption – but are not as low as the values for C_3 plants, with a mean of −12.5‰.

The distinctions between the isotope fractionations occurring in C_3 and C_4 plants is important because the $\delta^{13}C$ value of a human or animal is a reflection of the $\delta^{13}C$ values of the food that is consumed. A diet rich in maize can therefore be distinguished from one rich in C_3 plants by measuring the $\delta^{13}C$ values in human skeletons. This is precisely what was done by J.C. Vogel and N.J. van der Merwe in their study of human skeletons from the American northeast, published in 1977. They showed that the $\delta^{13}C$ values in skeletons from the Archaic (2500–2000 BC) and Early Woodland (400–100 BC) periods were consistent with the consumption of C_3 plants native to that region, but that Late Woodland (AD 1000–1300) and more recent skeletons had higher $\delta^{13}C$ values, consistent with the introduction of maize as a major part of the diet between 100 BC and AD 1000.

6.2.2 Carbon and nitrogen isotopes enable a marine diet to be distinguished

Photosynthesis in the oceans is carried out largely by organisms such as marine algae and phytoplankton, and is mostly via the C_3 pathway. Carbon isotopes undergo a two-stage fractionation process during photosynthesis in these organisms, exactly the same as in terrestrial C_3 plants. The starting point for photosynthesis in the oceans is not, however, atmospheric carbon dioxide, but dissolved bicarbonate, which is enriched in ^{13}C compared with the atmosphere and so has a more positive $\delta^{13}C$ value. The tissues of marine photosynthesizers therefore have higher $\delta^{13}C$ values than are found in terrestrial C_3 plants, the mean being approximately −20‰.

Other than seaweed, humans do not eat marine algae or phytoplankton, so how does the characteristic $\delta^{13}C$ value of these types of plant relate to biomolecular archaeology? The link is provided by the fish and marine invertebrates that are sometimes eaten extensively by those human populations living close to the sea. Many of these marine animals feed directly on the photosynthesizers, so their tissues take on $\delta^{13}C$ values close to those of the primary marine producers, usually around −16‰. The $\delta^{13}C$ values of humans whose diet has a substantial marine input reflects the values found in fish and shellfish, rather than those present in terrestrial vegetation and the animals that feed on that vegetation. A marine diet therefore leaves a distinctive $\delta^{13}C$ signature in a preserved skeleton.

The $\delta^{13}C$ value of marine animals, at −16‰, is close to that of maize and other terrestrial C_4 plants, which typically are close to −12.5‰. Bearing in mind that there can be quite a lot of variation around these means, there is a possibility of confusing a marine diet with one

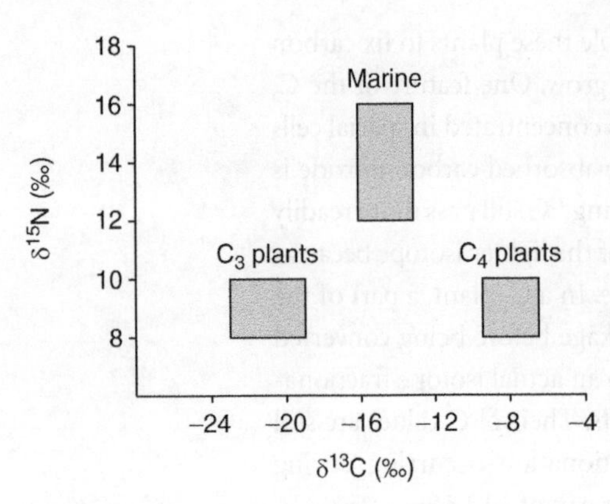

Figure 6.2 A $\delta^{13}C/\delta^{15}N$ plot showing the areas of the graph in which bone collagen values should fall for different types of diet.

rich in maize, and vice versa. Fortunately, measurements of nitrogen isotope ratios provide complementary information that gives greater precision to these dietary studies. Nitrogen has two stable isotopes, ^{14}N and ^{15}N, with the former predominant in the biosphere, making up 99.64% of all nitrogen atoms. Atmospheric nitrogen provides the standard for ^{14}N and ^{15}N measurements and so, by definition, has a $\delta^{15}N$ of 0 ‰. Nitrogen isotope fractionation in the biosphere is influenced in a complicated way by the balance between the microbial fixation of atmospheric nitrogen into inorganic and organic compounds, and its release again by denitrification. The important point is that the $\delta^{15}N$ values in marine environments are slightly more positive that those in terrestrial soils, mainly because a greater amount of denitrification occurs in the oceans. Coastal marine region have $\delta^{15}N$ values in the range 5–6 ‰, compared with 1–4 ‰ for terrestrial soils.

All terrestrial plants grown in a single area acquire similar $\delta^{15}N$ values, reflecting the value of the soil from which they acquire their fixed nitrogen. C_3 and C_4 plants are therefore indistinguishable when $\delta^{15}N$ is measured. Marine resources, on the other hand, have distinct $\delta^{15}N$ values compared with terrestrial plants. By plotting figures for $\delta^{13}C$ obtained from one or more skeletons against the $\delta^{15}N$ values for the same skeletons, clusters of points emerge, with C_3, C_4, and marine diets falling in different regions of the graph, enabling a clearer discrimination between the input of maize compared with marine animals (Figure 6.2). Note also that the $\delta^{15}N$ values of freshwaters are similar, though not identical, to those of the oceans, whereas freshwater $\delta^{13}C$ is closer to the terrestrial rather than marine figure (because photosynthesis in rivers is driven by atmospheric rather than dissolved carbon dioxide). It is therefore also possible to distinguish a diet rich in river fish from the $\delta^{13}C/\delta^{15}N$ plot.

6.2.3 Carbon and nitrogen isotope measurements enable carnivores to be distinguished from herbivores

Feeding experiments have shown that an animal's $\delta^{13}C$ and $\delta^{15}N$ values reflect the isotope ratios in the foods that have been eaten by that animal, but only when the animal as a whole is considered. Within the body, different tissues display different relationships with the dietary input. In large herbivorous mammals, for example, muscle proteins have a $\delta^{13}C$ value some 3 ‰ higher than the $\delta^{13}C$ content of the diet, and bone collagen is 2 ‰ higher again. Hence, if a large herbivore is grazing on C_3 plants, its diet will have a $\delta^{13}C$ of about −26.5 ‰, but its muscles will give a figure of −23.5 ‰ and it bone collagen a value of −21.5 ‰. The basis to this ^{13}C enrichment when going from dietary carbon to body carbon is not understood, but presumably resides in the way in which dietary amino acids are converted into protein. Similar effects are seen with nitrogen isotopes, the shift in $\delta^{15}N$ being +2–6 ‰ from diet to body protein, probably due to the excretion of nitrogen in urea, which is depleted in ^{15}N.

The critical point is that this isotope shift continues as one proceeds along a food chain. When a carnivore feeds on the muscle protein of a herbivore it ingests material with a $\delta^{13}C$

of approximately −23.5 ‰ (presuming the herbivore was consuming C_3 plants), and converts this into bone collagen and its own muscle protein with $\delta^{13}C$ values some 1–2 ‰ higher. Should a top carnivore consume that animal, then a similar shift will occur within its own tissues. The equivalent stepwise shift in $\delta^{15}N$ values is greater, around 3–4 ‰. Whether an animal is a herbivore or a carnivore can therefore be determined by measuring its $\delta^{13}C$ and $\delta^{15}N$ values (Figure 6.3).

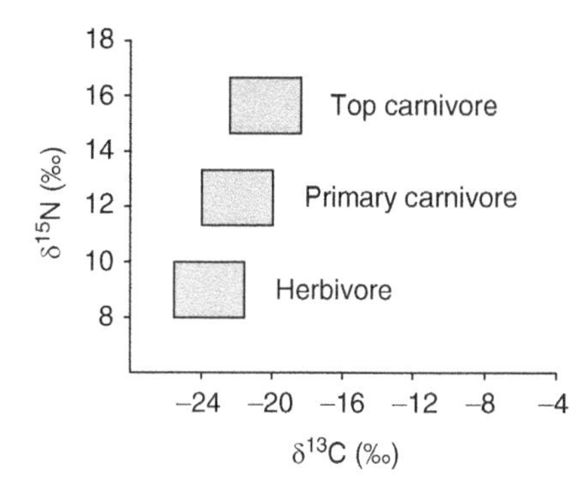

Figure 6.3 A $\delta^{13}C/\delta^{15}N$ plot showing the isotope shifts in bone collagen values that occur along a food chain.

Humans, of course, are omnivorous – we eat both plants and animals, but not always in equal proportions. Modern-day vegans have characteristic $\delta^{13}C$ and $\delta^{15}N$ values (determined from measurements made from hair keratin) reflecting their purely "herbivorous" diet. An increasing amount of meat in the human diet results in an equivalent increase in $\delta^{13}C$ and $\delta^{15}N$ values. Most land animals consumed by humans are primary rather than top carnivores, so the shift in isotope values relative to a plant diet is relatively small, but it is still sufficient to enable isotopic measurements taken from archaeological skeletons to distinguish individuals whose diets had a high meat content from those more dependent on plants, especially when $\delta^{15}N$ values are compared.

The shift in isotope ratios that occur along a food chain has one further, interesting application in biomolecular archaeology. When infants drink their mother's milk, they are, in effect, consuming a part of their parent. The infant is therefore a trophic level further along the food chain than the mother, and the infant's $\delta^{13}C$ and $\delta^{15}N$ values show an equivalent shift compared with the normal human figures. This shift quickly disappears once the infant is weaned and begins to consume regular foodstuffs including plants. The skeletons of infants who died while still being weaned can therefore be identified by $\delta^{13}C$ and $\delta^{15}N$ measurements, providing information on the length of the nursing period in past societies.

6.2.4 Strontium and oxygen isotopes can give information on human mobility

Stable isotope studies are not only used to study past human diets. They can also provide information on mobility, by identifying individuals whose isotopic compositions do not match that of the area in which their skeletons are found, and who are likely therefore to have migrated during their lifetime to the place at which they died. Strictly speaking, this type of research is not *biomolecular* archaeology, as the isotope ratios are measured not in biomolecules but in the inorganic fractions of bones and teeth, but these studies are increasingly carried out alongside genuine biomolecular examination of human and animal skeletons, and for this reason we will consider them briefly here.

Strontium isotopes are currently the most widely used in studies of human mobility. There are four natural isotopes of strontium, ^{84}Sr, ^{86}Sr, ^{87}Sr, and ^{88}Sr. The most abundant of these is ^{88}Sr, which makes up 82.58% of the total. All four of the isotopes are stable, but ^{87}Sr is formed by the radioactive decay of rubidium-87, and so increases in amount, very slowly as the half-life of ^{87}Rb is 4.9×10^{10} years. The amounts of ^{84}Sr, ^{86}Sr, and ^{88}Sr in geological formations are therefore fairly constant throughout the world, but the proportion of ^{87}Sr depends on the amount of ^{87}Rb that the rock contained when it was formed, and the age of the rock, which determines how much of that ^{87}Rb has decayed into ^{87}Sr. These geological variations are usually expressed as the $^{87}Sr/^{86}Sr$ ratio, which varies from 0.720 to 0.703 for terrestrial geosystems, the smaller values being consistent with the youngest volcanic formations.

Strontium has similar chemical properties to calcium and becomes incorporated, in small amounts, in calcium-containing structures such as the inorganic fractions of bones and teeth. Teeth are formed during childhood, their $^{87}Sr/^{86}Sr$ ratios reflecting the $^{87}Sr/^{86}Sr$ ratio of the geological region in which the child is living. After they are formed, there is no further exchange between the strontium in a tooth and that in the environment, so an adult skeleton whose teeth have $^{87}Sr/^{86}Sr$ ratios different to that of the geology underlying their place of death must have moved to that place since their infancy. Such movement should also be evident from comparisons between the $^{87}Sr/^{86}Sr$ ratios in the teeth and bones of a skeleton, as bones are continuously reformed over an individual's lifetime, which means that their $^{87}Sr/^{86}Sr$ content can change if the person moves to a region with a different geological signature.

Strontium isotope measurements can therefore identify individuals who have migrated into a new area during their lifetime, and has been particularly informative as a means of identifying incomers within a group of skeletons buried together (Section 11.3). Further information on the details of human migration can be obtained by comparing data from teeth and bones with geological data on $^{87}Sr/^{86}Sr$ ratios in different areas, obtained either by direct examination of rocks or by measuring the ^{87}Sr and ^{86}Sr contents of small animals indigenous to those regions. In some parts of the world, geological $^{87}Sr/^{86}Sr$ varies across quite limited geographical areas, so movements over relatively short distances can be detected. These studies are by no means restricted to questions of human mobility, and are also being applied to domestic animals in order to understand whether these were grazed close to or some distance from human settlements.

Similar information can be obtained from other elements whose isotopes have geographical variations and that become incorporated into bones and/or teeth, either as natural components or as environmental contaminants. Lead is one example of the latter category. After strontium, however, oxygen isotopes are the most widely used for this purpose. Oxygen has three stable isotopes, ^{16}O (by far the most abundant), ^{17}O, and ^{18}O. The $^{18}O/^{16}O$ ratio in rainwater varies in different regions according to latitude, altitude, and distance from the sea, and therefore displays geographical patterning, not as detailed as geological $^{87}Sr/^{86}Sr$ variations, but still sufficient to be of some value in studies of past human mobility. For this to be possible, the $^{18}O/^{16}O$ ratios in skeletal components such as the phosphates and carbonates in bones and tooth enamel must reflect the environmental values. This is the case because the oxygen present in these inorganic compounds is derived from the oxygen dissolved in the body's water content, which in turn is derived from ingested water and food. Analysis of $^{18}O/^{16}O$ ratios in bones and teeth can therefore be used in the same way as $^{87}Sr/^{86}Sr$ studies to identify individuals who have moved during their lifetimes. The oxygen isotope approach becomes less secure the further one goes back in time, because environmental hydrology is not unchanging, which means that the $^{18}O/^{16}O$ value for rainfall in a particular region in the past might not be the same as it is today, but for archaeological studies a reasonable degree of consistency between past and present can be assumed.

6.3 Practical Aspects of Stable Isotope Studies

Now we must examine the methodology used to obtain the stable isotope ratios that underlie the dietary and mobility studies outlined above.

Stable isotopes of carbon and nitrogen in bulk materials such as collagen and bioapatite are quantified by **isotope ratio mass spectrometry** (**IRMS**), the slight mass differences between, for example, ^{12}C and ^{13}C being sufficient to give the ionized versions of these isotopes different mass-to-charge ratios, enabling them to be separated in the mass spectrometer (Section 4.2). A magnetic sector mass spectrometer, rather than the quadrupole system, is generally used, and the ionized isotopes are carried through the machine in a stream of helium gas. The starting material must first be converted to pure carbon dioxide and nitrogen, which is achieved by combustion at 1020°C in a chamber containing copper, nickel, and platinum wires, followed by passage over a second set of copper wires at 600°C to reduce residual nitrogen oxides to nitrogen. The carbon dioxide and nitrogen are then subjected to electron ionization before passage through the mass spectrometer. Ions of the different isotopes deflect to different extents in the magnetic sector, so each one can be directed to a different **Faraday collector**, which catches the ions, generating a current which is measured in order to quantify the amount of each isotope in the starting sample (Figure 6.4).

When used for single compound studies, the basic IRMS format must be modified by addition of a preliminary gas chromatography stage in order to separate the individual compounds within what might be a quite complex starting mixture. The purified compounds that emerge from the GC sequentially enter the combustion chamber of the IRMS machine, which means that the isotopic composition of each can be individually measured.

Oxygen and strontium isotope ratios are also measured by mass spectrometry. Oxygen isotopes are released from carbonates by acid digestion, and from phosphates by pyrolysis, and then analyzed by IRMS. Strontium isotopes are ionized either by heating on a metal filament (**thermal ionization mass spectrometry, TIMS**) or by plasma exposure (**plasma ionization mass spectrometry, PIMS**; and **inductively coupled plasma mass spectrometry, ICP-MS**). Currently, a major difficulty is the need to use chemical methods to purify the strontium from the sample prior to ionization. With geological material that is relatively rich in strontium, a laser beam is commonly used to take samples, as this procedure – called **laser ablation** – generates ions directly and so obviates the need for chemical purification. Laser ablation is increasingly used with archaeological material such as teeth. It is ideal for taking measurements from individual enamel layers, and also has the benefit that the tooth is not entirely destroyed in the process. Laser ablation can be coupled directly with IRMS, so the material released from the tooth is directly fed into the analyzer.

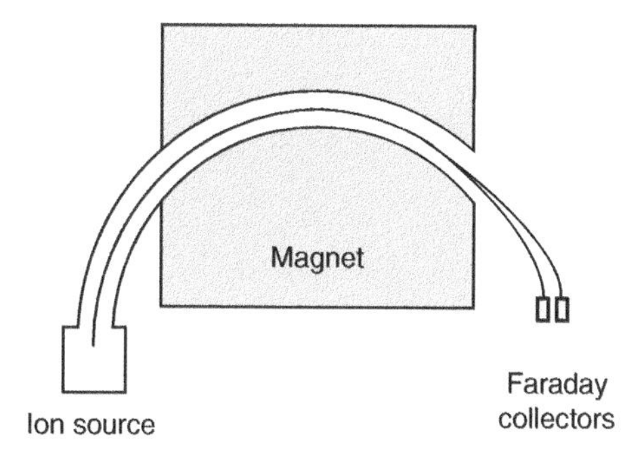

Figure 6.4 Isotope ratio mass spectrometry.

Further Reading

Bentley, R.A. (2006) Strontium isotopes from the earth to the archaeological skeleton: a review. *Journal of Archaeological Method and Theory*, 13, 135–87.

Bocherens, H. & Drucker, D. (2003) Trophic level isotopic enrichment of carbon and nitrogen in bone collagen: case studies from recent and ancient terrestrial ecosystems. *International Journal of Osteoarchaeology*, 13, 46–53.

Copeland, S.R., Sponheimer, M. & le Roux, P.J., *et al.* (2008) Strontium isotope ratios (87Sr/86Sr) of tooth enamel: a comparison of solution and laser ablation multicollector inductively coupled plasma mass spectrometry methods. *Rapid Communications in Mass Spectrometry*, 22, 3187–94.

Hedges, R.E.M. & Reynard, L.M. (2007) Nitrogen isotopes and the trophic level of humans in archaeology. *Journal of Archaeological Science*, 34, 1240–51.

Lee-Thorp, J.A. (2008) On isotopes and old bones. *Archaeometry*, 50, 925–50.

Schoeninger, M.J. & Moore, K. (1992) Bone stable isotope studies in archaeology. *Journal of World Prehistory*, 6, 247–96.

Vogel, J.C. & van der Merwe, N.J. (1977) Isotopic evidence for early maize cultivation in New York State. *American Antiquity*, 42, 238–42. [One of the earliest applications of stable isotope analysis in archaeology.]

PART II

PRESERVATION AND DECAY OF BIOMOLECULES IN ARCHAEOLOGICAL SPECIMENS

7

Sources of Ancient Biomolecules

Now that we are familiar with the different types of ancient biomolecule we must examine the materials from which those biomolecules are obtained. The most important of these are the bones and teeth of humans and animals, these being the types of biological material most frequently recovered by archaeological excavation. But we must also consider specimens such as mummies and bog bodies, which contain preserved soft tissue as well as skeletal remains, and hair and **coprolites** (preserved excrement), both of which are important sources of human and animal biomolecules. Our survey must also include plants, which are preserved in the archaeological record mainly as charred, desiccated, or waterlogged remains, often but not always of seeds. Invertebrates are also important in archaeology, discarded shells being good indicators of a marine or freshwater diet and the identities of land snails present at a site providing information on climate, but invertebrate remains have not yet been exploited as sources of ancient biomolecules, and we can therefore exclude them from our study.

When we consider each type of remain, we must remember that the material from which ancient biomolecules are extracted is not the original biological structure, but the preserved version of that structure, which almost certainly has undergone transformation since it was alive. We must therefore understand not just the nature of the living material but the changes that might have occurred to that material since its death. These changes can be grouped into three categories:

- Changes occurring soon after death due to human activity, including cremation or mummification of human and animal remains.
- Changes due to **diagenesis**, the general breakdown and decay of dead biological material over time.
- Changes occurring after excavation, due to continued decay during storage or resulting from treatments applied by curators in attempts to preserve the material. Post-excavational

 © 2011 Terry Brown & Keri Brown

change is particularly important, as biomolecular archaeologists often work with material that has been stored in museums or elsewhere for years or decades.

Clearly, we must deal with a range of materials and a number of different issues. To simplify our task we will focus in this chapter specifically on the biological materials. The related question of how the biomolecules themselves change and decay over time will be the subject of the next chapter.

7.1 Bones and Teeth

The preserved bones and teeth of animals and humans are by far the commonest types of biological material recovered by archaeological excavation. Partly for this reason, bone and teeth are also the types of material most frequently studied by biomolecular archaeologists. Bone and teeth are the most important sources of ancient DNA, the demonstration in 1989 that DNA is preserved in at least some archaeological skeletons being the major stimulus in development of ancient DNA research from a backwater into the mainstream of archaeology, zoology, and paleontology. Bone also contains proteins, such as collagen and osteocalcin, which can be studied in their own right or, along with the mineral component, used as sources of stable isotopes for dietary studies. Stable isotopes are also obtained from teeth, and teeth contain the strontium isotopes that are used to provide information on human and animal mobility.

7.1.1 The structure of living bone

Bone is a living tissue that constantly undergoes remodelling during the lifetime of the organism. The cells responsible for the synthesis and resorption of bone are called **osteoblasts** and **osteoclasts**, respectively. A third type of cell, **osteocytes**, are osteoblasts that have become embedded in the bone that they have formed, but which remain alive and continue to contribute to turnover of the bone matrix.

Osteoblasts secrete proteins including collagen, along with calcium, magnesium, and phosphate ions, which combine to form the inorganic component of bone. This inorganic component is, in chemical terms, a crystalline hydroxyapatite carbonate mineral called dahllite, more commonly referred to simply as **bioapatite**. The organic component, which makes up about 20% of the dry weight of a bone, largely comprises collagen molecules that have combined into triple helical structures called **tropocollagen**, which in turn combine to form **collagen fibrils**. These fibrils are embedded within the bioapatite matrix, which also permeates between the individual tropocollagen units of a fibril, giving an overall structure that combines flexibility with rigidity.

Collagen fibrils can be organized in different ways to give three distinct types of bone microstructure:

- **Woven bone**, in which the fibrils takes up a random arrangement. Woven bone is relatively weak but can be synthesized rapidly and so is often the first type of bone to be made at points of bone growth or repair.
- **Lamellar bone** contains aligned collagen fibrils and is much harder than woven bone, which it usually replaces as the bone structure matures. Lamellar bone is usually organized into

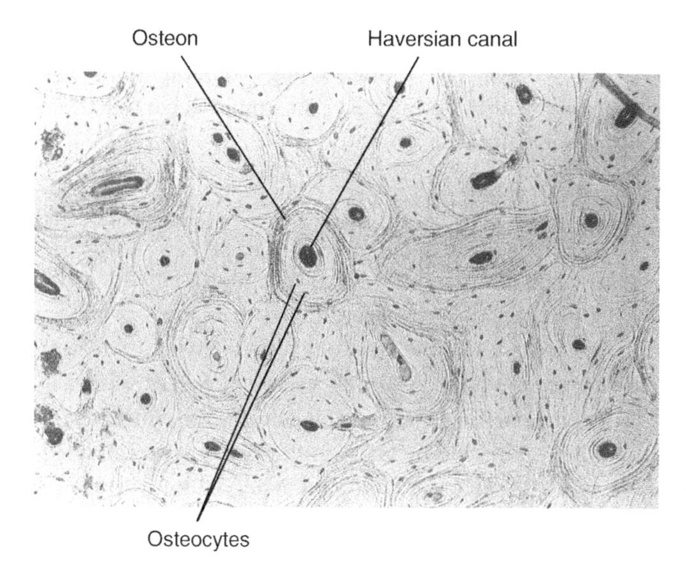

Figure 7.1 The typical appearance of well-preserved lamellar bone. The micrograph shows a cross-section with individual osteons clearly visible, each approximately 0.2 mm in diameter. The histology would be graded as "5" in the index described in Table 7.1. Reprinted from Robert Hedges, Andrew Millard, and Alistair Pike, "Measurements and relationships of diagenetic alteration of bone from three archaeological sites," *Journal of Archaeological Science*, 22, 201–9, copyright 1995, with permission from Elsevier and the authors.

Table 7.1 The Oxford Histological Index of archaeological bone decay.

Index	*% intact bone*	*Description*
0	< 5	No original features identifiable, other than Haversian canals
1	<15	Small areas of well-preserved bone present, or some lamellar structure preserved by pattern of destructive foci
2	<50	Clear lamellate structure preserved between destructive foci
3	>50	Clear preservation of some osteocyte lacunae
4	>85	Only minor amounts of destructive foci, otherwise generally well preserved
5	>95	Very well preserved, virtually indistinguishable from fresh bone

cylindrical **osteons**, each one a few mm in length and made up of concentric layers of bone surrounding a central **Haversian canal** containing blood vessels and nerves (Figure 7.1).

• **Parallel-fibered bone** has an intermediate arrangement between woven and lamellar. It is the least frequent type and is found mainly on new bone surfaces.

At the macroscopic level, two types of bone are recognized. The first is **compact** or **cortical bone**, as shown in Figure 7.1, which is made up of parallel bundles of osteons and has great strength. Compact bone makes up the shafts of the long bones in the arms and legs, as well as the outer parts of ribs and wrist and ankle bones. Layers of compact bone also form the various components of the skull. The second macroscopic type is **cancellous** or **trabecular bone**, often called "**spongy**," which has a much less rigid organization, being less dense and permeated with channels. Spongy bone is found in the central parts of long bones, forming the bone marrow in which blood cells are made, and is also present within the bones of the skull. Vertebrae and the hip bones have a larger spongy component, though this component is denser than that found in bone marrow. The compact part of these "irregular" bones is confined to a thin outer surface.

7.1.2 The decay processes for bone after death are complex

The diagenetic changes occurring in bones after death are complex and dependent on the burial environment, in particular the presence of water. Even the water aspect is complex, modelling studies showing that the movement of water through a bone leads to different decay processes than those occurring when a bone is simply soaked in water. Neither is there any reason to suppose that the environment will remain static during the entire period between burial and excavation, adding another layer of complexity. This multifaceted nature of diagenesis applies not just to bones but to all biological materials, but we tend to recognize it more clearly with bones because the most detailed attempts to understand diagenesis have been made with this material.

Arguably the most important cause of bone decay in the period immediately following death is attack by microorganisms, both bacteria and fungi. Microbial activity leads to changes in bone histology that are clearly visible when a transverse section is examined with a light microscope, the first signs of attack being the presence of small holes in the bone structure. These holes or "destructive foci" are not all uniform in appearance and different types of foci are associated with different burial environments. Whether or not they are caused by different species of microorganism is not known. Continued microbial attack leads to an increasing amount of structural damage, culminating in badly decayed bones lacking almost all of their original histological features (Figure 7.2). Microbial attack has been observed in some bones after just a few months of burial, but in other cases bones buried for over 30 years have displayed very little of this kind of decay. These differences might not be entirely due to the burial environment but instead might reflect the nature of the burial. Burial of an entire animal or human body, as opposed to butchered segments, exposes the skeleton to microorganisms released from the guts, which are thought to promote bone decay. There might also be differences between butchered bones buried with attached flesh and those from which the flesh has been removed. Experiments with animal bones suggest that ones that are partially cooked before burial undergo more rapid microbial attack than those that are buried without this treatment.

These considerations prompt the question of what exactly is being broken down by the microorganisms present in a bone. The destructive foci indicate that the inorganic component is being broken down in some way, but this is almost certainly not the only effect that microorganisms have on a bone. To survive, bacteria and fungi must obtain nutrients and it will be the organic fraction of the bone that satisfies this requirement. Collagen, however, is a difficult compound to degrade, so some type of pretreatment such as partial cooking might promote microbial attack by partially breaking down the bone collagen, or making it more accessible by freeing parts of the collagen fibrils from the inorganic component of

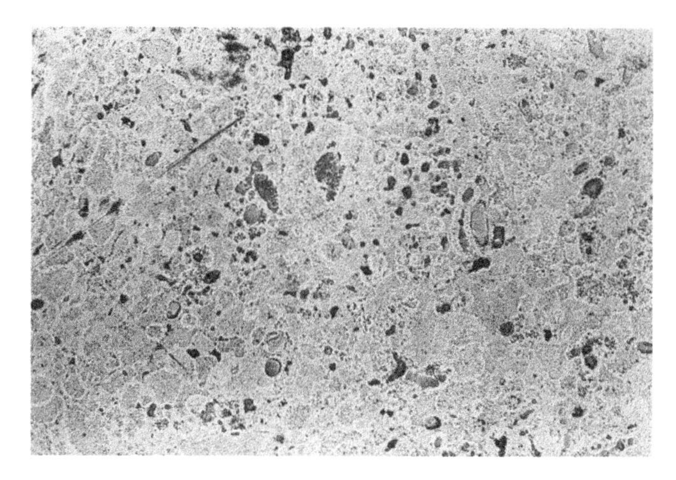

Figure 7.2 The typical appearance of badly degraded lamellar bone, histology index "0." The micrograph is at the same magnification as Figure 7.1. Reprinted from Robert Hedges, Andrew Millard, and Alistair Pike, "Measurements and relationships of diagenetic alteration of bone from three archaeological sites," *Journal of Archaeological Science*, 22, 201–9, copyright 1995, with permission from Elsevier and the authors.

the bone. Microbial attack can certainly result in considerable depletion of the organic content of a bone, and the process is sufficiently rapid to reach completion after a few hundred years.

Non-biological processes can also contribute to loss of collagen, and will act on any that remains after microbial decay, and can also result in rapid decay of collagen in bones that are not subject to attack by microorganisms. These chemical decay reactions are water driven and, like all chemical reactions, act more quickly at higher temperatures. Other processes are thought to prevent or at least delay the chemical breakdown of collagen. These include the formation of a complex between collagen and humic acids, which can move from the soil into buried bones.

The inorganic matrix of bone also undergoes chemical change, additional to the focal destruction caused by microbial attack. These changes include dissolution of the bone matrix in water, which is more rapid in acidic environments and when the bone is constantly exposed to fresh sources of water. If the water is not recharged, then its content of calcium and phosphate ions eventually reaches equilibrium with that of bone and no further dissolution occurs. There can also be uptake of ions from the groundwater, possibly resulting in substantial changes in the inorganic content of the bone. One important consequence of uptake is that certain ions naturally present in the bone might become exchanged with ions of the same element present in the burial environment. This has been shown to happen with strontium, which means that the characteristic strontium isotope ratio present in a bone at the time of death will change over time due to the diagenetic exchange between bone and environmental strontium ions. This change must be accounted for when strontium isotope ratios from bones are used in studies of human and animal mobility (Section 6.2).

7.1.3 Methods that enable the extent of bone diagenesis to be measured are being sought

Faced with such a complex process, is it possible to devise any methods for estimating the amount of diagenesis that has occurred within a bone? Various methods have been devised to address this question, but with most of these it is debatable if diagenesis as a whole is being measured, or just some component of it.

The simplest way of measuring bone diagenesis is to examine thin transverse sections with the light microscope in order to assess the amount of structural damage that has occurred. To aid this type of analysis, a histological index has been devised (Table 7.1) to grade the extent of destruction between the two extremes illustrated in Figures 7.1 and 7.2. This index is largely based on the nature and extent of the destructive foci that are present, and so is essentially a measure of microbiological attack rather than the overall degree of diagenesis. The amount of organic decay that has occurred can be estimated by assaying the collagen content of the bone, by measuring the weight loss that occurs when the mineral component is dissolved away by soaking in 0.6M hydrochloric acid. Collagen makes up approximately 23% of the dry weight of a living bone, so values less than this indicate increasing organic destruction. Diagenesis is also associated with an increase in the **crystallinity** of the bone, probably due to loss of the smallest bioapatite crystals along with recrystallization of the less small crystals to produce even larger structures. Both factors lead to a greater crystalline content that can be measured by X-ray diffraction analysis or infrared spectroscopy.

The methods that we have considered so far give good indications of the extent to which particular components of the diagenetic process have proceeded to completion, but do not give an overall measure of diagenesis as a whole. One parameter that might be used to obtain a more complete picture is bone porosity. About 12% of living compact bone is made up of pore spaces, but this value increases during diagenesis as the bone is gradually destroyed. An important observation is that different kinds of diagenetic process produce pores of different sizes. Degradation of the collagen fibrils, for example, results in pores between 0.01 and 0.1 μm in diameter, microbiological attack produces slightly larger pores up to 8.5 μm, and destruction of the mineral component gives pores up to 70 μm in diameter. By measuring porosity it is therefore possible to assess both the overall degree of diagenesis and the relative contributions of organic, microbial, and mineral decay.

How is bone porosity measured? The technique called **mercury intrusion porosimetry** enables the total volume of pores of different sizes to be measured in materials ranging from concrete to fine pharmaceutical powders and has also been applied with success to archaeological bones. The principle of the method is the inability of a non-wetting liquid such as mercury to penetrate into the pores of a material unless it is forced in by physical pressure. A porosimeter comprises a bath of mercury, into which the test material is placed, linked to a device that gradually increases the hydraulic pressure acting on the mercury. As the pressure increases the mercury is squeezed into ever smaller pores, resulting in a gradual decrease in its volume. By measuring the volume change at different pressures the volume of pores in different size ranges can be estimated (Figure 7.3). Mercury intrusion porosimetry can therefore provide a measurement of the overall porosity of a bone as well as individual estimates of the relative contributions made by pores of the different diameters associated with organic, microbial, and mineral decay.

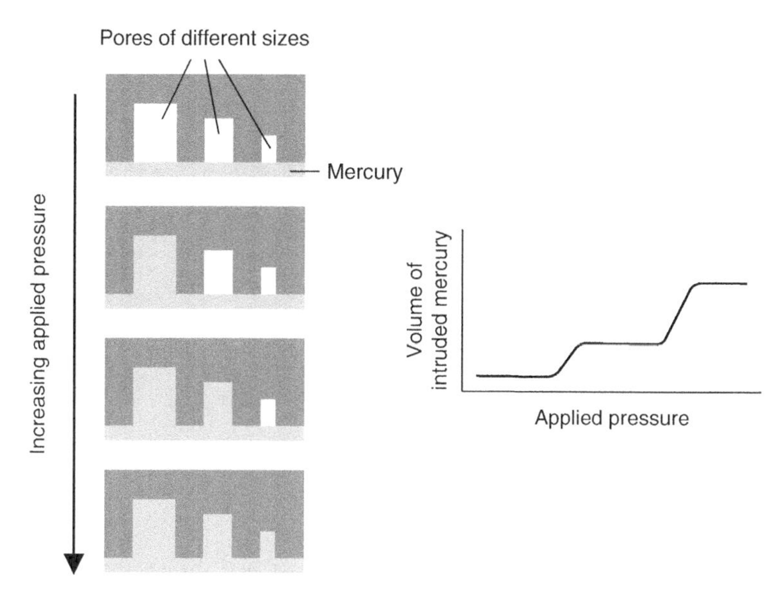

Figure 7.3 Mercury intrusion porosimetry. The drawing on the left shows a material with pores of three different sizes. As the applied pressure increases, mercury intrudes into pores of decreasing size. The graph shows the relationship between applied pressure and the amount of mercury that is intruded into the material. From this graph, the sizes of the pores can be deduced.

7.1.4 Cremation, but not cooking, results in extensive changes in bone structure

So far we have only considered bones that have not been subjected to any kind of harsh treatment prior to burial. This is the case for many bones recovered from archaeological sites, but not all. In particular, cremation was an important mode of disposal for human bodies in the past, as it is today, and in some instances was accompanied by cremation of animals as part of the death ritual. Cooking of animal bones was also commonplace. But what effects do these processes have on the structure of the material that is recovered hundreds or thousands of years later?

Cremation can result in quite substantial changes in bone structure. Prehistoric cremations were not always as efficient as those carried out today, but experiments have shown that temperatures in excess of 1000°C can be achieved within fairly unsophisticated pyres. Such high temperatures result in both dehydration and oxidation of the bone material. This leads to the bones shrinking by up to 25% and cracking in typical ways such as along the shafts of the long bones, which means that cremated material is often recovered as fragments with only the smaller bones remaining intact (Figure 7.4). These morphological changes are associated with alterations in the crystal structure of the mineral matrix, the fusion of individual crystals giving an increased crystallinity, similar to the processes occurring during diagenesis of bones that have not been subject to cremation.

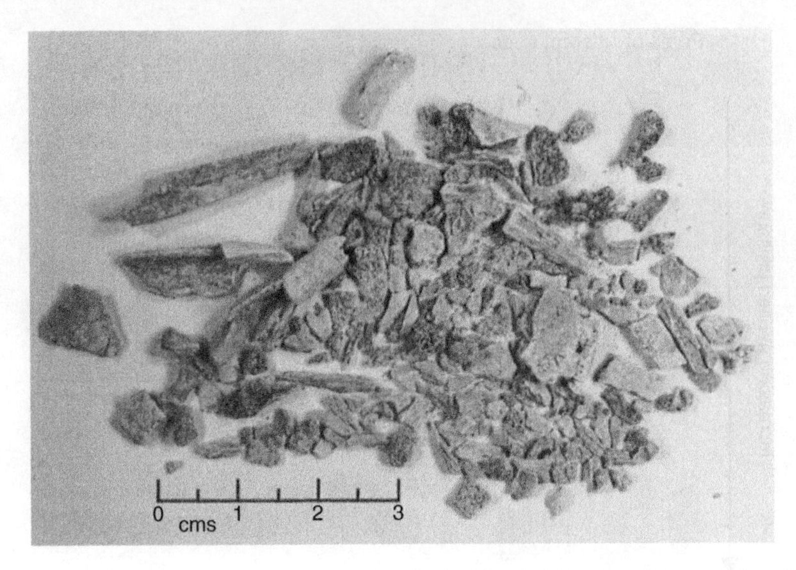

Figure 7.4 Typical appearance of cremated archaeological bone. This material comes from Bedd Branwen, an Early Bronze Age cemetery cairn in Anglesey, North Wales, excavated by Frances Lynch. Reprinted from Keri Brown, Kerry O'Donoghue, and Terence Brown (1995), "DNA in cremated bones from an Early Bronze Age cemetery cairn," *International Journal of Osteoarchaeology*, 5, 181–7, with permission from John Wiley & Sons, and the authors.

Bones change color during cremation, becoming brown or black at temperatures of 300°C or so, and at higher temperatures returning to grey or white as the material becomes completely oxidized. The blackening of bone indicates that most of the organic material has been destroyed, leaving a carbon shell, and chemical analyses show that all or most of the collagen has degraded by this stage. Although there have been reports of ancient DNA being recovered from cremated bone, DNA is much more heat-labile than collagen and it is difficult to believe that any could remain in bones that have been subjected to temperatures sufficient to cause blackening. It is clear, however, that prehistoric cremations were not always entirely efficient, with single assemblages of bone including material of different colors and displaying different degrees of morphological change. Experiments have shown that during cremation the temperature inside the corpse might not have exceeded 300°C. If not all parts of a cremation reach a particularly high temperature, then some biomolecules might persist in the material after cooling.

Cooking of bones presents a different set of problems. Cooking rarely involves heating to the temperatures associated with cremations – such high temperatures would clearly spoil the food. The chemical changes that take place in bones during cooking are similar to those occurring over a much longer period during the diagenesis of uncooked bones. It can therefore be very difficult to distinguish cooked bones in the archaeological record, as they simply look like bones that have undergone the normal process of diagenesis. This inability to identify cooked bone limits the amount of information that can be obtained about the types of food being eaten at an archaeological site. Recently it has been shown that heating bone to temperatures used in cooking results in gradual demineralization of collagen fibrils, which take on an unpacked appearance that is observable by electron microscopy. Demineralization and unpacking of collagen fibrils also occurs during diagenesis, but it is suggested that at a single site, where all the buried bones are subject to similar environmental conditions and hence equivalent diagenetic regimes, any bones that have been cooked can be distinguished, as these display a greater degree of collagen unpacking.

7.1.5 Bone may continue to deteriorate after excavation

Finally, we must consider the changes that might occur within an archaeological bone after excavation. A few decades ago, archaeologists and museum curators tended to place greater importance on cultural remains – pottery, metalwork, and suchlike – rather than skeletons, and bones collected prior to the 1960s were not always conserved as well as they might have been. Biomolecular deterioration in material excavated in the past has been demonstrated in convincing fashion by comparisons between the ancient DNA contents of two parts of a 3200-year-old aurochs bone, one part excavated in 1947 and the other in 2004. When examined in 2005, no DNA could be detected in the 1947 segment, whereas the 2004 fragment yielded positive results with a range of PCRs. It was estimated that DNA degradation within the 1947 fragment had increased seventyfold after its excavation.

In recent years, the curation and conservation of archaeological bones has become a highly professional discipline, but the effectiveness of the strategies that have been adopted has been measured largely by their ability to preserve the overall morphological appearance of a bone, rather than by their effects on the biomolecular content. It is common practice to wash bones immediately after excavation, to remove dirt, which can potentially lead to some biomolecules and ions being lost if these are not bound within the bone matrix in an insoluble form. There is also a possibility that washing increases the chances that a bone becomes contaminated with modern DNA. Fortunately, more and more excavators are becoming aware of the requirements of biomolecular analysis and are setting aside some bones for this purpose prior to washing. Additionally, many biomolecular archaeologists now direct their research specifically at freshly excavated material, so that potential problems associated with curation are circumvented.

Although not a major problem for archaeological material, bones that have adhering soft tissue or grease are usually treated in some way to clean them and hence prevent subsequent attack by microorganisms and other pests. The procedures that are used are necessarily harsh and can include boiling the bones for short periods, treating with enzymes designed to break down the protein content of adhering flesh, or soaking in organic solvents for periods of days to remove grease. The combination of these procedures would be expected to have a detrimental, one might say disastrous, consequence for the biomolecular content of a bone. If a bone is fragile then a consolidant might be used to stabilize it. Most consolidants are polymeric compounds that soak into the bone and then harden, improving its structural integrity. A wide range of chemicals are used, including acrylic, polyvinyl, and cellulose resins, none of which would be expected to have any damaging effect on the biomolecules within a bone. Consolidants can, however, hinder recovery of those biomolecules by making it less easy to break the bone up and to make a soluble extract.

Although conservation practices can have a detrimental effect on the biomolecular content of a bone, these problems pale into insignificance compared with the post-excavation procedure that is most damaging to biomolecular, and indeed all archaeological, research. This is reburial. Few archaeologists dispute the imperative for reburial of skeletal material that can be clearly identified as ancestral to a group of living people. Less easy to accept is the reburial of British and European skeletons due to nebulous concepts of "ownership" promoted by groups whose cultural identities lie in New Ageism or, at best, 19th century romanticism.

7.1.6 Teeth are more stable than bones

Teeth are the second major hard tissue present in vertebrates and commonly recovered at archaeological sites. Strictly speaking, teeth are not part of the skeleton, but form a separate component of the body called the dentition.

Like bones, teeth are made up primarily of collagen and bioapatite, but structurally they are quite different to bones. Each tooth comprises a central **pulp cavity** surrounded by **dentine** which in turn is coated with **enamel** and **cementum**. The enamel covers that part of the tooth that is exposed within the mouth, and the cementum covers the part that is attached to jaw. The pulp cavity contains living cells, blood vessels, and nerves and connects directly with the soft tissue of the jaw via a canal present at the tip of each root. Being soft tissue, the pulp usually degrades soon after death and is not present in archaeological remains. The other three components have significant collagen and mineral contents and are much more resistant to deterioration. Dentine is made up of collagen fibrils mineralized in bioapatite and arranged as tubular structures, between 1.0 and 2.5 μm in diameter, radiating out from the center of the tooth. The organic collagen component makes up about 20% of living dentine, with the mineral part contributing some 70%, and the rest water. Cementum has a slightly higher organic content, approximately 35%, again mainly collagen, with 45% bioapatite and 20% water. Enamel is virtually entirely made up of bioapatite, with less than 5% organic content. The latter does not include collagen, but instead is made up of special enamel proteins such as **amelogenin**, which we will meet again in Section 10.3, as its gene is used in a DNA test for identifying the sex of a skeleton.

Compared with bone, there have been relatively few studies of the diagenesis of teeth. Most of the detailed studies that have been carried out have focussed on uptake and exchange of ions between the mineral fraction of teeth and the burial environment, the isotopic signature of preserved teeth being important not only in archaeological research but in studies of past climates. This type of work has shown that there is minimal exchange between tooth and environmental strontium during diagenesis, lending support to the use of dental strontium in studies of human and animal mobility. It has generally been assumed that the enamel and cementum layers are so solid that microorganisms are unable to penetrate into a tooth, although microbial invasion does occur through the openings at the tips of the roots. Despite the relative lack of work on tooth diagenesis it is clear that teeth degrade more slowly than bone, although burial conditions that result in complete destruction of bone, such as acidic environments, will destroy teeth also.

As well as their use in isotope studies, teeth are attractive sources of ancient DNA, partly because of their relatively good preservation but also because their outer surfaces can be rigorously cleaned to remove contaminating modern DNA, which is less easy to do with bone because the surface is porous. Samples for DNA analysis can be taken from the interior of a tooth by drilling through the root canal which, if carried out carefully, results in minimal damage to the overall morphology of the tooth. The use of laser ablation (Section 6.3) similarly enables samples for isotopic analysis to be taken without damaging the tooth.

These non-destructive methods are important because morphological examination of teeth can provide information relevant to studies of kinship and diet. Small inherited structural variations can be used to infer if groups of skeletons might be related (Section 11.1), and microwear patterns can indicate the types of food that were eaten (Section 12.1). This information would be lost if teeth were damaged when used in biomolecular research.

7.2 Vertebrate Soft Tissues

Bones and teeth are by no means the only remains of humans and other vertebrates that occur in the archaeological record. Soft tissues are also preserved under some circumstances. Most notable are mummified remains, which include both the artificial **mummies** resulting from the burial practices of ancient Egypt and elsewhere, and mummies that arise naturally if the body lies in a cold desiccating environment. Immersion in the cold and oxygen free environment of a sphagnum bog also leads to preservation of soft tissues, giving rise to a **bog body**, examples of which have been found in Denmark and other parts of northern Europe.

In addition to mummies and bog bodies, two other types of soft tissue remains are important in biomolecular archaeology. The first is hair, which arguably is not a type of soft tissue but which we will consider in this section. The second is **coprolites**, fossilized excrement, which contain biomolecules derived both from the organism responsible for the deposit and the animals and plants recently consumed by that organism.

7.2.1 Mummification results in preservation of soft tissues

Mummification is an alternative diagenetic route for human and animal remains, which results in preservation not only of the skeleton but also of some or all of the soft tissues. Mummification can arise naturally under certain environmental conditions or can be induced by artificial treatment of the body soon after death.

Under most circumstances, the soft tissues of a body are lost in the days and months after death. The process of decomposition is complex, and includes autolysis of tissues by chemicals and enzymes released from the dead cells, further breakdown of cellular components by the microflora present in the digestive tract, and attack by external organisms such as environmental bacteria, fungi and insects. Removal of large chunks of flesh by carnivores is not usually looked on as part of this process, but can of course occur if the body is accessible to this type of attack. For the body to become mummified, these normal decomposition processes must be retarded or prevented altogether.

In the natural environment, mummification is usually associated with dry and/or cold conditions. If the body dries out then autolysis is prevented and the action of microbial and insect decomposers is greatly retarded. The best known of these natural mummies are the "frozen" mammoths of Siberia and the **Tyrolean iceman**. The mammoths are thought to have become frozen shortly after death, possibly after drowning by falling through the ice on the surface of a lake. Subsequently their bodies became encased in permafrost until erosion eventually exposed their remains on the tundra surface. Almost 50 of these, between 20,000–60,000 years in age, are known to science; many others have probably been discovered in the past but only their ivory collected.

The iceman, sometimes called Ötzi as he was found in the Ötstal region of the Alps on the border between Austria and Italy, is one of the most famous of all archaeological remains. He dates to about 3300 BC and was preserved along with many items of clothing and other artifacts, which have revealed a great deal about life in Copper Age Europe. Originally it was thought that Ötzi died from exposure on the high mountains, even though his clothing should have been sufficient to protect him from the cold. More recently, the discovery of an arrow head in his left shoulder has suggested a more sinister end. His body became mummified in the cold conditions on the mountain, probably under a thin layer of snow that protected him from animal predation, and possibly aided by a dry, desiccating wind during the days after his death. His body then became trapped in a glacier until its discovery in 1991. He is one of the most complete and best preserved of all natural mummies that have been found, anatomical changes being restricted to shrinkage and displacement of some of his organs, and the stomach contents sufficiently well preserved to enable identification of the food that he ate prior to his death. A greater amount of decomposition is evident at the microscopic level, the cellular structure of individual tissues being indistinct with no cell nuclei visible, though with well-preserved collagen fibrils and fat deposits.

The iceman is the most famous, oldest, and arguably best preserved of the natural human mummies that have been discovered (Table 7.2). Equally evocative, however, is the **Siberian ice maiden**, a female frozen mummy from the 5th century BC, found in one of a number of tombs belonging to the Pazyryk culture. Her tomb probably became flooded soon after her burial, the water freezing and remaining as permafrost until excavation in 1993. The ice maiden was heavily tatooed and buried with rich trappings, suggesting that she held high status in her community. Other, much younger frozen mummies, from 500 to 600 years ago, have been discovered in Greenland and British Columbia in North America. Mummies of Inca children, again dating to 500–600 years ago have been found on mountaintops in the Andes, thought to be sacrificial victims.

7.2.2 Artificial mummification was not restricted to ancient Egypt

The distinction between natural and artificial mummification becomes blurred at the edges. Mummification in ancient Egypt began with simple burial in the sand, probably as a convenient means of disposing of the dead rather than with any aim of preserving the body. Natural mummification is promoted by the dry, desiccating conditions in the hot desert and it is possible that the ancient Egyptians "discovered" mummification by recognizing that bodies buried in this way do not decompose. During the 4th millennium BC,

Table 7.2 Examples of natural mummification.

Cause of mummification	*Examples*	*Date*
Freezing	Tyrolean iceman	3300 BC
	Siberian ice maiden	5th century BC
	Greenland mummies	14–15th centuries AD
Desiccation	Spirit Cave, Nevada	7400 BC
	Urumchi mummies	c.2000–1000 BC
	Maronite mummies, Lebanon	AD 1283
	Inca sacrificial children	14th–15th centuries AD
Preservation in a bog	Koelberg Woman	c.6000 BC
	Windover brains	c.6000 BC
	Lindow Man	1st century AD
Enclosure in salt	Iranian salt mummy	5th century AD
Enclosure in tightly sealed coffin	Chinese Marquise of Tui	1st century BC

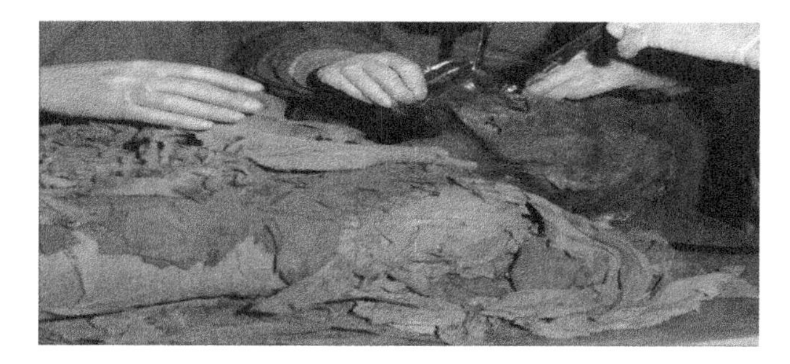

Figure 7.5 An Egyptian mummy. This mummy, in the Leeds Museum, UK, is a man named Natsef-Amun who was a priest and scribe in the temple of Karnak, Thebes, during the reign of Ramesses XI, Dynasty 20, 1099–1069 BC. He was in his mid-40s at the time of his death. Image kindly provided by the KNH Centre for Biomedical Egyptology, University of Manchester.

the burial practices changed, with elite individuals being placed in tombs rather than being buried in the sand. For a culture that believed in life after death the subsequent rapid decomposition of the bodies of their rulers would have been quite a concern. And so artificial methods of reproducing the natural events occurring in the sand burials were developed (Figure 7.5).

Egyptian mummification was based around the use of **natron**, a mixture of salts, mainly sodium carbonate and sodium bicarbonate, found in dried lake beds in some parts of Egypt. Natron is a very effective desiccant and dries out the body more quickly than burial in the sand, and also has a relatively high pH of 9–10, further inhibiting microbial decay. The earliest techniques involved simply drying out the body with natron, but by 1000 BC the procedure had become much more sophisticated, with most of the internal organs removed and stored in jars. The exceptions were the heart, which was often left in place, and the brain, which was thrown away because it was thought to be useless. Inert materials would be placed in the body to retain its shape and make it suitable for reuse by the deceased in the afterlife. The overall degree of soft tissue preservation in Egyptian mummies is similar to that seen in the iceman, histological examination of rehydrated samples revealing that the cellular structure has largely broken down but that tissues with a high collagen content remain intact.

Table 7.3 Examples of artificial mummification.

Type	*Period*
Egyptian mummies	c.3500–200 BC
Chinchorro mummies, South America	c.5000–2000 BC
Guanche mummies, Canary Islands	Uncertain, pre AD 1400
Kabayon mummies, Philippines	2000 BC?
Pazyryk mummies, Scythia	8th–4th centuries BC

In popular culture, artificial mummification is most closely associated with ancient Egypt, but this does not mean that it is unknown in other parts of the world. At different periods, mummification has been practiced in Asia, the Pacific islands, the Canary islands, and in South America (Table 7.3). The most significant of these non-Egyptian practices was that of the Chinchorro people of South America, who lived in the region now marked by the border between Chile and Peru. The Chinchorros began artificial mummification of their dead around 5000 BC, earlier than the Egyptians, and virtually everyone, even unborn infants, were treated in this way. At least two different mummification procedures were used at different periods, the objective appearing to be not preservation of the body as such, but recreation of an effigy of the living person. Hence, in the "black mummification" process, the body was taken apart and most of the internal tissues stripped from the skeleton. The parts were then reassembled with the help of sticks and clay packing, and covered with the original skin or with skin from an animal such as sea lion. Great care was often taken to reproduce the facial features of the living person. Although fascinating from the anthropological viewpoint, the lack of preserved soft tissues except for skin (which might not be the skin of the deceased) means that the Chinchorro mummies are less valuable in biomolecular archaeology.

7.2.3 Biomolecular preservation in mummified remains

Intuitively we would expect mummified remains to be good sources of ancient biomolecules. Although the processes responsible for biomolecular degradation after death have never been studied, we can assume that the enzymatic and chemical activities associated with autolysis are a major cause of breakdown of biomolecules. The reduction in autolysis that occurs as a result of mummification should therefore promote survival of biomolecules. We might also predict that desiccation of the body will help to preserve its DNA and triacylglycerol contents because it is well known that hydrolysis is one of the main drivers in breakdown of these biomolecules (Chapter 8). These predictions are not entirely borne out by the studies that have been made of biomolecular preservation in mummified remains, and the general assumption that every mummy will be a goldmine for the biomolecular archaeologist is erroneous.

Thermal history is an important factor in biomolecular decay, so mummies that have been preserved in very cold conditions are likely to display better preservation than those from warmer environments. With frozen mammoths, DNA preservation is so good that substantial parts of the nuclear genome have been sequenced, but the DNA for this work was extracted not from the preserved tissues but from the bones within the specimens and from hairs. The mitochondrial genome of the iceman, on the other hand, has been obtained from soft tissue extracts. Quantitative PCR of these extracts suggested that the DNA content of the iceman's tissues was similar to that of the mammoth bones. As the iceman is substantially younger than the mammoths, the implication from this observation might be that DNA preservation in mummified soft tissue is no better than that in the bones of the same specimen. The iceman's skin has also been examined, showing that although its gross structure is well preserved, proteins other than collagen have broken down, and the triacylglycerols and phospholipids have been hydrolyzed to fatty acids. We will look in more detail at the breakdown pathways for different types of biomolecule in Chapter 8: here it is sufficient to note that the degree of preservation in mummified tissues is not as great as we might have assumed.

With Egyptian mummies, there have been numerous reports of the successful extraction of ancient DNA and detection of proteins via immunological and other methods, but very few rigorous studies aimed at understanding the general biomolecular content. The similarities between the histological appearance of tissue from Egyptian mummies and that of the iceman suggest that the protein and lipid contents might be degraded to an equivalent extent. As well as its own biomolecules, most Egyptian mummies also contain compounds of biological origin from the embalming fluids and other preparations applied during the mummification process. Identification of the resins, oils, waxes, and spices used in these preparations is an important part of biomolecular archaeology in its own right, but their presence as contaminants in extracts also has the potential to interfere with studies directed at a mummy's endogenous biomolecules.

Additional practical problems have to be taken into account when mummies are studied. Mummified remains include many of the most valuable bioarchaeological treasures, and destructive sampling, as is needed for most types of biomolecular study, has to be carefully justified. In the early days of scientific Egyptology, researchers had to make do with pieces that had literally dropped off the mummy they wished to study. Scientific study is nowadays valued more highly, but still the samples taken for biomolecular research are often not the ones that would be chosen if there were no other considerations. It is also clear that some mummies have deteriorated since their collection. Natural mummification is not an irreversible process, and exposure of the body to moisture or heat will cause partial rehydration and the possibility of gradual decomposition. For some mummies, notably the Siberian ice maiden, substantial deterioration is thought to have occurred since excavation, and even for the most carefully curated specimens such as the iceman concerns have been raised about the onset of decay. This is less of a problem for Egyptian mummies as the embalming process continues to preserve the specimen even today, but some have still suffered as a result of deficiencies in the curation practices adopted during the earlier decades of the last century.

7.2.4 Bog bodies are special types of mummy

The strict definition of a mummy is a specimen in which some of the soft tissues are preserved. This means that bog bodies are a type of mummy, although we will treat them as a separate category (as do most archaeologists) because they are quite different from other types of mummy in terms of their biomolecular preservation.

Almost 1000 partial or complete bog bodies have been found in Denmark, Sweden, northern Germany, the Netherlands, Britain, and Ireland. Most are from the last five centuries BC but the oldest, the Koelberg Woman from Denmark, dates to about 6000 BC. Some show evidence of execution or ritual killing, as is the case with Lindow Man, a British bog body from the 1st century AD, who appears to have had his throat cut and head fractured as well as being strangled prior to being placed in the bog. Others may simply have fallen into the bog and drowned, the weight of the bog subsequently causing some fracturing of the bones, implying a more violent death than actually occurred.

Once inside the bog, the body becomes preserved through a combination of acidity, low oxygen content, and the presence of specific compounds that promote tanning of the skin and inhibit microbial action. The acidity is due to the presence of large quantities of humic acid, released from the degradation of plant material. To survive as a bog body, it has been suggested that the body must enter the bog in winter so the acids can soak into the body before microbial activity increases during the warmer summer months. The anaerobic conditions arise because most bogs have poor drainage so there is little replenishment of the water at the bottom of the bog. The body therefore resides in a cold, acidic, oxygen-free environment, but these conditions on their own are probably insufficient to lead to preservation as a bog body. Almost all of these remains have been found in **sphagnum bogs**, ones whose peat is derived largely from sphagnum moss. The decay products of this moss include sphagnum acid and a polysaccharide, called sphagnan, which may have tanning and antimicrobial properties that play a central role in the preservation process that leads to the bog body.

The conditions in the bog lead to loss of the teeth and bones, whose mineral contents dissolve in the bog acid over a period of a few hundred years. As with other types of mummy, the cellular structure within the skin and soft tissues is indistinct, but fibrous proteins such as collagen and hair keratin are well preserved. Other proteins, such as blood hemoglobin and cellular enzymes, are much more unstable in acid and hence degrade. DNA is also very acid labile and ancient DNA has never been detected in any bog body. Lipids, on the other hand, survive remarkably well in the skin and muscular tissue. As with other types of mummy, phospholipids and, to a lesser extent, triacylglycerols break down to their constituent fatty acids, but these are resistant to further decay. Cholesterol and other sterols also appear to survive with little alteration.

Although intact bog bodies have only been found in northern Europe, soft tissue remains have also been discovered in a few other bogs around the world. One such site is the Windover bog in Florida, where the acidic bog lies on an alkaline substratum, which means that at the bottom of the bog the pH is approximately neutral. Windover was used as a burial site by the paleoindians living there 8000 years ago, and the skeletons of at least 40 individuals have been recovered. Some of the skulls contain preserved brain material which, although shrunken, retains the anatomical features of the original tissue and has some preserved cellular structure. In the early 1990s, there were several reports of ancient DNA recovery from examples of these brains, but attempts to replicate this work have been unsuccessful.

7.2.5 Hair is important in DNA and stable isotope studies

Hair is a characteristic feature of all mammals and is one of the parts of the body that is most resistant to decay. Attached hair is frequently found on natural mummies, and also on artificial ones, if it was not removed when the body was prepared for burial. Hair is also sometimes found attached to the skull of a body that is otherwise completely skeletonized. It is quite possible that detached human hair is present at many archaeological sites, not only associated with burials, but is simply overlooked during excavation. In those cases where hair has been searched for, for example in cave sites, it has often been found.

Hair has a more complex structure than is often appreciated. Most people are familiar with the lengthwise division into the shaft or hair fiber, which protrudes from the skin and is the visible component, and the root or follicle which is embedded in the dermis. At the base of the root are a small number of living cells, comprising the dermal papilla, whose division results in elongation of the fiber and growth of the hair. The new cells that bud off from the dermal papilla form the structure of the hair shaft. These cells become filled with the fibrous protein **keratin**, and with pigments that give the hair its natural color. In cross-section, the hair fiber is made up of three concentric layers, referred to as the medulla (in the middle of the hair), the cortex, and the outer cuticle. The cortex contains densely packed keratinized cells and usually makes up the greatest part of the bulk of the hair. The medulla also contains keratinized cells, but these are less densely packed and possibly separated by air spaces. This part of the shaft helps provide the hair with its insulating properties. The cuticle is just a single layer of cells which controls the water content of the hair and gives it most of its mechanical strength. The hairs of different mammals all have the same structure, the major differences between species being the relative thicknesses of the hairs on different parts of the body. Humans have a similar density of hair follicles on all parts of their body as other mammals, but most of the follicles produce very fine downy hair. On the other hand, the hair on our heads and on masculine faces is thicker than the hair on the heads and faces of other mammals.

The textile and cosmetics industries have carried out extensive research into the changes that occur during processing and aging of animal hair and as a result of cosmetic treatment of human hair. It is not clear to what extent this work is relevant to the changes that hair undergoes in an archaeological setting, and very few studies have addressed this particular issue. Hair from mummies retains much of its cross-sectional structure, though usually the outlines of individual cells are no longer visible. In more extreme cases, the keratin fibers become unpacked and the melanin pigment granules aggregate into larger structures. Although looked on as relatively resistant to decay, hair can decompose over a period of months under some conditions, probably when the burial environment contains fungi or bacteria able to degrade keratin. Such microorganisms are not ubiquitous in the biosphere but a number of different species are known and these have a wide geographical distribution.

As the hair fiber has a cellular structure, it is no surprise that living hair contains DNA. Both nuclear and mitochondrial DNA can be obtained from the root of the hair, and both are also easily recoverable from the shaft. The presence of DNA in shed and cut hair, even after a few years of storage, has been one of the pillars on which modern forensic biology has been built since the development of genetic profiling. It has also been known since the 1980s that hair from archaeological specimens often contains DNA, but hair has only

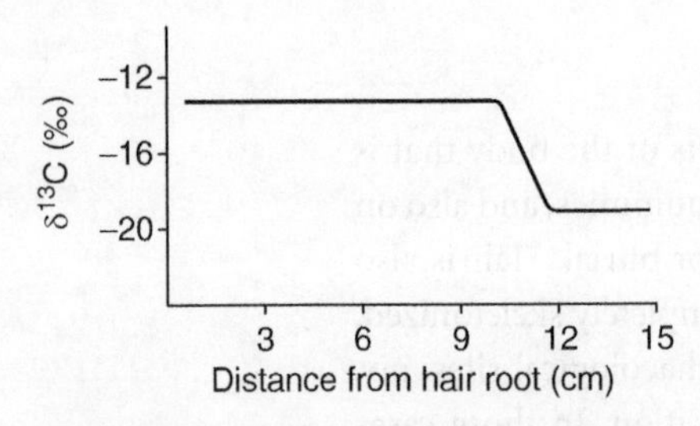

Figure 7.6 Graph showing a sudden change in δ^{13}C values along a hair shaft, suggesting that this person's diet shifted to one rich in a C_4 plant such as maize about one year before the hair was shed. This dietary change resulted in an increased δ^{13}C value for the region of the hair closest to the root. This result is similar to that obtained when the hair of an Inca child mummy from an Andean mountaintop was examined, suggesting that this child's diet underwent a dramatic improvement about one year before her sacrifice.

infrequently been exploited in this way because usually other parts of the body, such as soft tissues, bones and teeth, are also available from the same specimen. The alternative sources are usually present in larger amounts than the hair, and can be sampled with less visible damage to the specimen. It has been suggested that the ancient DNA in hairs of frozen mammoths has undergone less chemical degradation than that in bones, and so gives more accurate nucleotide sequences, but this report has been disputed and remains unproven.

The second application of hair in biomolecular archaeology is as a source of stable isotopes. The δ^{13}C and δ^{15}N ratios in hair protein are influenced by diet in the same way as the ratios in bones and teeth, which means that hair can be used along with the hard tissues in dietary studies. Hair has the additional advantage that it provides a time course for the period immediately before the death of the individual, or the time when the hair was shed. The isotope ratios in segments closest to the root reflect the very recent diet, and those further along the shaft relate to earlier periods when those parts of the hair were being formed. Seasonal variations and sudden changes in diet can therefore be discerned by segmental analysis along a hair (Figure 7.6).

7.2.6 Biomolecular archaeologists are becoming increasing interested in coprolites

Coprolites are the final type of animal remain that we will consider. The archaeological study of coprolites has an interesting history. Originally discarded as being no value, interest in coprolites began in the 1960s due almost entirely to the efforts of a single person, Eric Callen, who sadly died in the field in 1970. Today it is recognized that coprolites provide a record of diet and disease, and specialists in their study are located throughout the world.

Coprolites are frequently preserved in dry climates, especially in caves where deposits may cover large areas and reach significant depths. Features such as size, shape, color after treatment with certain chemicals and, reputedly, smell after rehydration enable human coprolites to be distinguished from those of animals. Although there are controversies over the use of coprolites to indicate the extent of meat in the diet, plant remains such as seeds, pollen, and phytoliths can be identified by visual and microscopic examination and used in dietary reconstruction. There have been several reports of endoparasite remains, giving indications of the health of the human or animal responsible for the deposit.

All types of biomolecule appear to be preserved in coprolites. Steroids have been detected in specimens up to 2500 years old, with sufficiently good chemical preservation for sex-specific hormones such as testosterone to be typed, enabling the sex of the originator to be identified. A range of lipid biomarkers for dietary plants have been identified in much older ground sloth coprolites, thought to be 11,000 years in age. Immunological tests have detected proteins with claims that some of these derive from animals that were eaten, but these results suffer from questions regarding specificity, as is the case with all uses of immunological detection in analysis of ancient proteins. The greatest success has been in DNA analysis, which has been used both in reconstruction of the plant and animal components of the diet and to identify the mitochondrial DNA haplotype of the originator. Although not strictly archaeology, one interesting variation has been to type the plant DNA in animal coprolites from different layers of a single cave, with the resulting data used to reconstruct changes in the local vegetation over a period of time. The reason why such well-preserved biomolecules survive in coprolites of quite substantial age has not yet been asked and little is known about the diagenetic processes acting within this type of material.

7.3 Plant Remains

Plant remains are important in archaeology as indicators of past subsistence patterns. At pre-Neolithic sites they provide information on the wild resources that were utilized, and at agricultural sites they can tell us exactly what types of crop were grown at particular times. Intact plants are only rarely recovered, largely because the non-nutritious parts of the plant were discarded or not collected at all and so had little opportunity of being preserved. Seeds are common, especially for cereals such as wheat, barley, and rice because for these the seeds (the grain) are the part that was processed and eaten. If these cereal seeds come from stores that have been abandoned, then they may be mixed in with fragments of chaff, but individual seeds that were lost during processing are also frequently found. Maize seeds are usually still attached to their cob, and sometimes the cobs on their own are recovered, having being discarded during food processing. Fruit seeds that have passed through the digestive system unharmed can be recovered from latrines and similar settings, and intact nuts are also found at some sites.

In this section, we will focus on the macrofossil remains of plants used as food, most of which are preserved in the archaeological record as desiccated, charred, or waterlogged material. We will not cover the microfossils of these plants, such as starch grains, phytoliths, and pollen, as we dealt with the relevant information concerning these in Section 5.2. Neither will we look at plant residues in cooking and storage vessels, or products such as tars, oils, and resins, not because these are unimportant in biomolecular archaeology, but because the correct place to study them is in later chapters on prehistoric diet (Chapter 12) and technology (Chapter 14).

7.3.1 Desiccated remains are most suitable for biomolecular study

In some parts of the world, plant remains are found as desiccated specimens that have undergone little visible change even though they may be several thousand years old (Figure 7.7). Maize cobs, with or without attached kernels, are often found in this form from cold dry sites at high altitude in the Andes, but perhaps the most spectacular preservation is that found in the Old World, at Qasr Ibrim, on the upper Nile near the boundaries of modern Egypt and Sudan. Qasr Ibrim, which lies in an essentially rainless environment, was occupied from the 1st millennium BC to the 18th century AD, and has been excavated since the early 1960s by the Egyptian Exploration Society. A wide range of desiccated plant

Figure 7.7 Desiccated plant remains. Top: maize cob from the lower Ica valley, south coastal Peru, dating to the Late Nasca period, about AD 750. Bottom: bouquet of sorghum of intermediate race bicolor from Qasr Ibrim, directly dated to AD 420–640, scale shows cm. Top image kindly provided by Claudia Grimaldo and David Beresford-Jones; bottom image kindly provided by Peter Rowley-Conwy.

remains have been discovered, from different periods, including cereals such as barley, wheat, and sorghum, legumes such as beans, and crops grown for economic purposes such as cotton and radishes, the latter a source of oil.

Desiccated remains, especially seeds, display remarkable preservation of biomolecules. This was first established with 1500-year-old radish seeds from Qasr Ibrim. which were shown to have fatty acid and sterol profiles very similar to those of modern radish seeds (Figure 7.8), the only significant differences being that the triacylglycerols had become hydrolyzed and some of the polyunsaturated fatty acids had broken down. DNA was present in sufficiently large amounts for nucleotides and bases to be detectable by gas chromatography-mass spectrometry, desiccated seeds being the only type of archaeological material from which ancient DNA has been detected directly by chemical means rather than by PCR amplification. The nucleic acid bases that were detected included uracil, which can arise as a breakdown product of cytosine in DNA, but which was present in such large amounts as to suggest that RNA was also present in these seeds.

The presence of ancient DNA, both nuclear and chloroplast, in most desiccated seeds up to 5000 years in age is now well established, and the RNA contents of these seeds is beginning to be exploited, especially with maize kernels. With modern material, RNA typing (**transcriptomics**) is used to study the extent to which individual genes are expressed in a particular tissue at a particular time. This in turn enables the protein content of the tissue to be inferred. Comparing ancient RNA in archaeological maize kernels with RNA in kernels from modern elite cultivars might therefore indicate how the protein content of kernels has evolved in response to human selection for nutritionally better varieties.

Why are desiccated seeds such good sources of ancient biomolecules? The answer almost certainly lies with the natural role of the seed, which is to preserve its biochemical content during a period of dormancy, which can last several years and involve exposure to harsh environmental conditions. As the seed approaches dormancy, its water content drops, but in a controlled fashion that does not cause damage to cells and avoids release of chemicals and enzymes that might damage its biomolecules. The dormant seed can therefore be looked on as a natural structure for biomolecular preservation. Exactly how the biomolecules are prevented from breaking down is not understood, but it is clear that the process does not require constant maintenance, because the dormant seed is metabolically inactive. This means that if dormancy is extended to such an extent that the

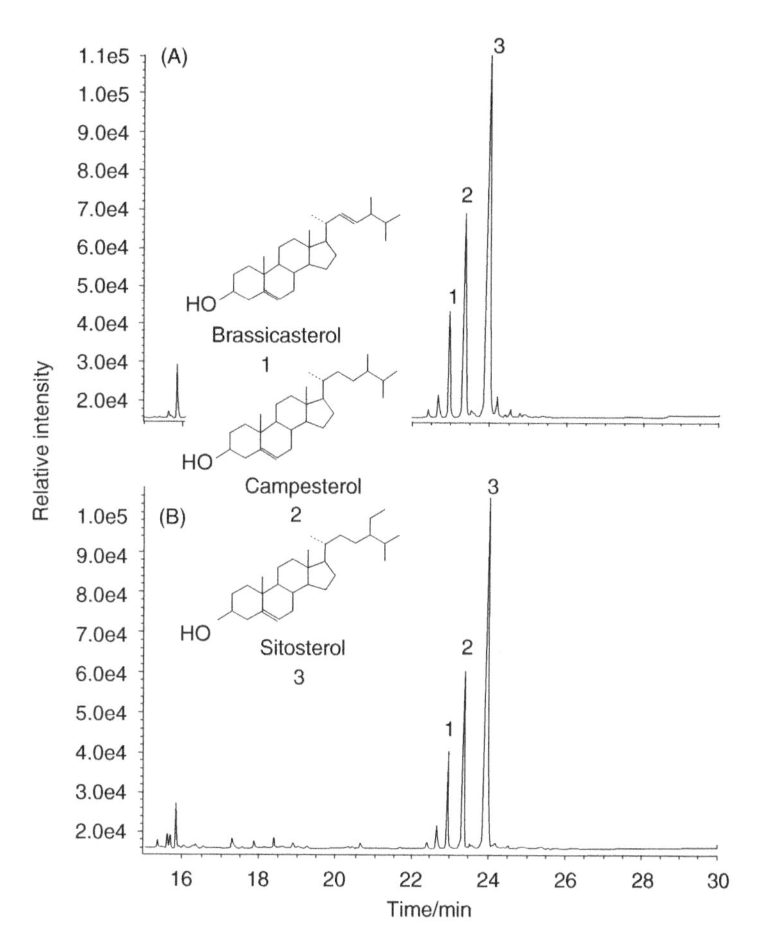

Figure 7.8 Partial gas chromatogram showing the similarity between the sterol composition of (A) modern radish seeds and (B) 1500-year-old radish seeds from Qasr Ibrim. The structures of the compounds represented by each peak are shown. Reprinted from Kerry O'Donoghue, Alan Clapham, Richard Evershed, and Terence Brown (1996), Remarkable preservation of biomolecules in ancient radish seeds, *Proceedings of the Royal Society of London, Series B*, 263, 541–7.

seed loses its ability to germinate and eventually dies, the protective mechanisms might continue to operate, at least in the short term. The outcome would be the relatively good biomolecular preservation that we observe with desiccated seeds from archaeological sites. The prediction that desiccated structures other than seeds might have less well preserved biomolecules is borne out, as ancient DNA is not so frequently recovered from remains such as husks and leaves.

In addition to these archaeological specimens, there are several kinds of desiccated plant material, younger in age, that have historic significance. The most abundant of these are herbarium specimens, which date from the 17th century to the present day, and are widespread in collections around the world. Herbarium material is preserved by drying, and in many respects is equivalent to the desiccated remains that we have already

Figure 7.9 A corn dolly from the Museum of English Rural Life, Reading. Image courtesy of Diane Lister.

considered. Nuclear and chloroplast DNA has been isolated from wheat grains from 19th century herbarium collections, and it seems likely that DNA is widespread in this type of material. Whether biomolecules other than DNA are present is not known. DNA has also been detected in seeds from corn dollies, decorative structures made of wheat, barley and other dried plants, some of which are several hundred years old (Figure 7.9). The cereal plants used to make some of the older corn dollies might come from varieties that are now extinct, hence the interest in studying them, but the dollies themselves are such valuable objects that destruction of even small pieces needs stringent justification. The roofs and walls of some old buildings also contain cereal plants used as thatch, insulation, and to add strength to mud bricks and daub. So far this type of material has not yielded ancient DNA, possibly because it has been exposed to the vicissitudes of climate and, although dried when first used, has subsequently gone through cycles of partial rehydration which might result in DNA degradation.

7.3.2 Some charred and waterlogged plant remains contain ancient DNA

Desiccated plant remains are only found in those parts of the world where the environmental conditions are conducive for this type of preservation. The vast majority of archaeological plant specimens, including virtually all from Europe, have been preserved by burning, resulting in oxidized remains that archaeologists have traditionally referred to as carbonized, but which are more correctly called charred as it is clear that not all of the organic material has been completely converted into carbon. Charred seeds are typically black with a shiny "clinkered" surface, though some can be dark brown (Figure 7.10). Charred remains are frequently recovered as individual seeds, which probably became burnt by falling into the fire while being cooked. Occasionally much larger quantities are found in granaries or other seed stores that caught fire and were abandoned by their owners.

Ancient DNA was first detected in charred wheat seeds in the early 1990s. Initially the results appeared dubious as the total organic content of this type of material is very low and

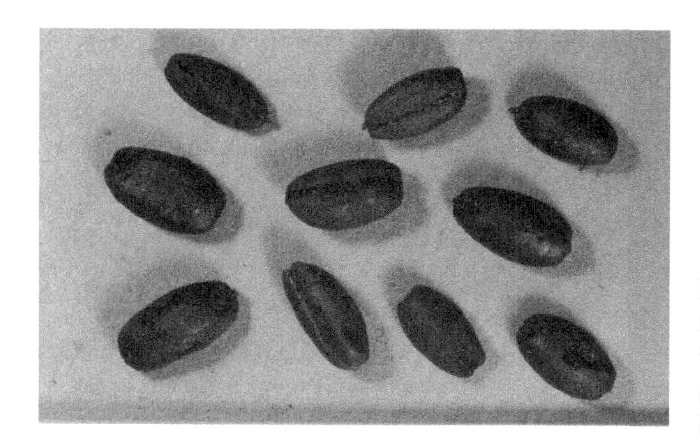

Figure 7.10 Charred grains of spelt wheat from Assiros, Greece, approximately 3000 years old. Image kindly provided by Glynis Jones.

other biomolecules are clearly not present, but similar detections have since been made with charred cereal grains of various kinds and with charred maize kernels and cobs. DNA from charred material typically has undergone a great deal of damage but short nucleotide sequences have been obtained, sufficient to identify short tandem repeat alleles in some instances. Many charred remains contain no detectable DNA at all, and even in a single assemblage there are often many seeds that contain no DNA and only a few that do.

Artificial charring experiments have been carried out to try to rationalize the presence of DNA in material that has been heated to temperatures sufficient to cause charring. This work has shown that if wheat seeds are heated above 225°C then all the DNA breaks down during the first 4 hours, but at lower temperatures the rate of DNA decay gradually slows to such an extent that some is still present after 5 hours. The rate of decay is slower when the amount of oxygen is limited, which might explain the presence of DNA in some grains recovered from burnt stores, such as at Assiros in Greece and Danebury, England. In these stores, grains were kept in sealed pots or clay lined pits, in which a small amount of germination would take place, reducing the oxygen level to a point where further germination was inhibited and the seeds remained dormant over winter. Destruction of these stores by fire would therefore result in seed charring in the low oxygen conditions shown to reduce DNA degradation. Simulations in which grain stores have deliberately been burnt down have shown that although parts of the structure reach very high temperatures, in excess of 600°C, at a depth of 0.6 m immediately below the conflagration the temperature does not get above 200°C. The presence of DNA in charred grains from storage pits can therefore perhaps be understood, even if its survival in material from other sources remains unexplained.

Seeds and other plant remains can also be recovered from waterlogged environments, such as the sediment at the bottom of a lake, or simply from part of a site that is below the water table. Although many biomolecules, including DNA, are susceptible to hydrolysis, the anaerobic conditions associated with waterlogging might be expected to retard their degradation. In some cases, seeds obtained from waterlogged locations had previously been charred, examples being cereal grains from Swiss lake villages, where houses were built on stilts above the surface of the edge of a lake, and burnt grains were likely to fall from the cooking area into the water. Ancient DNA has been found in several sets of such grains. Plant remains that have been preserved solely by waterlogging have not been studied so extensively, but protein profiling of extracts of grape seeds from a waterlogged medieval site in York, England, suggests that partially degraded proteins may be present in some of these.

Further Reading

Brown, K.A., O'Donoghue, K. & Brown, T.A. (1995) DNA in cremated bones from an Early Bronze Age cemetery cairn. *International Journal of Osteoarchaeology*, 5, 181–7.

Bryant, V.M. & Dean, G.W. (2006) Archaeological coprolite science: the legacy of Eric O. Callen (1912–70). *Palaeogeography, Palaeoclimatology, Palaeoecology*, 237, 51–66.

David, A.R. & Tapp, E. (eds.) (1992) *The Mummy's Tale: The Scientific and Medical Investigation of Natsef-Amun, Priest in the Temple of Karnak*. Michael O'Mara Books, London.

Evershed, R.P. (1990) Lipids from samples of skin from seven Dutch bog bodies: preliminary report. *Archaeometry*, 32, 139–53.

Hedges, R.E.M., Millard, A.R. & Pike, A.W.G. (1995) Measurements and relationships of diagenetic alteration of bone from three archaeological sites. *Journal of Archaeological Science*, 22, 201–9. [Describes the Oxford Histological Index.]

Koon, H.E.C., O'Connor, T.P. & Collins, M.J. (2010) Sorting the butchered from the boiled. *Journal of Archaeological Science*, 37, 62–9. [Distinguishing between cooked and burnt bone.]

Lister, D.L., Bower, M.A., Howe, C.J. & Jones, M.K. (2008) Extraction and amplification of nuclear DNA from herbarium specimens of emmer wheat: a method for assessing DNA preservation by maximum amplicon length recovery. *Taxon*, 57, 254–8.

Lynnerup, N. (2007) Mummies. *Yearbook of Physical Anthropology*, 50, 162–90.

Nielsen-Marsh, C.M. & Hedges, R.E.M. (2007) Bone porosity and the use of mercury intrusion porosimetry in bone diagenesis studies. *Archaeometry*, 41, 165–74.

O'Donoghue, K., Clapham, A., Evershed, R.P. & Brown, T.A. (1996) Remarkable preservation of biomolecules in ancient radish seeds. *Proceedings of the Royal Society of London series B*, 263, 541–7. [Preservation in desiccated seeds from Qasr Ibrim.]

Pruvost, M., Schwarz, R. & Correia, V.B., *et al.* (2007) Freshly excavated fossil bones are best for amplification of ancient DNA. *Proceedings of the National Academy of Sciences USA*, 104, 739–44.

Reiche, I., Vignaud, C. & Menu, M. (2002) The crystallinity of ancient bone and dentine: new insights by transmission electron microscopy. *Archaeometry*, 44, 447–59.

Smith, C.I., Nielsen-Marsh, C.M., Jans, M.M.E. & Collins, M.J. (2007) Bone diagenesis in the European Holocene I: patterns and mechanisms. *Journal of Archaeological Science*, 34, 1485–93.

Threadgold, J. & Brown, T.A. (2003) Degradation of DNA in artificially charred wheat seeds. *Journal of Archaeological Science*, 30, 1067–76.

Turner-Walker, G. (2008) The chemical and microbial degradation of bones and teeth. In *Advances in Human Palaeopathology* (ed. Pinhasi, R. & Mays, S.), 3–29. John Wiley and Sons, Chichester.

Wilson, A.S., Taylor, T. & Ceruti, M.C., *et al.* (2007) Stable isotope and DNA evidence for the ritual sequences in Inca child sacrifice. *Proceedings of the National Acadamy of Sciences USA*, 104, 16456–61. [Segmental analysis of hair reveals a change in diet one year before sacrifice.]

Wilson, J., Hardy, K. & Allen, R., *et al.* (2010) Automated classification of starch granules using supervised pattern recognition of morphological properties. *Journal of Archaeological Science*, 37, 594–604.

8

Degradation of Ancient Biomolecules

An understanding of the degradation of ancient biomolecules is an essential complement to the use of these molecules to address archaeological questions. At the most basic level, a knowledge of decay rates enables material that is too old to contain preserved biomolecules to be identified, so that research effort can be directed at younger specimens more likely to yield useful information. For some types of biomolecule, especially lipids, study of the degradation pathways has identified breakdown products that can be used as markers for the parent compounds, which may no longer be present in a specimen. With DNA, degradation can lead to changes in the nucleotide bases, which result in errors in the DNA sequence that is obtained after PCR. An understanding of the nucleotide changes enables the errors to be recognized and taken into account when the sequence data are analyzed. With collagen, which contributes to the structural integrity of the hard and soft tissues of animals, a knowledge of the decay pathway has an additional level of importance as a means of understanding the diagenetic breakdown of bones and mummified remains.

8.1 Complications in the Study of Biomolecular Degradation

Considerable progress has been made in understanding how DNA, protein, lipids, and, to a lesser extent, carbohydrates, degrade in archaeological specimens, but this progress has not been easy to achieve. This is because of the complexity of the decay processes and because of limitations in the methods available for their study. Before surveying what we currently know about degradation of the individual types of ancient biomolecule, we must first look at the general problems inherent in this area of research.

 © 2011 Terry Brown & Keri Brown

8.1.1 A variety of factors influence the decay of an ancient biomolecule

The main complication in understanding the degradation of a biomolecule is the variety of different factors that can influence the process. These can be grouped under three broad headings: autolysis, environmental factors, and microbial attack.

Autolysis occurs during the period immediately after death when breakdown of internal controls within cells can lead to biomolecules being attacked by endogenous enzymes from which they are protected when the cell is alive. Enzymes which digest DNA, proteins, lipids, and carbohydrates are contained within the lysosomes of living animal cells, small membrane bound vacuoles distributed throughout the cell's cytoplasm. In plant cells, an equivalent role is played by the larger vacuole in the center of the cell. The enzymes in lysosomes and vacuoles are used to turn over cellular biomolecules, and possibly to break down exogenous molecules for nutritional purposes and to help digest invading bacteria. The important point is that all of these processes are carefully controlled so that the cell's endogenous biomolecules are unaffected, which is why the enzymes are packaged away from the rest of the cell components. After death, the cell's membranes start to break down and the digestive enzymes are released from the lysosomes and vacuoles. The resulting activity is capable of completely degrading the DNA, protein, lipid, and carbohydrate in the cell, and ancient biomolecules probably only survive when the autolytic processes are halted before reaching completion, for example if the tissue becomes desiccated. As well as the enzymatic activity during autolysis, the breakdown of the cell's structure will generate reactive chemicals such as peroxides, which may also contribute to destruction of biomolecules.

Those biomolecules that survive the initial, rapid degradative activity occurring during autolysis have the chance of being preserved in the archaeological record. They then become subject to the slower but nonetheless incessant decay promoted by environmental factors, both chemical and physical. The chemical factors are primarily water and oxygen, both of which are highly reactive compounds participating in, respectively, hydrolytic and oxidative reactions. It is also possible that chemicals released by decay of one biomolecule can react with and promote the breakdown of a second type of biomolecule. The physical factors include the various forms of radiation – cosmic, ultraviolet, and geological – which have a significant impact on the integrity of some biomolecules in living cells, and equally important effects after death. Heat is also an important physical factor that influences biomolecular decay, not by initiating its own specific chemical reactions but by increasing the rate of the reactions promoted by all the other factors. The relative impacts of these different chemical and physical factors on the overall decay of the biomolecules in a specimen will depend on the nature of that specimen and the environmental conditions to which it is exposed. Desiccated remains, for example, might be expected to suffer less from hydrolytic attack, ones buried in an anoxic environment will be less subject to oxidative reactions, and specimens in permafrost should undergo relatively slow biomolecular decay due to the low temperature.

Microbial activity can be looked on as a special type of environmental factor. We have already learnt about the important impact that bacteria and fungi have on the overall diagenesis of bones and other bioarchaeological specimens (Chapter 7). Microorganisms are initially attracted by the organic material within biological remains, which provide a source of energy and nutrients. To utilize this material, a microbe must be able to secrete one or more enzymes able to break down the biomolecule into smaller units that can be absorbed. These enzymes are similar to those that are released by the cell itself during autolysis, and have the same effect, although much more slowly. The overall impact that microbial attack has on the biomolecular content of an archaeological specimen depends on the nature of the microflora in the environment in which the specimen in buried. Microbes that secrete extracellular nucleases and/or lipases are relatively common in the biosphere, but ones able to digest other biomolecules such as collagen and keratin are much rarer.

8.1.2 There are limitations to the approaches available for studying degradation

There are two ways of studying the degradation pathways of an ancient biomolecule. The first is to carry out experiments in the laboratory, using standard chemistry procedures, and through these to identify the effects of hydrolysis, oxidation, and other known factors, chemical and physical, on the breakdown of the molecule. This approach has two major strengths. First, because it follows the principles of experimental chemistry the resulting information is precise and can be made very detailed. Second, rate constants can be determined for each of the various reactions that contribute to decay, enabling the length of time that the biomolecule can survive in the archaeological record to be predicted.

This *in vitro* chemical approach also has two limitations. The first of these is that it is unable to take a great deal of account of decay arising from enzymatic activity occurring during autolysis or as a result of microbial activity. The *in vitro* approach therefore provides incomplete information, as not all of the factors influencing decay are included. The second limitation relates to the difficulty in extrapolating from the relatively simple environment of a chemical solution to the much greater complexity of the environment within an archaeological specimen. In archaeological remains, rather than being in solution, preserved biomolecules could be absorbed to mineral surfaces or aggregated with other compounds. These complex associations might influence the nature and rate of the degradation reactions caused by the environmental factors and microorganisms responsible for decay.

The second way of studying the degradation pathways of an ancient biomolecule is to make direct observations of the decay products that are present in archaeological materials of different types and different ages. This approach avoids the limitations of *in vitro* chemical study because they identify the decay products that are actually formed, and so do not exclude any part of the process and do not involve extrapolations from aqueous reactions to those occurring in the complex environment in which the biomolecules are degrading. But direct observation suffers from its own problems. The most important of these is the detection limit of the methods used to identify the decay products. Individual compounds might be present in very low quantities, so low that they cannot be detected. So again the information obtained is incomplete as some products are missed. Even if all decay products can be identified, it can be difficult to use this information to deduce the decay pathway that gave rise to them. If the decay pathway itself is not understood, then the products present in one type of material, subject to its own particular preservation conditions, cannot be used to predict the degree or nature of preservation of the biomolecule in a different type of material.

The information from chemical studies must therefore be combined with direct identification of decay products in archaeological remains in order to develop an overall understanding of the degradation pathways for ancient biomolecules. In the next sections, we will study what these complementary approaches have so far told us about the decay of DNA, protein, lipid, and carbohydrates.

8.2 Degradation of Ancient DNA

Understanding how DNA degrades is important not only in biomolecular archaeology. The environmental factors that, in the long term, lead to DNA breakdown in archaeological specimens are the same factors that give rise to mutations and other types of DNA damage in living cells. DNA degradation is therefore relevant to mutagenesis, cancer, and other central areas of biological research, and extensive work on DNA breakdown has been carried out in this context since the 1950s. The available information was brought together in a seminal paper "Instability and decay of the primary structure of DNA," published in *Nature* by Tomas Lindahl in 1993. This paper was greatly influential in ancient DNA research, as it showed that reports, common at that time, of DNA in million-year-old fossils could not be correct, as these contravened the known kinetics of DNA breakdown.

The complement to chemical study of DNA breakdown – direct observation of decay products in archaeological remains – has been much less successful because of the small amounts of DNA that are preserved. Only with desiccated plant remains has there been convincing detection of ancient DNA by a direct analytical method – gas chromatography-mass spectrometry – rather than indirectly by PCR amplification, and in that study only the original nucleic acid bases could be detected rather than any decay products. During the last few years this problem has partially been circumvented by the use of next generation sequencing and novel variants of PCR, which have shown that only one of the predicted nucleotide base decay products is abundant in ancient DNA. We will examine that work and its implications at the end of this section. First we will concentrate on the most important type of decay process affecting ancient DNA, the strand breakage caused by reaction with water.

8.2.1 Hydrolysis causes breakage of polynucleotide strands

A characteristic feature of ancient DNA is the limited length of the PCR products that can be obtained after amplification. With even the best preserved specimens it is rarely possible to obtain products longer than 300 bp (Figure 8.1). This is often taken as an indication that the ancient DNA molecules are themselves equally short, but the assumption is only partly correct. Breakage of polynucleotides is certainly an important part of DNA degradation, but PCR product length is also affected by the presence of chemical modifications that block progress of the *Taq* DNA polymerase and prevent all of an ancient DNA molecule

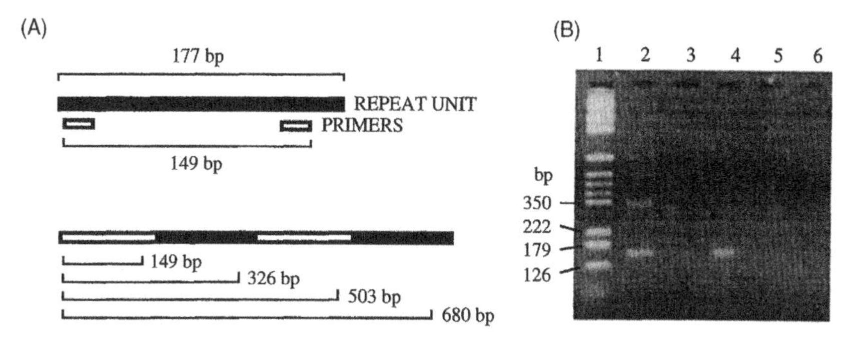

Figure 8.1 Illustration of the short lengths of ancient DNA molecules in even the best preserved archaeological specimens. DNA has been extracted from the desiccated radish seeds whose sterol profiles were shown in Figure 7.8. PCRs were directed at a repeat unit, as shown in part (A), with expected products of 149, 326, 503, and 680 bp. Part (B) shows the resulting gel. In lane 2, PCR with modern radish DNA has given a ladder of products, but in lane 4 PCR with DNA from the archaeological seeds has given only the shortest of these products. Lane 1 contains size markers. Reprinted from Kerry O'Donoghue, Alan Clapham, Richard Evershed, and Terence Brown (1996), Remarkable preservation of biomolecules in ancient radish seeds, *Proceedings of the Royal Society of London, Series B*, 263, 541–7.

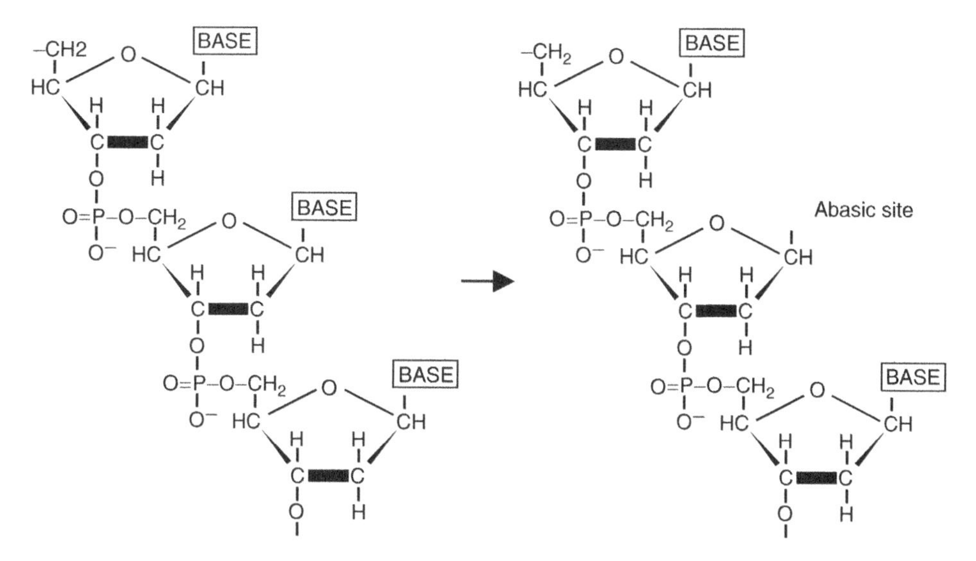

Figure 8.2 Water-induced cleavage of a β-*N*-glycosidic bond creating an unstable abasic site in a polynucleotide.

from being copied. We will examine the nature of these **blocking lesions** in a moment: first we will consider how the polynucleotide breaks arise.

Breakage of polynucleotides is caused indirectly by water, which attacks the β-*N*-glycosidic bond linking the base to the sugar component of the nucleotide (Figure 8.2). Hydrolysis of this bond releases the base, with G and A nucleotides about 20 times more

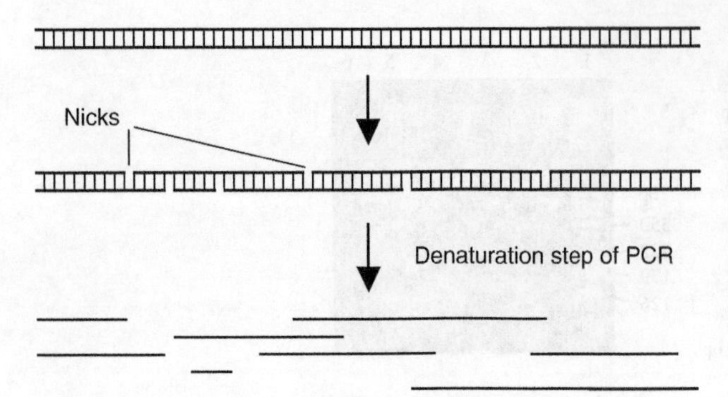

Figure 8.3 A double-stranded molecule that contains nicks will remain intact until the denaturation step of a PCR.

susceptible to attack than Cs and Ts. Breakage of a β-*N*-glycosidic bond does not immediately cause cleavage of the polynucleotide, but this occurs soon afterwards because the sugar that is left at the **abasic site** is prone to a chemical rearrangement that will break the strand in two. The continual hydrolysis of β-*N*-glycosidic bonds over time therefore results in gradual fragmentation of a polynucleotide.

It is important to remember that cellular DNA molecules are double stranded, which means that cleavage of the individual polynucleotides will not immediately result in breakage of the DNA molecule as a whole. Instead a series of nicks will appear in either strand of the molecule. If a pair of nicks occurs at roughly adjacent positions in the two polynucleotides, giving a region held together by just a few base pairs, then the molecule might break simply because these few base pairs are not strong enough to hold it together. But it is quite possible that some ancient DNA molecules are several kilobases in length, though containing frequent nicks in both strands. When amplified by PCR, these molecules would only yield short products because the initial denaturation step of the reaction will cause the molecules to fall apart into their constituent polynucleotides (Figure 8.3). Recognition of this fact has prompted attempts to repair nicks in ancient DNA molecules by treating with a DNA ligase, possibly in conjunction with a DNA polymerase to fill in any short gaps that might be present. The hope is that this treatment might result in longer templates and hence longer PCR products, but none of the attempts that have been reported so far has been conspicuously successful.

8.2.2 Blocking lesions can arise in various ways

A blocking lesion is any type of modification to a polynucleotide that prevents progression of a DNA polymerase along the strand. In living cells, blocking lesions prevent DNA replication and can lead to death of the cell, but most are repaired before they are able to have such dire consequences. After death, the repair mechanisms cease and blocking lesions gradually accumulate in DNA molecules. Their presence in ancient DNA reduces the lengths of the PCR products that can be obtained, because they hinder the progress of *Taq* polymerase along the initial templates that are being copied.

The abasic sites that result from hydrolytic cleavage of β-*N*-glycosidic bonds are blocking lesions, but as these quickly become converted into strand breaks they are not common in ancient DNA molecules. The most important blocking lesions in ancient DNA are probably those caused by oxidation of the purine and pyrimidine bases or of the ribose sugar. Oxygen itself (O_2) is not particularly reactive, but ionizing radiation such as cosmic rays and geological radiation generates derivatives, including hydrogen peroxide (H_2O_2), super-oxide ($\cdot O_2$), and free hydroxyl radicals ($\cdot OH$), which are more powerful oxidants. Oxidation of the bases can give rise to a variety of products, the commonest reaction being replacement of one or more double bonds with single bonds, the chemical process called ring saturation. In some cases, replacement of a double bond gives an unstable structure and the purine or pyrimidine ring breaks open. Many of these base oxidation products are known to occur in living cells, and most or all are likely to be present in ancient DNA also. Oxidation of the sugar predominantly affects the bond linking the 2′ and 3′ carbons, which breaks, opening up the ribose ring but possibly not causing cleavage of the polynucleotide.

Oxidation can also cause larger scale alterations to DNA molecules, including dimerization of purines on opposite strands. The resulting crosslink is an effective blocking agent. Ultraviolet radiation can form links between purines on the same strand, and other types of crosslink can be formed as a result of the **Maillard reaction**, which occurs between sugars and amino acids, and hence can link together polynucleotides and peptides. Even if such a reaction does not cause a crosslink, the presence of a peptide attached to a polynucleotide is likely to act as a blocking lesion in its own right. Maillard products are commonly formed during the decomposition of biological material, so are likely to formed in most archaeological specimens at some stage during their diagenesis. Thiazolium compounds such as *N*-phenacylthiazolium bromide (PTB) are able to break the sugar-protein link in Maillard products, and it has been claimed that treatment of ancient DNA extracts with PTB increases PCR success by removing blocking peptides from polynucleotide chains. It is equally possible that the added PTB breaks down free Maillard products that copurify with DNA, these products interfering with the PCR by inhibiting the action of the polymerase in some way, without actually forming blocking lesions.

8.2.3 Breaks and blocking lesions have different effects on PCR

The combination of polynucleotide breakage and accumulation of blocking lesions limits the lengths of PCR products that can be obtained from ancient DNA. Over time, the frequency of strand breaks and blocking lesions will increase until a point is reached where no PCR products can be obtained. Inability to obtain PCR products is often equated with absence of DNA, but in many cases DNA will still be present, but too damaged to work as a template for PCR. Even in a specimen that gives PCR products the bulk of the ancient DNA that is present might fall into this overly damaged category. This has been shown by a PCR technique called single primer extension (SPEX). In the first step of SPEX, the ancient DNA molecules are copied by extension of a single primer, this extension continuing until the polymerase reaches a strand break or a blocking lesion (Figure 8.4). This is exactly the same as what happens in the first cycle of a standard PCR (Figure 2.12). The difference is that the SPEX primer has a biotin label and so the extended strands can be purified by binding to beads coated with streptavidin. A poly(C) sequence – a series of

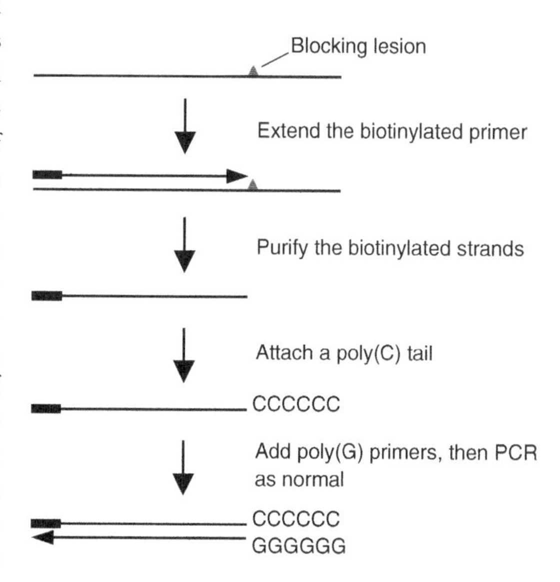

Figure 8.4 Outline of the key steps of the single primer extension (SPEX) method.

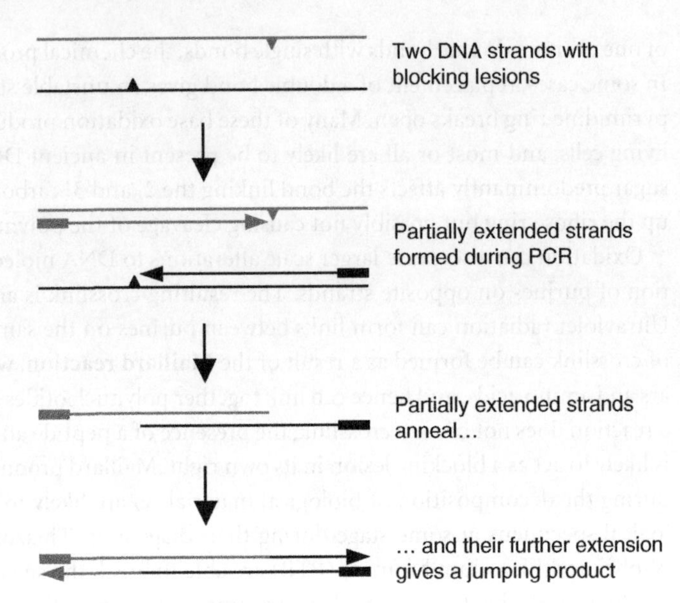

Figure 8.5 One of the ways in which hybrid PCR products can be synthesized by template switching.

C nucleotides – is then synthesized at the 3′ end of each strand, this sequence subsequently acting as the annealing site for a second primer. The strands can now be amplified by PCR, enabling their lengths to be worked out. In most specimens that have been studied by SPEX, the vast majority of the extended strands are less than 30 nucleotides in length. Those long enough to have generated a product in a standard PCR form a very small proportion of the total.

Despite the presence of strand breaks and blocking lesions, it is sometimes possible to obtain relatively long PCR products from highly damaged ancient DNA preparations. This is because of **template switching**, more commonly called **jumping PCR**, in which the short, partially extended strands that are synthesized during the first cycle of a PCR with damaged ancient DNA act as primers in the second and later cycles. These partially extended strands could prime onto other ancient DNA molecules, or possibly onto one another (Figure 8.5). In either case, further extension of these strands, possibly after three or more rounds of "jumping," might eventually lead to a full-length product that could then be amplified in the normal fashion during the remaining steps of the PCR.

At first glance, jumping PCR might appear to be beneficial as a means of generating long PCR products from ancient DNA molecules that are heavily damaged. In some circumstances, this is the case. For example, if an extract from a human bone contains ancient mitochondrial DNA and is not contaminated with any modern mitochondrial DNA, then jumping PCR could result in synthesis of long products whose sequences are entirely representative of the ancient DNA. Imagine, however, the result if modern contamination is also present, and the contaminating mitochondrial DNA has a different sequence to the ancient version (i.e. they are different haplotypes). Now it is possible for jumping to occur between the ancient and modern molecules, creating hybrid products whose sequences are part one haplotype and part another. Unwary researchers might think they have found a new haplotype in their archaeological specimen, and become excited, but all they have done is generate an artifact. Jumping PCR has been shown to be a common occurrence during PCR of DNA from many types of archaeological specimen.

8.2.4 Miscoding lesions result in errors in ancient DNA sequences

We have seen how many types of chemical modification occurring during DNA degradation can block copying of ancient DNA molecules during PCR. A second and more insidious type of nucleotide modification is one that does not block the DNA polymerase, but causes it to misread the DNA sequence. These modifications, called **miscoding lesions**, have been extensively studied in living cells because they can lead to mutations when genomic DNA is replicated.

The most common miscoding lesion occurring in ancient DNA is caused by loss of the amino ($-NH_2$) group from cytosine bases (Figure 8.6A). This is called deamination and is another of the hydrolytic effects that water has on a DNA molecule. The deaminated form of cytosine is uracil, one of the bases normally found in RNA and which base pairs with adenine. Figure 8.6B takes us through a PCR with an ancient DNA molecule that contains a deaminated cytosine. In the first cycle of the PCR, the uracil pairs with an adenine, which in turn pairs with thymine when it is copied in the next cycle. This means that when the PCR product is cloned and sequenced, the deamination leads to the C being read as a T, causing an error in the DNA sequence. Note that as well as C→T errors, cytosine deamination can also lead to a G being read as an A, if the other strand of the PCR product is the one that is sequenced.

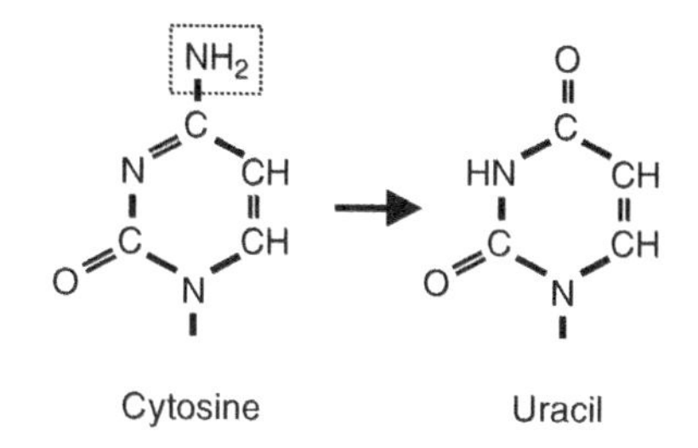

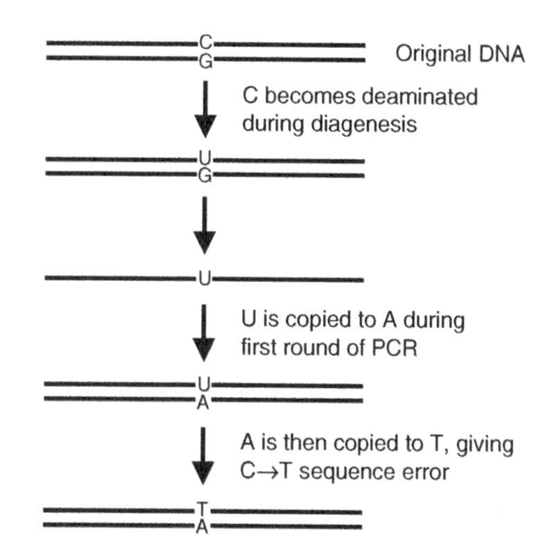

Figure 8.6 Cytosine deamination and generation of a C→T sequence error. In (A), the amino group that is lost during deamination is boxed.

In living cells, 10–30% of the cytosine bases in nuclear DNA (and rather fewer in mitochondrial DNA) are naturally modified by addition of a methyl ($-CH_3$) group to carbon number 5, giving 5-methylcytosine. This modification has a number of functions, including silencing genes that must be kept inactive in a particular tissue. 5-methylcytosine base pairs with guanine and so does not act as a miscoding lesion, but is even more susceptible to deamination, giving rise to thymine. Once again, if this occurs in ancient DNA, then the sequence that is obtained has a C→T or G→A error. The accumulation of cytosine and 5-methylcytosine deamination over time can lead to a substantial number of miscoding lesions being present in some ancient preparations, these having an important impact on the accuracy of the DNA sequences that are obtained (Figure 8.7).

Original sequence

```
AAGTACAGCAATCAACCCTCAACTATCACACATCAACTGCAACTCCAAAGCCACCCCT
```

Clone sequences

```
AAGTACAGCAATCAACCCTCAATTATCACACATCAACTGCAACTCCAAAGCCACCCCT
AAGTACAGCAATCAACCCTCAACTATCACACATCAACTGCAACTCCAAAGCCACCCCT
AAGTACAGTAATCAACCCTCAACTATCACACATCAACTGCAACTCCAAAGCCACCCCT
AAGTACAGCAATCAACCCTCAACTATCACACATCAACTGCAACTCCAAAGCCACCCCT
AAGTACAGCAATCAACCCTCAACTATCACACATTAACTGCAACTCCAAAGCCACCCCT
AAGTACAGCAATCAACCCTCAACTATCACACATCAACTGCAACTCCAAAGCCACCCCT
AAGTACAGCAATCAACCCTCAACTATCACACATCAACTGCAACTTTAAAGCCACCCCT
AAGTACAGCAATCAACCCTCAACTATCACACATCAACTGCAACTCCAAAGCCACCCCT
AAGTACAGCAATCAACCCTCAACTATCACACATCAACTGCAACTCCAAAGCCACCCCT
```

Figure 8.7 Multiple alignment of the sequences of 10 clones of a PCR product. C→T sequence errors caused by deamination of cytosines in the ancient DNA molecules are highlighted.

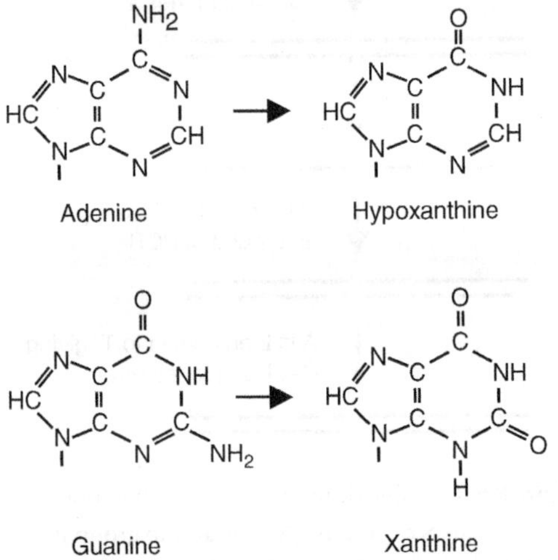

Adenine → Hypoxanthine

Guanine → Xanthine

Figure 8.8 Deamination of adenine to hypoxanthine, and of guanine to xanthine.

Other nucleotide bases undergo deamination, but the effects on ancient DNA sequences are less significant. Adenine is deaminated to hypoxanthine (Figure 8.8), the latter preferring to base pair with cytosine rather than thymine, giving an A→G miscoding lesion. The rate of adenine deamination is approximately 2–3% that of cytosine deamination, so these A→G lesions are much less common in ancient DNA. Guanine can also be deaminated but the product, xanthine, still base pairs with cytosine and so is not miscoding. Thymine cannot be deaminated, as it has no amino group.

8.3 Degradation of Ancient Proteins

We know a great deal about some aspects of protein degradation, and much less about others. Our knowledge is strongest with regard to the degradation of collagen, which has been studied in experimental systems and by examination of the structure of partially degraded collagen in bones and other archaeological specimens. We also understand a great deal about the breakdown of collagen by digestive enzymes secreted by certain types of microorganism. We know rather less about the degradation of keratin, but what we do know includes important information on the effects of microbial activity on this protein. Our main area of ignorance lies with the nature and rate of the degradation processes for non-structural proteins such as hemoglobin and enzymes. There have been no experimental studies of the environmental breakdown of these proteins, and we are forced to use information from more general areas of protein chemistry to deduce the possible decay pathways. The resulting deductions do not always appear to agree with the surprisingly good preservation of hemoglobin and enzymes in some types of specimen. We will therefore begin this section on solid ground by studying the chemical and enzymatic degradation of collagen and keratin before moving on to the more difficult issues regarding non-structural proteins.

8.3.1 Collagen breaks down by polypeptide cleavage and loss of the resulting fragments

Humans, and other vertebrates, synthesize several different types of collagen. The individual polypeptides are approximately 1000 amino acids in length, made up largely of a repeating sequence that is either glycine–X– proline or glycine–Y–hydroxyproline, where X and Y can be any amino acid. Hydroxyproline is a derivative of proline that is made by post-translational modification (Figure 8.9). The repeat pattern gives the collagen polypeptide a left-handed helical structure, quite different to the right-handed α-helix commonly found in other proteins. The conformation of this helix is such that three collagen polypeptides can wind around each other to form the triple helical structure called tropocollagen. Tropocollagen has considerable tensile strength due to the tightness of the helix and the presence of cross-linking covalent bonds that form between lysines in the individual polypeptides. The collagen polypeptides are also attached to one another by hydrogen bonds, which individually are weak but, because there are so many of them, make a major contribution to the strength and stability of the triple helix. Tropocollagen molecules align and become linked to one another by more lysine–lysine bonds, forming the collagen fibrils that can be seen by electron microscopy.

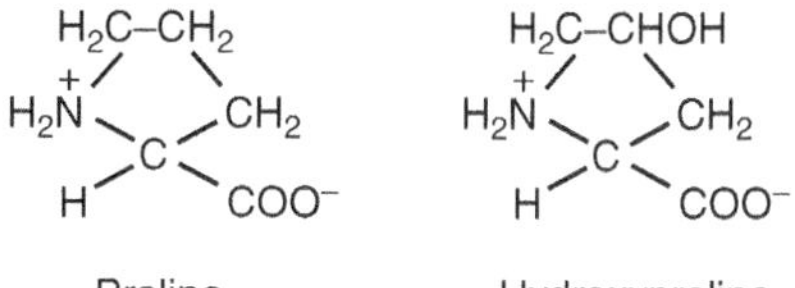

Proline | Hydroxyproline

Figure 8.9 The structures of proline and hydroxyproline.

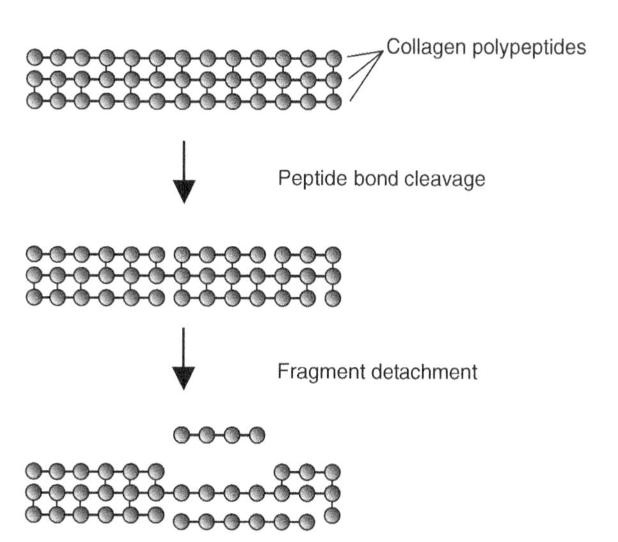

Figure 8.10 Model for collagen degradation. Three collagen polypeptides are shown, linked to one another by intrastrand hydrogen bonds. Hydrolysis of peptide bonds gives fragments that detach from the fibril. The ease with which a fragment detaches depends in the number of hydrogen bonds linking it to the adjacent polypeptide.

In aqueous solutions, collagen fibrils break down by a two-step process (Figure 8.10). The first step involves hydrolysis of peptide bonds, causing the collagen polypeptides to fragment into shorter units. This step is very dependent on pH, the rate of bond hydrolysis being 10 times more rapid at pH 1 (highly acid), and 100 times more rapid at pH 12 (alkaline pH), compared with the rate at neutral pH. In the second step, fragments detach from the fibril and are lost in the surrounding solution. The rate at which fragments are lost depends on their length, or more specifically on the number of hydrogen bonds holding the fragment to the neighboring collagen polypeptides. Shorter fragments will be held in place by fewer hydrogen bonds, and hence are more likely to dissociate from the fibril. The overall effect of peptide bond hydrolysis and loss of fragments is a gradual unpacking of the fibril, converting collagen into gelatin.

Non-mineralized collagen fibrils, such as those in connective tissue and skin, are less resistant to degradation than the mineralized fibrils in bone. Gelatinization can be induced by heating unmineralized collagen to 70°C, but does not occur in bones below 150°C. The mineral component of bone therefore provides a protective environment that enables collagen to survive for much longer than is possible in soft tissues, unless the latter are preserved by artificial or natural mummification. Microscopic examination of bones of different ages show that breakdown of collagen fibrils begins within the first few years of burial of a bone, but is limited to small-scale swelling and unravelling. The process is so slow that even after 150,000 years it is still possible for a bone to retain the majority of its original collagen content. Models have been proposed for the protective effect of bone mineral, these suggesting that the bone matrix decreases the rate of hydrolysis by preventing water molecules from accessing the collagen fibril, and holds broken fragments in place, slowing down their dissociation from the fibril.

As well as the chemical pathways described above, collagen can also be degraded by enzymes secreted by various species of bacteria and fungi. Some of these are true **collagenases**, specific for this protein, and others are non-specific proteases which act on a range of proteins including collagen. Both types of enzyme attack peptide bonds in order to release polypeptide fragments that can be absorbed and digested by the microorganism. Collagenases, being adapted to the unusual and compact structure of collagen fibrils, degrade the protein more quickly than the non-specific proteases, the latter often being unable to attack a fibril until it has been partially unpacked by the action of collagenases or by chemical decay. Many microbial species are known to synthesize collagen-degrading enzymes, and several of these are common in soil, examples being members of the genera *Clostridium*, *Streptomyces*, and *Achromobacter*. They would have a much more dramatic effect on bone survival were it not for one important factor – the secreted enzymes are too large to penetrate the mineral matrix of intact or partially degraded bone. Initially they are able only to degrade collagen exposed on the surface of a bone, and then are ineffective until some part of the mineral component has been broken down so additional collagen is exposed. This exposure may occur within the destructive foci thought to be caused by microbial decay of bone (Figure 7.2), or it may require a more general dissolution of the mineral matrix in acidic groundwater.

8.3.2 Much less is known about the degradation pathways for other ancient proteins

Collagen is the only protein whose decay pathway in archaeological remains can be described with any conviction. With keratin, the second most prevalent structural protein possessed by mammals, found mainly in hair, there have been fewer direct studies of degradation, but enough is known to suggest that the decay pathway is similar to that of collagen. Like collagen, keratin is a fibrous protein, each keratin polypeptide forming a long α-helix, pairs of which are able to wind around one another to produce a super-helix. This structure is then cross-linked by disulfide bridges that form between cysteine amino acids in adjacent polypeptides. Keratin super-helices combine into protofilaments, which in turn align to give the keratin fibers found in hair cells. Chemical degradation of keratin involves a combination of oxidation to break the disulfide bridges and hydrolysis to cleave peptide bonds. There are also keratinolytic fungi that secrete enzymes able to degrade keratin, again by attacking both the disulfide and peptide bonds. Removal of the disulfide bridges causes the fibre to unpack, increasing its susceptibility to attack by non-specific microbial proteases.

Other proteins that have been detected in archaeological remains have a **globular** rather than fibrous structure. This means that different regions of their polypeptide chains take up different secondary structures, the whole folded into a tertiary structure comprising a variety of α-helices and β-sheets, together with less organized regions (Figure 3.4). The tertiary structure is stabilized by a variety of non-covalent bonds (e.g. hydrogen bonds, electrostatic interactions, and hydrophobic forces) and, with some proteins, disulfide bridges between cysteines in different parts of the polypeptide chain. Globular proteins are therefore less compact than fibrous ones, and hence more susceptible to hydrolytic attack. Once a few peptide bones have been broken, the resulting fragments are likely to dissociate relatively rapidly, unless by chance they are held in place by disulfide bonds. Hence we might imagine that globular proteins would be relatively unstable after death of the organism, certainly less stable than collagen.

It is therefore surprising how many globular proteins have been reported in archaeological remains, especially bone, and how remarkably intact many of these appear to be. A few enzymes, such as alkaline phosphatase and super-oxide dismutase, have been identified in mummified remains and/or bones, and in some cases these enzymes are still capable of catalyzing the biochemical reactions they carry out in living tissue. To still be active in this way, not only must the protein retain its overall tertiary structure without deletion of peptide fragments, but it must also retain at its active site the exact conformation of chemical groups needed to bring the substrates together in the orientation needed for the enzymatic reaction to occur. Non-enzymatic blood proteins, such as hemoglobin and albumin, have also been detected in bones, using immunological methods, which, as described in Section 3.3, depend on the target protein retaining its overall three-dimensional structure and still having on its surface the same pattern of chemical groups as displayed by the original protein. Confidence in the authenticity of the detections of globular proteins that have been reported over the years would be greatly enhanced if experimental studies or models were available to show how these proteins can survive, apparently intact, in archaeological remains.

8.3.3 Amino acid racemization is an important accompaniment to protein degradation

So far we have looked only at the degradation pathways for entire proteins. We must also consider one important change that individual amino acids within a polypeptide chain undergo as the protein as a whole degrades. This process, called **amino acid racemization**, is important because it has been used as a means of dating archaeological and paleontological materials, and also because it has been suggested that the degree of racemization gives a convenient indication of the overall degree of biomolecular degradation that has occurred in a specimen.

An amino acid comprises a central carbon atom attached to four different chemical groups: a hydrogen atom, a carboxyl group, an amino group, and the R group that is different in each amino acid. This means that the central carbon is chiral and there are two enantiomers for each amino acid (Figure 8.11). This is true for every amino acid except glycine, whose R group is a hydrogen atom and hence does not have a chiral carbon. In nature, all

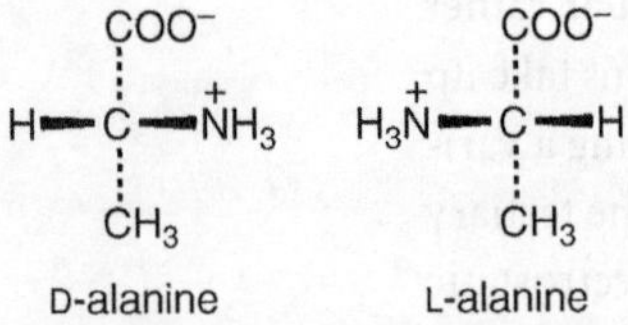

Figure 8.11 The two enantiomers of alanine. Compare with Figure 5.2, where the arrangement of atoms around a chiral carbon is shown.

amino acids exist as the L-enantiomer, except a few in specialized structures such as bacterial cell walls. In chemical solutions, on the other hand, the L- and D-enantiomers of an amino acid are in equilibrium, this equilibrium maintained by the interconversion of one enantiomer into the other. This interconversion is called racemization.

The biochemical processes occurring in living tissues maintain cellular amino acids in their L-forms. After death, however, these biochemical processes cease and racemization is able to proceed unhindered. Over time, the amounts of the L- and D-enantiomers reach equilibrium. This is not as simple a process as might be imagined, because the rate of racemization is different for each amino acid, and is affected by a wide range of environmental parameters, including temperature, pH, water content, ionic strength, and the presence of buffering compounds. The rate is also affected by the location of an amino acid within its protein, amino acids at the ends of a polypeptide and on the surface of a protein undergoing more rapid racemization, and can be retarded by absorption of a protein onto a solid surface, as might happen in a bone.

Despite the number of variables affecting the process, racemization of aspartic acid has been proposed as a dating tool. This method got off to an inauspicious start when it provided a date of 48,000 years for a set of paleoindian skeletons from California. As this placed the remains some 36,000 years before the time when human are thought to have arrived in the Americas, it caused quite a stir. Subsequently, radiocarbon dating was applied to the same specimens, giving a more realistic date of 5000–6000 years ago. Some archaeologists would argue that amino acid racemization dating is hopelessly unreliable, but this view can only be supported for bone, which is subject to inaccuracies caused by contamination with amino acids from the environment. These amino acids will already be in racemic equilibrium and hence will dilute the L-enantiomer content of the specimen. Closed systems, such as mollusc shells, appear to give much more accurate results when dated by this method.

The degree of amino acid racemization has also been proposed as a general indicator of the amount of biomolecular preservation in a specimen. The assumption is that specimens in which amino acid enantiomers are in equilibrium will have undergone so much biochemical diagenesis that their biomolecules will be heavily degraded. A **biochemical proxy** of this kind would be particularly useful in ancient DNA research, as a means of helping to validate supposed detections of ancient DNA in a specimen. The accuracy of amino acid racemization as a biochemical proxy has been tested by comparing racemization values with actual ancient DNA contents for a range of specimens. Initial results suggested that there was a good correlation between the two, but more recent studies have questioned this. Once again, the possible contamination of bone with exogenous amino acids is a major problem.

8.4 Degradation of Ancient Lipids

Lipids are a diverse group of compounds and as such it is not possible to make general statements about their degradation pathways in archaeological remains. Some lipids are very stable and can be detected in unchanged form in artifacts several thousand years in age.

Examples are the waxes: not just beeswax, but also the types found on the surfaces of the leaves of plants, the latter sometimes used as biomarkers for the types of vegetables processed in cooking vessels. These long-chain alcohol compounds have few reactive groups (Figure 4.4) and so are resistant to attack by chemicals and enzymes secreted by microorganisms. Other lipids break down to relatively stable end products, whose identities enable the original presence of the parent molecule to be inferred.

Rather than attempting a broad discussion of the possible breakdown pathways for lipids as a whole, we will focus on the degradation issues that are relevant to two of the most important applications of ancient lipids in biomolecular archaeology. These applications are the detection of residues of fats and oils in cooking pots and storage vessels, and the use of sterols as biomarkers for enrichment of agricultural soils. Together, these two case studies illustrate the range of issues that must be considered when ancient lipids are analyzed.

8.4.1 Fats and oils degrade to glycerol and free fatty acids

Identification of the fats or oils absorbed into the surfaces of cooking pots or present as residues in storage vessels is often used to infer the usage of those implements. Fats and oils are triglycerides, compounds made up of three fatty acids attached to a single molecule of glycerol (Figure 4.2). The breakdown of triglycerides has been studied extensively by the food industry, as part of broader research into the transformations that lipids undergo during cooking and other kinds of food processing. This work has shown that the major degradation pathway for triglycerides involves hydrolytic cleavage of the ester bonds that link the fatty acid chains to the glycerol component. Examination of intermediates suggests that the fatty acids are cleaved one at a time, though in no particular order. The end products of the pathway are glycerol and the free fatty acids.

In Chapter 4, we learnt that there are two different types of fatty acid, saturated and unsaturated. In a saturated fatty acid, every carbon in the chain is attached to hydrogen atoms and each pair of adjacent carbons is linked by a single bond. Saturated fatty acids are relatively stable and do not undergo additional change under normal cooking conditions. Unsaturated fatty acids, on the other hand, have at least one double bond in the carbon chain, providing opportunities for further degradative reactions of various kinds. These reactions include oxidation of a methylene ($-CH_2-$) located between two unsaturated carbons, which gives rise to a highly reactive group called a **free radical**. This group can attack other lipids, giving a range of breakdown products, including compounds classified as ketones, ketoacids, and dialdehydes. The accumulation of these products in foods leads to rancidity, which in effect is a warning signal to humans that the food is unpalatable, as many of the compounds are toxic. Unsaturated fatty acids are difficult to detect in the archaeological record because these breakdown products are non-specific, the same ones being obtained from different fatty acids and some also being possible breakdown products of other types of lipid. They cannot therefore be used to infer the presence of a particular fatty acid. This can be a problem when trying to detect plant oils in the archaeological record, as most of these are polyunsaturated triglycerides. Two minor decay pathways do, however, yield compounds that can act as biomarkers for unsaturated fatty acids. The first of these pathways involves conversion of the double C=C bond in a monounsaturated

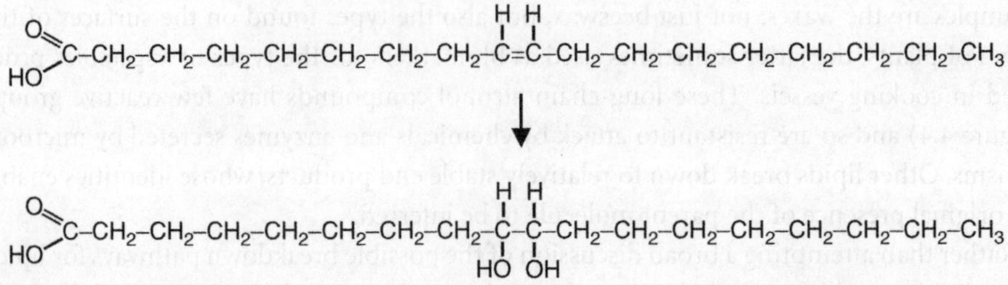

$$HO(O=)C-CH_2-CH_2-CH_2-CH_2-CH_2-CH_2-CH_2-CH=CH-CH_2-CH_2-CH_2-CH_2-CH_2-CH_2-CH_2-CH_3$$

$$\downarrow$$

$$HO(O=)C-CH_2-CH_2-CH_2-CH_2-CH_2-CH_2-CH_2-CH(OH)-CH(OH)-CH_2-CH_2-CH_2-CH_2-CH_2-CH_2-CH_2-CH_3$$

Figure 8.12 Conversion of a monosaturated fatty acid to a more stable dihydroxy fatty acid whose structure indicates the position of the double bond in the parent compound.

fatty acid to a single bond, accompanied by addition of hydroxyl (–OH) groups to the adjacent carbons (Figure 8.12). The resulting dihydroxy fatty acid is relatively stable and its precise structure enables the parent fatty acid to be identified. The second informative breakdown pathway converts polyunsaturated fatty acids into stable cyclic compounds called ω-(*o*-alkylphenyl)alkanoic acids. These are formed only at temperatures in excess of 270°C, and a single fatty acid can give rise to a variety of products. Detection of ω-(*o*-alkylphenyl)alkanoic acids is therefore an indication that a vessel has been used to process a polyunsaturated oil, although the identity of that oil cannot be deduced.

Most animal fats are saturated, so their fatty acid breakdown products are more stable. Fats from animals such as cattle and pigs have characteristic triglyceride profiles, and often the derivative fatty acids are sufficiently well preserved in potsherds for distinctions between vessels used to cook meats from these animals to be made. Individual pots were probably used many times, so animal fats that became absorbed in their surfaces would have been subjected to repeated rounds of heating and cooling, Under these relatively harsh conditions, the carboxyl groups of saturated fatty acids can react with one another, yielding long-chain ketones of the general formula $CH_3(CH_2)_mCO(CH_2)_nCH_3$. This reaction occurs at temperatures above 300°C and is possibly promoted by conditions within the matrix of a clay pot. Although puzzling when they were first identified, these compounds are now looked on as reliable biomarkers for the saturated fatty acids from which they are derived.

8.4.2 Decay products of cholesterol are biomarkers for fecal material

Our second case study concerns the way in which an understanding of the decay pathways for cholesterol enables the presence of fecal material to be detected, which is often used to indicate that a soil has been enriched by manuring (Section 14.1). Cholesterol is the most abundant animal sterol and is an important component of cell membranes. Like all sterols, a cholesterol molecule comprises four linked hydrocarbon rings, three of these containing six carbons each and one containing five carbons, with additional groups attached at various points on these rings (Figure 4.10). Sterols are relatively stable compounds and their diagenetic changes during an archaeological timescale are restricted to small structural alterations. For cholesterol, the commonest diagenetic change is conversion of the double

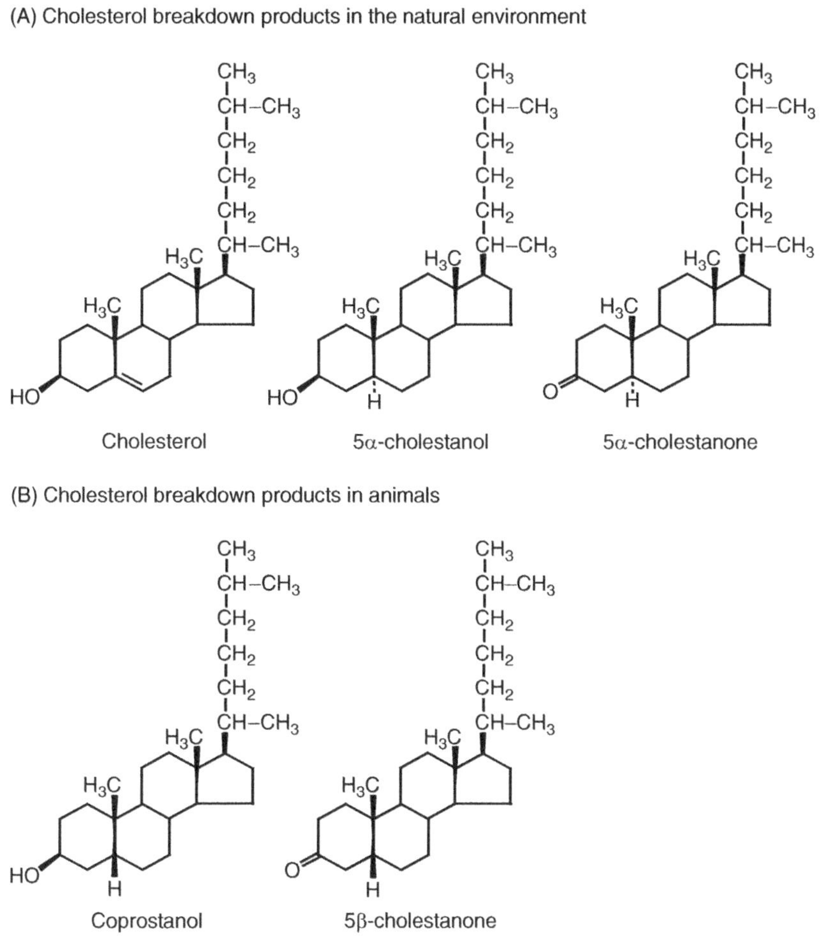

Figure 8.13 Cholesterol breakdown products.

bond between carbons 5 and 6 to a single bond, leading to addition of a hydrogen atom to carbon 5 (Figure 8.13A). This carbon is asymmetric, so the hydrogen can be added in two different orientations: the product of cholesterol decay in the natural environment has the hydrogen in the α orientation and is called 5α-cholestanol. This compound might, in turn, be converted to 5α-cholestanone by replacement of the hydroxyl group (–OH) attached to carbon 3 with a carbonyl group (=O).

5α-cholestanol and 5α-cholestanone are therefore diagenetic biomarkers for cholesterol, and their detection indicates the past presence of this compound. They are not, however, biomarkers for fecal material, as cholesterol that is ingested in the diet follows a slightly different decay pathway. Within the intestines of humans and most other animals, cholesterol is attacked by the gut microflora, again resulting in conversion of the C_5–C_6 double bond into a single bond, but this time adding the hydrogen group to carbon 5 in the β orientation. This gives rise to 5β-cholestanol, which is more commonly called **coprostanol** (Figure 8.13B). After excretion, coprostanol can undergo further diagenetic

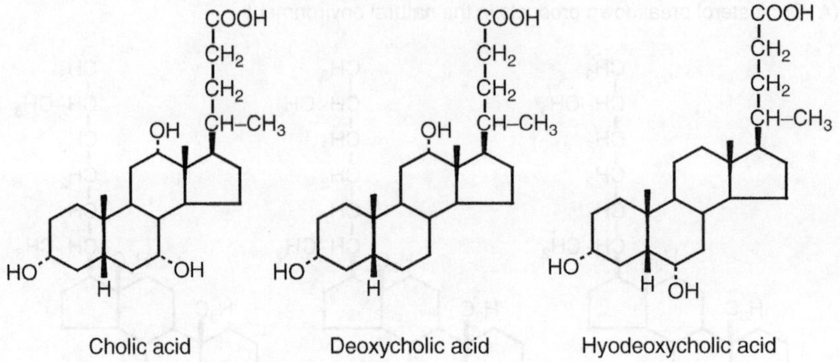

Figure 8.14 Bile acids. In these drawings, the positions of bonds linking methyl groups to the sterol ring are marked, but the methyl groups themselves are omitted.

changes to give compounds such as 5β-cholestanone. Coprostanol and its derivatives are therefore alternative biomarkers for cholesterol, but more importantly are biomarkers for fecal material, as this particular decay pathway for cholesterol occurs only in animal intestines.

Understanding the differences between the decay pathways for cholesterol that has been deposited directly in a soil (e.g. from a dead body) and cholesterol that has passed through an animal gut therefore enables fecal material to be detected. The analysis can be taken even further to enable the type of animal whose feces are present to be identified. There are two ways of doing this. The first is based on the amount of plant material in the diet. Plants contain their own specific types of sterol, such as sitosterol and campesterol. These compounds also undergo microbially induced transformations in animal intestines, giving rise to their 5β derivatives. Their presence in relatively large amounts suggests that the feces are from ruminants such as cattle.

The second way to infer the species of origin brings us to a third transformation undergone by cholesterol. An animal's own cholesterol is excreted after oxidation in the liver as a derivative called a **bile acid**, which is transferred via the bile duct to the intestine and thence into the feces. Bile acids are a family of compounds, differing from one another in the positions of hydroxyl groups (–OH) that are attached to the sterol ring (Figure 8.14). Deoxycholic acid, for example, has a hydroxyl group at carbon 12 and is present in human and cattle feces, but not in that of pigs. Hyodeoxycholic acid is the dominant bile acid in pigs, and cholic acid is present in human but not cattle feces. Typing the bile acids in soils can therefore identify the origin of any fecal material that is present.

8.5 Degradation of Ancient Carbohydrates

The only ancient carbohydrate that has so far been utilized in biomolecular archaeology is starch. Starch grains derived from plants have been recovered from a variety of archaeological settings, including sediments, the surfaces of implements used in food processing, from charred residues within cooking vessels, and even from the stomach contents of bog bodies and the calculus present on the teeth of human skeletons. In order to identify the species from which archaeological grains are derived, comparisons are made with modern reference material, so any diagenetic alterations to the size and/or morphology or the archaeological grains are likely to confound the analysis. We must therefore consider how starch is degraded in the environment.

8.5.1 Enzymes that degrade starch grains are common in the natural environment

Chemical hydrolysis of the glycosidic bonds in starch breaks both amylose and amylopectin down into their constituent glucose units, but this reaction is very slow except at high temperatures and in the presence of acid. In the natural environment, the chemical reaction is usually not a factor in starch degradation because many of the microorganisms and other creatures living in soil possess the enzymes needed to break starch down much more quickly, and some of these organisms scavenge through soil in search of starch grains which they use as an energy source. This activity might be less in certain types of soil than in others, and has been shown to decrease with depth, but should still be sufficient to turn over most of the grains that are deposited in soils by plant decomposition. Most of the starch content of freshly fallen spruce and aspen leaves is degraded within a few months, and experiments have shown that breakdown can occur more quickly than this if starch is added directly to a soil. Bacteria and fungi account for a large part of this activity, but extensive starch decomposition is also carried out by insects and other small animals, by earthworms, and in the rhizosphere – the surface area occupied by plant roots.

The pathway for starch grain decomposition in the natural environment has never been studied, but we can assume that it is similar to the processes by which grains are utilized as an energy source in plant tissues. These are slightly different for transitory and stored starch, but both involve a range of enzymes acting in a coordinated fashion to break the polysaccharide chain into shorter units and to convert the final monosaccharides into carbon dioxide and water with release of energy. The main enzymes responsible for cleaving the ($\alpha 1 \rightarrow 4$) glycosidic bonds in amylose and amylopectin are the amylases. Most organisms synthesize one or more α-amylases, which cut these bonds at random, gradually breaking the starch polymers into smaller units and eventually into maltose (two linked glucose units), maltotriose (three units), and some free glucose. Bacteria, fungi, and plants also possess β-amylases, which make more specific cleavages at every second glycosidic bond from a non-reducing end, and hence release maltose units. A third type of enzyme, the α-glucosidases, convert maltose units into glucose. None of these enzymes is able to attack the ($\alpha 1 \rightarrow 6$) bonds that form the branch points in amylopectin molecules (Figure 5.9). Instead, this reaction is carried out by a debranching enzyme such as pullulanase or limit dextrinase, pullulan and limit dextrin being different names for the branched oligosaccharides resulting from extensive cleavage of ($\alpha 1 \rightarrow 4$) bonds.

Studies of starch grain breakdown in plant tissues have shown that the morphologies of the grains change during their utilization, and not just through a general decrease in size. In at least some plants, the initial stages of decomposition involve alterations to the shape of the grains, with pits and holes appearing on their surfaces. This is worrying from an archaeological perspective as it suggests that partial enzymatic breakdown of a grain might change its morphology in a way that could lead to misidentification. It has also been suggested that some types of grain are relatively resistant to attack by scavengers, and hence more likely to survive in the archaeological record. Again this is worrying because it means that identification of starch grains in archaeological soils might give an incomplete picture of the types of plants present at a particular site. Another problem is that smaller grains are readily moved around within soils through the effects of groundwater, so their presence in a certain stratigraphic layer cannot be used as a secure indication of their presence in the associated archaeological context.

8.5.2 Why are starch grains preserved at all?

Starch degrading enzymes are so ubiquitous in the natural environment that we might wonder how any starch grains are able to survive into the archaeological record. Part of the answer probably lies with the huge number of grains that are deposited from plant debris, which in some environments might simply overwhelm the degradative ability of the soil flora and fauna. Of all the billions that are present, a few individual grains might survive.

Survival might be aided by the microenvironment around individual grains or groups of grains, which might protect those grains either by making them less accessible to the degradative enzymes or by inhibiting the activities of those enzymes. An example of inaccessibility has been discovered in grasslands, where fungal utilization of starch grains results in formation of soil aggregates of 250 μm and greater in diameter. The grains in these aggregates are rapidly degraded, but neither fungi nor bacteria are able to access grains that remain in smaller soil particles alongside these aggregates. Those grains can only be degraded after the soil has been reworked, either by later fungal growth or by physical disruption. The second possibility, localized inhibition of enzyme activity, has been proposed for soils containing heavy metals such as lead and aluminum. Heavy metals inhibit many types of enzyme, with α-amylases particularly susceptible. The presence of clay particles may also reduce enzyme activity by absorbing the enzyme molecules and/or the starch grains so that the two are not brought together.

It has even been suggested that starch grains present in archaeological soils have not been preserved in that context but derive from artifacts to which they were originally attached. This assumes that some types of artifact surface provide a protective environment for starch grain survival, which is itself a difficult topic. One might imagine that unless an artifact such as a processing or cooking implement is located in an area free from soil, then the grains will still be susceptible to enzyme degradation. An argument frequently made for survival of biomolecules on stone implements is that the artifact surface includes pits and cracks within which the biomolecule can become trapped and that are too large to be penetrated by scavenging microorganisms. Empirical evidence to support this hypothesis for the survival of starch or any other biomolecule on artifact surfaces has yet to be obtained.

Further Reading

Bethel, P.H., Ottaway, J., Campbell, G., Goad, L.J. & Evershed, R.P. (1994) The study of molecular markers of human activity: the use of coprostanol in the soil as an indicator of human faecal material. *Journal of Archaeological Science*, 21, 619–32.

Bouwman, A.S., Chilvers, E.R., Brown, K.A. & Brown, T.A. (2006) Identification of the authentic ancient DNA sequence in a human bone contaminated with modern DNA. *American Journal of Physical Anthropology*, 131, 428–31. [Includes examples of jumping PCR.]

Brotherton, P., Endicott, P. & Sanchez, J.J., *et al.* (2007) Novel high-resolution characterization of ancient DNA reveals C>U-type base modification events as the sole cause of *post mortem* miscoding lesions. *Nucleic Acids Research*, 35, 5717–28.

Collins, M.J., Penkman, K.E.H. & Rohland, N., *et al.* (2009) Is amino acid racemization a useful tool for screening for ancient DNA in bone? *Proceedings of the Royal Society Series B*, 276, 2971–7.

Collins, M.J., Riley, M.S., Child, A.M. & Turner-Walker, G. (1993) A basic mathematical simulation of the chemical degradation of ancient collagen. *Journal of Archaeological Science*, 22, 175–83.

Evershed, R.P. (2008) Organic residue analysis in archaeology: the archaeological biomarker revolution. *Archaeometry*, 50, 895–924.

Evershed, R.P., Copley, M.S., Dickson, L. & Hansel, F.A. (2005) Experimental evidence for the processing of marine animal products and other commodities containing polyunsaturated fatty acids in pottery vessels. *Archaeometry*, 50, 101–13. [Alkanoic acids as biomarkers for polyunsaturated fatty acids.]

Haslam, M. (2004) The decomposition of starch grains in soils: implications for archaeological residue analysis. *Journal of Archaeological Science*, 31, 1715–34.

Lindahl, T. (1993) Instability and decay of the primary structure of DNA. *Nature*, 362, 709–15.

Raven, A.M., van Bergen, P.F., Stott, A.W., Dudd, S.N. & Evershed, R.P. (1997) Formation of long-chain ketones in archaeological pottery vessels by pyrolysis of acyl lipids. *Journal of Analytical and Applied Pyrolysis*, 40–41, 267–85.

Regert, M., Bland, H.A., Dudd, S.N., van Bergen, P.F. & Evershed, R.P. (1998) Free and bound fatty acid oxidation products in archaeological ceramic vessels. *Proceedings of the Royal Society of London series B* 265, 2027–32. [Identification of monounsaturated fatty acids from their decay products.]

Turner-Walker, G. (2008) The chemical and microbial degradation of bones and teeth. In *Advances in Human Palaeopathology* (ed. Pinhasi, R. & Mays, S.), 3–29. John Wiley and Sons, Chichester.

9

The Technical Challenges of Biomolecular Archaeology

All areas of research present a technical challenge, and biomolecular archaeology is no different in this regard from any other scientific discipline. The study of ancient biomolecules has, however, thrown up a greater than average number of controversies over the years. Some of these have been prompted by a desire among some researchers to discover the oldest preserved biomolecules, as was the case with ancient DNA in the early 1990s. During this period there was a spate of reports of DNA in million-year-old fossils, despite a clear realization that the kinetics of DNA decay provided little opportunity for survival of intact polynucleotides in specimens of more than about 100,000 years in age. The reports included a 17–20 million-year-old fossilized *Magnolia* leaf, 80 million-year-old dinosaur bones, and a variety of amber specimens culminating with a 120–135-million-year-old weevil, the paper describing this result coincidentally published in *Nature* in the same week in 1993 that the film *Jurassic Park* was released in the USA. The quest for the "oldest in the world" has been a peculiar feature of ancient biomolecules research, and not restricted just to DNA. In recent years, there have been reports of preserved proteins in dinosaur bones, these reports heavily criticized as being unproven and equally energetically defended as being authentic.

The multimillion-year-old DNA sequences are generally looked on as being artifacts resulting from contamination of the specimens with modern DNA, though this has been proven only with the dinosaur bones. These yielded sequences from the mitochondrial cytochrome b gene that did not match any that were known at that time, and were therefore thought to be genuine dinosaur DNA. It was subsequently discovered that they were human sequences, not of the human mitochondrial genome itself, but of mitochondrial DNA fragments that had become inserted in the nuclear genome and then, over time, accumulated mutations that disguised their origin. The fact that nuclear DNA contains degraded mitochondrial inserts was well known among geneticists but had to be rediscovered by the

Biomolecular Archaeology: An Introduction, by Terry Brown and Keri Brown © 2011 Terry Brown & Keri Brown

ancient DNA community before the "dinosaur" sequences could be discounted. This controversy was very instructive, as the original work had been done extremely carefully, with precautions taken to avoid contaminating the bone samples with human or any other type of DNA, but still contamination had occurred.

Contamination presents a major problem for ancient DNA research because of the extreme sensitivity of PCR, which when fully optimized is capable of amplifying a single contaminating DNA molecule, and even when not optimized can routinely give a product if fewer than 100 molecules are present. The smallest amount of modern DNA contamination can therefore complicate and possibly invalidate an ancient DNA project. Contamination with modern biomolecules is not such a substantial problem when ancient proteins, lipids, and carbohydrates are studied, because the methods used to detect these molecules are not as exquisitely sensitive as PCR. Small amounts of contamination that fall below the detection limits of the analytical procedures do not interfere with the studies being carried out on any ancient biomolecules that are present. This is not to say that contamination is irrelevant with molecules other than DNA. Fingerprints, for example, contain detectable levels of fatty acids, cholesterol, and waxes, and care must therefore be taken when handling artifacts whose ancient lipid contents are to be examined. When starch grains are studied, it is vital that material from one layer in an archaeological soil does not contaminate a second layer, as the resulting loss of stratigraphic information can lead to erroneous conclusions regarding the age of the grains that are discovered. However, these types of contamination do not cause such great problems as those that bedevil ancient DNA research, because the measures needed to prevent them fall into the category of "good scientific practice," and do not approach the complexity of the more exceptional precautions that have to be taken when ancient DNA is being studied. Most of this chapter is therefore devoted to ancient DNA, but there are also important technical challenges, though of a different nature, in work with all types of ancient biomolecule, as we will see at the end of the chapter.

9.1 Problems Caused by Modern DNA Contamination

Modern DNA contamination is such a problem when human specimens are studied that it has been suggested that there is no point in carrying out work on ancient human DNA, as it is never possible to be certain that the results are authentic. The second part of that statement – that one can never be *certain* that ancient human DNA results are authentic – is probably correct, but if a sufficiently strict technical regime is followed, then it is possible to ensure that there is a *high possibility* that genuine ancient human sequences can be identified. This regime is based on practical measures that minimize the amount of contamination that can occur, control procedures that are designed to recognize contaminating DNA sequences when these are obtained, and an assessment of the likelihood that ancient DNA has survived in the material being studied.

It is sometimes thought that a strict anti-contamination regime is needed only when human specimens are being studied, the likeliest source of contamination being people who have handled or worked with the specimen. As we will see, this view is erroneous and

Table 9.1 Stages in the history of an archaeological specimen when contamination with non-endogenous DNA could occur.

Stage	*Nature of contamination*
During burial	Contamination resulting from handling during burial rites
When buried	Movement of DNA between buried specimens
During and after excavation	Handling by excavators, curators, osteologists, ancient DNA researchers and other specialists
In the ancient DNA lab	Cross-contamination with amplicons from previous PCRs
In the ancient DNA lab	Use of contaminated plasticware or reagents

contamination is an equally difficult problem when ancient DNA is extracted from animal or plant material, or when pathogen DNA is studied in human specimens.

9.1.1 There are five possible sources of DNA contamination

There are five ways in which a specimen, DNA extract, or PCR can become contaminated with non-endogenous DNA (Table 9.1). Two of these involve events that might have occured before the specimen is excavated. The first of these concerns burial practices, which could result in contamination with non-endogenous human DNA if they involve extensive handling of the remains. If the burial rites involve any practices that expose the remains to human blood, urine, or feces, then contamination is almost certain. There is no solution to this problem; in practical terms all that can be done is to take into account what may have happened to the remains before inhumation, and to assign appropriate confidence limits to the nucleotide sequences that are obtained. The second pre-excavation source of contamination is the movement of DNA within the burial environment, possibly from one specimen into the surrounding matrix and then into other specimens. Under certain conditions, it seems quite reasonable to expect a degree of leaching of DNA from biological remains into the surrounding soil, and if a number of different specimens are interred together, then exchange of DNA cannot be discounted. This could be a major problem in mixed deposits of animal bones. Again there is no solution to the problem, although experiments could be designed to assess to what degree this type of cross-contamination occurs in different burial environments. This has rarely been attempted, though in Section 15.2 we will look at one project in which the soil adjacent to human bones was tested for the presence of ancient DNA, with none being detected.

Handling of specimens during and after excavation could result in contamination with modern human DNA from dead skin cells, sweat, saliva, and suchlike. The ease with which this type of contamination can occur is clear from the ability of forensic scientists to obtain both mitochondrial and nuclear DNA from fingerprints. Handling is looked on as the primary way by which specimens can become contaminated with modern human DNA, but it is important to realize that specimens, DNA extracts, and PCRs can also become contaminated with amplicons from previous PCR experiments. This type of contamination arises because the simple process of opening a test tube in which a PCR has been carried out generates an aerosol containing DNA amplicons from that reaction. These molecules can remain airborne for long periods, or can land on laboratory surfaces, equipment, and personnel, and via any of those routes could subsequently be transferred to new PCRs. This is the major source of contamination when animal, plant, or pathogen DNA is being studied.

Finally the plasticware and reagents used in the extraction and amplification of DNA might be contaminated with DNA from various sources. Human DNA contamination has been reported in the small plastic test tubes in which PCRs are carried out, and cattle DNA is sometimes present in preparations of bovine serum albumin (BSA), a protein that is sometimes added to PCRs because it can absorb inhibitory compounds and hence aid amplification of the target sequence. Ironically BSA has been shown to be particularly effective in absorbing inhibitors from animal bones.

Of these five contamination sources, the first two – handling prior to burial and movement of DNA in the burial environment – cannot be prevented and must simply be taken into account as possibilities when ancient DNA results are analyzed. The final of the five sources – plasticware and reagents – is easily avoided by careful sourcing of these materials. The greatest problems in ancient DNA research are caused by contamination by handling and from "old" amplicons. We will therefore look more closely at how these two types of contamination arise and can be prevented.

9.1.2 Handling contaminates specimens with modern human DNA

The ease with which archaeological specimens can become contaminated by handling became clear in the early 1990s when animal bones from museum collections were first shown to contain human DNA. The problem was brought into sharp focus in 1995 when it was realized that the DNA sequences obtained from the 80 million-year-old dinosaur bones were modern human contaminants. As the genuine sequence of the dinosaur DNA was unknown, the extracts from these bones had been tested with PCRs that used non-specific primers able to amplify a range of mammalian, reptilian, and avian mitochondrial DNAs, including human sequences. Careful precautions were therefore taken to avoid these bones coming into contact with human DNA, but contamination still occurred.

Contamination of animal and other non-human specimens with modern human DNA is only a problem in those unusual cases when the contamination might be mistaken for the genuine ancient DNA. Usually when non-human material is studied, the PCRs are specific for the organism being studied and will not amplify human DNA, so the problem does not arise. With human specimens, the situation is much more difficult, as the contaminating DNA will be amplified along with any ancient DNA that is present. A positive PCR cannot therefore be taken as evidence that ancient DNA is present, as the product might be derived wholly from the contamination. Even after the amplicon has been cloned and sequenced, it might not be possible to identify the genuine ancient DNA. If all the sequences are the same, then they may all be contaminants. If two sets of sequences are present then both may be contaminants, and even if one set is ancient DNA, then

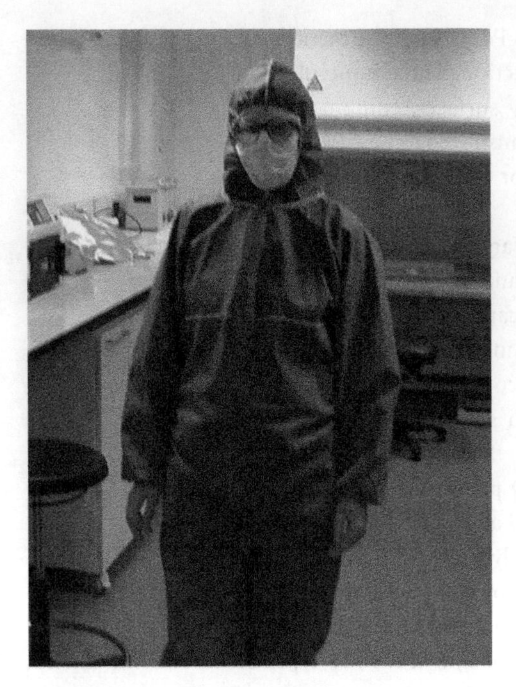

Figure 9.1 Biomolecular archaeologist wearing the appropriate protective clothing for working with ancient DNA.

distinguishing which one is genuine and which contaminating can be difficult, if not impossible. It is perhaps not surprising that in the early 2000s there were suggestions that work on human ancient DNA should be abandoned, as it was not possible to be certain that the results were correct.

How can we solve or circumvent the problems caused by modern contamination of human specimens? Experiments have shown that if a specimen, such as a bone, is handled, then it will almost inevitably become contaminated with modern human DNA. Even if sterile gloves are worn, contamination can still occur from skin flakes from the hair, face, and other exposed areas, and probably also from exhaled breath from the mouth and nose. In the laboratory, researchers working with human specimens wear full forensic clothing when preparing extracts and setting up PCRs, including a facemask, hair net, and glasses (Figure 9.1). Tests have shown that this level of protection does prevent contamination, but that small errors such as leaving a wrist exposed can jeopardize the containment and significantly increase the likelihood that modern DNA finds it way into the PCRs. Such precautions must be taken with all human specimens – not just skeletons – as it is equally easy to contaminate a soft tissue sample such as one taken from a mummy.

Although forensic clothing can be worn in the laboratory, it is not usually feasible to dress up in this way in the field when specimens are being excavated. Many field archaeologists are aware of the need to minimize their contact with human bones and teeth destined for DNA analysis, and so excavate the skeletons while wearing gloves and a facemask and immediately package the samples in sterile bags (Box). Unfortunately, these practices have been adopted only in recent years, and most of the human material in museums and other collections is heavily contaminated, often from multiple sources.

9.1.3 Methods are needed for removing or identifying modern human DNA

We must accept that most human specimens, especially bones and teeth, have been contaminated by at least one person. Methods are therefore needed for removing this contamination, or recognizing modern human DNA sequences when these are obtained, so that genuine ancient sequences can be identified. Such methods have been devised, but are not foolproof.

Methods for removing modern contamination from a bone or tooth are based on the assumption that this contamination is largely on the surface of the specimen. The surfaces of bones and teeth are therefore washed with 3% sodium hypochlorite – a diluted bleach solution – or exposed to ultraviolet radiation prior to DNA extraction. Bleach is a strong oxidizing agent and so causes blocking lesions and crosslinks between DNA molecules; and ultraviolet radiation has the same effect by forming bonds between adjacent nucleotide

Box Precautions that should be taken by field archaeologists when handling specimens destined from DNA analysis.

1 Wear clean gloves when excavating and handling material for DNA analysis. Disposable medical latex gloves are ideal, as the outer surfaces of these are sterile. Rubber gloves are acceptable, provided they are rinsed thoroughly with clean water after being put on. Beware of contaminating the gloves you are wearing by, for example, scratching your skin or touching your hair. Change gloves for each new sample. Beware of a gap between gloves and sleeve edge. Prevent exposure of skin by taping the ends of the gloves to your sleeves if necessary.
2 Cover nose and mouth with a surgical mask or at least a clean scarf. Keep hair tied back or covered by a shower cap.
3 Do not smoke.
4 Keep sample dry at all costs. If it is already wet, then place it on a clean surface and let it dry thoroughly.
5 Do not allow the sample to come into contact with the ground once it has been excavated.
6 Store the sample in a clean (preferably sterile), dry, airtight container. A screw-cap bottle or container is ideal. Do not store in plastic bags, as these encourage the growth of microorganisms.
7 Store samples in a cool, dark place; ideally a fridge. Longer-term storage requires a −20°C freezer. Keep out of direct sunlight.
8 Only a few grams are needed for DNA analysis, ideally from two or three different bones in the same individual (in separate bags). Any part of the skeleton is suitable, including teeth. Use clean, dry metal tweezers or similar implements – sterilize these by flaming with a cigarette lighter frequently, and always between taking samples from different individuals.
9 Make a note of soil conditions around each bone sample – type of soil, pH, presence of other objects such as metal, dry or wet soil conditions, etc.
10 Finally, all archaeologists involved in the excavation of human remains should also have their DNA sampled (usually a saliva swab – non-invasive). This is to eliminate sources of contamination.

bases, especially pairs of thymines (Section 8.2). It is generally believed that these treatments alter the surface DNA to the extent that it cannot be copied at all during a PCR, but this might not be the case for all the contaminating molecules, particularly those that have become partially absorbed into the bone or tooth matrix. If this is the case, then bleach or ultraviolet treatment may create molecules that, although containing some blocking lesions, are still able to participate in the PCR. This would exacerbate the contamination problem, as these molecules would be copied into short products that could promote jumping PCR (Figure 8.5). For bones, an alternative to chemical treatment and irradiation is to remove and discard the outer surface, by scraping with a sterile scalpel blade, prior to preparation of the DNA extract. Tests in which bones are deliberately handled have shown that most of the contaminating DNA is contained within the outer 2 mm, and that abrasion of the surface to this depth is able to remove the contamination. There have been worries that washing a bone, which is routinely done after excavation to remove soil and other debris, might cause contaminating DNA to move deeper into the matrix, but experiments have suggested that this is not the case.

Chemical treatment, ultraviolet irradiation, and surface abrasion can reduce the amount of contaminating DNA on the surface of a specimen, and in some cases remove it completely, but it is dangerous to assume that a PCR product that is subsequently obtained derives from ancient DNA. Contamination is such an insidious problem that additional checks must be made before the product is considered genuine. These checks take two forms. First, genuine ancient DNA is expected to display at least some miscoding lesions resulting from chemical damage occurring as the polynucleotides degrade (Section 8.2). These lesions are revealed when cloned sequences are examined, with sporadic C→T and G→A errors that occur irregularly within the sequences being particularly diagnostic (Figure 8.7). A difficulty here is that contaminating DNA from dead skin cells might itself have undergone some chemical degradation and so might not be free from these miscoding lesions.

The second check is to make comparisons between the "ancient" DNA sequences and the corresponding sequences of all laboratory members who have come into contact with the specimen. PCRs should therefore be directed at the mitochondrial DNA hypervariable regions, even if the main aim of the project is to study nuclear DNA, and any "ancient" sequence that matches a haplotype possessed by a lab member treated with suspicion. Ideally, this comparison should be extended to excavators, osteoarchaeologists, and others who have handled the specimen before it reached the safe confines of the laboratory. For this reason, many researchers now work only with freshly excavated material for which this chain of contact can be established, and aim to minimize the number of individuals who handle the specimen, to reduce the possibility that a genuine ancient DNA result is discarded because of a chance match between the haplotype of the specimen and that of one of the potential contaminators.

9.1.4 Contamination with amplicons from previous PCRs is a problem with all types of archaeological specimen

Contamination by handling is specifically a problem when human specimens are studied. The second serious form of contamination affects all ancient DNA research, with human, animal, and plant material equally susceptible. This is the contamination that occurs when airborne amplicons from previous PCRs become deposited on a specimen or find their way into extracts as these are being processed. The possible extent of this contamination cannot be overemphasized. A completed PCR can contain 10^{11} copies of the amplicon in a volume of 10 μl, so a single aerosol droplet of 1 nl derived from the PCR can carry ten million amplicons. In contrast, a gram of human or animal bone might contain fewer than 1000 copies of this sequence in the form of ancient DNA.

Contamination of a new bone extract with just a single aerosol droplet will completely obscure the ancient DNA, and a PCR prepared with this extract will give products that derive wholly from the contaminants rather than the ancient DNA.

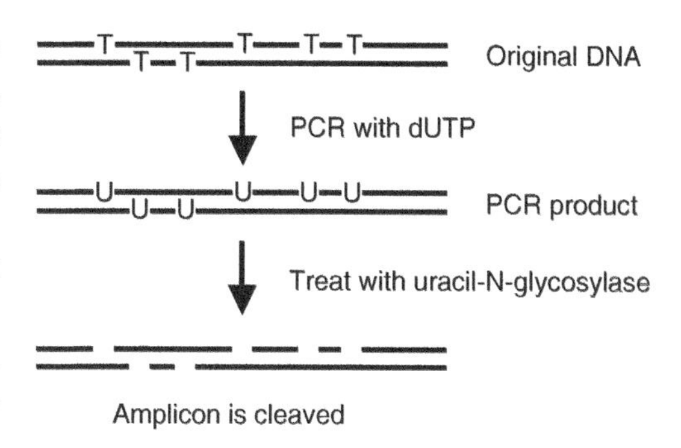

Figure 9.2 The principle behind the use of uracil-N-glycosylase to prevent amplicon cross-contamination.

Amplicon cross-contamination is more difficult to detect than contamination that arises from handling. If a contaminating amplicon is the product of a previous ancient DNA PCR, then it will display miscoding lesions and appear, once again, to be genuine ancient DNA. If it is a human DNA amplicon, then comparisons with the sequences of lab workers and excavators will not reveal that it is a contaminant, because it is not derived from any of these individuals. Neither is it sufficient to rely on negative controls to reveal this type of contamination, because it is quite possible for the sample to become contaminated, while the negative controls remain unaffected. Often, amplicon cross-contamination is only recognized when an unusual DNA sequence, obtained in a previous experiment, reappears when later specimens are examined. This means that amplicon cross-contamination can be detected only when the amplicon displays variability within the species being studied. If the amplicon sequence is invariant, then there is no way of knowing if it is genuine or the result of cross-contamination. This is a particular problem with the standard PCRs used to screen human remains for the presence of *Mycobacterium tuberculosis*, the bacterium that causes tuberculosis (Section 15.2). These PCRs target sequences that are thought to be specific for this bacterium, but that display no variability within the species. Amplicons that result from genuine ancient tuberculosis DNA cannot therefore be distinguished from cross-contaminants, and results obtained with these PCRs are always insecure.

As amplicon cross-contamination is so difficult to recognize, and sometimes impossible to detect, the only way of dealing with it is to prevent its occurring in the first place. Various methods have been proposed to achieve this end, the most useful being the inclusion of the modified nucleotide 2′-deoxyuridine 5′-triphosphate (dUTP) instead of dTTP in the PCR. This does not affect DNA amplification as dUTP has the same base pairing properties as dTTP, and so can be incorporated into the new strands without affecting the sequence that is eventually obtained. Importantly polynucleotides containing dUTP can be cleaved by treatment with the enzyme uracil-N-glycosylase (Figure 9.2). The regime is therefore to use dUTP in all PCRs, and to treat new ancient DNA extracts with uracil-N-glycosylase so that contaminating amplicons are cleaved. The enzyme is heat labile and so is inactivated during the first cycle of the PCR set up with the new extract, and so cannot attack the new amplicons resulting from this PCR.

9.1.5 Criteria of authenticity must be followed in all ancient DNA research

The problems caused by contamination, whether resulting from handling or previous amplicons, have led to the proposal of nine "criteria of authenticity," to be followed in all ancient DNA projects (Table 9.2). These criteria have three aims:

Table 9.2 The original "criteria of authenticity" for ancient DNA research.

Precaution	*Details*
Physically isolated work areas	Ancient DNA research must be carried out in a dedicated, isolated laboratory
Control amplifications	Multiple extraction and PCR controls
Appropriate molecular behavior	Amplification products greater than 500 bp should be rare or never obtained; mitochondrial DNA should be amplifiable if single copy nuclear or pathogen DNA is detected; sequences should make phylogenetic sense
Reproducibility	Results should be reproducible with different PCR systems
Cloning	Sequences of cloned PCR products should be obtained to assess endogenous vs. exogenous sequences, miscoding lesions, etc.
Independent replication	Results should be replicated in a second laboratory
Biochemical preservation	The presence of ancient DNA implies that there has been only limited biochemical diagenesis, and this should be reflected by the degree of preservation of other biomolecules
Quantification	Ancient DNA copy numbers should be >1000, as lower numbers are difficult to distinguish from sporadic contamination
Associated remains	Work with human specimens should be accompanied by studies of faunal remains, with which contamination is less of a problem; if the human material contains ancient DNA, then ancient DNA should also be present in the associated faunal remains

- **To prevent contamination**, by isolating the specimen within laboratories used for no purpose other than the preparation of ancient DNA extracts and PCRs. Airborne contamination is prevented by equipping these labs with an ultrafiltered air supply that maintains positive displacement pressure, so that a minimal amount of unfiltered air enters the room when the door into the lab is opened. Within the lab, workers wear forensic clothing, as described above, to prevent contamination by handling. When not in use, all work surfaces are exposed to ultraviolet radiation in order to crosslink any DNA that is present.
- **To recognize contamination when it occurs**. This is achieved partly by accompanying all ancient DNA extractions with an "extraction blank" in which the starting material is omitted. If the PCR set up with the solution resulting from this blank gives a product, then the plasticware or reagents used in the extraction or in the PCR have become contaminated. Contamination is also suspected if positive results are given with control PCRs designed to yield amplicons greater than 300 bp in length, as authentic ancient DNA is usually shorter than this due to the polynucleotide cleavage that occurs during DNA degradation (Section 8.2). Replication of results, both in the same lab and by a second independent lab, give additional confidence that an ancient DNA detection is real, and examination of the variations displayed by sequences obtained after cloning enables miscoding lesions to be identified.

• **To assess whether the overall degree of preservation of the specimen is compatible with the presence of ancient DNA.** If a specimen has undergone extensive physical degradation, and/or other types of biomolecules are absent, then that specimen is unlikely to contain ancient DNA. Similarly if animal bones buried alongside human ones do not contain ancient DNA, then any positive results obtained with the human bones should be treated with suspicion.

Although these criteria provide only circumstantial evidence that an ancient DNA detection is genuine, adherence to them reduces the possibility that contamination will be mistakenly identified as authentic ancient DNA. Unfortunately, despite being proposed as long ago as 2000, only a minority of ancient DNA projects have attempted to meet all nine criteria. In some cases, this is understandable as it is not always possible to meet every one of them. For example, associated remains can be examined only if these exist, and tests for the presence of other biomolecules is not possible if only limited amounts of material are available for study. But in some cases, more dubious reasons appear to underlie a failure to adhere to these criteria. There is sometimes an assumption that these measures are really necessary only when human specimens are being studied. This assumption is based on the mistaken belief that handling is the only cause of contamination, whereas in reality amplicon cross-contamination is equally likely and more difficult to detect. In some projects, a subset of the criteria have been adopted, with no explanation as to why others have been omitted. In this regard, it would be fair to say that the ancient DNA research community has not yet fully met the technical challenges of their discipline.

9.2 Problems Caused by Overinterpretation of Data

Although contamination with modern biomolecules is primarily a problem for studies of ancient DNA, research with ancient proteins, lipids and carbohydrates can still be controversial. In most cases, these disputes center on data interpretation, a problem that can arise with ancient DNA also. There are a number of case studies that we could choose to illustrate the dangers of data overinterpretation, but we will focus on just one, the perplexing questions that surround the supposed detections of preserved hemoglobin molecules on prehistoric stone implements.

9.2.1 The "blood on stone tools" controversy illustrates the dangers of data overinterpretation

The detection of blood residues on stone tools holds a fascination for the general public and has considerable archaeological importance as a means of understanding the functions of individual implements. If it were possible to move beyond simple detection of blood residues to identification of the species from which the blood was derived, then novel information could be obtained on prehistoric diets.

The first apparent breakthroughs in studying blood residues on stone tools were reported in 1983. This research, carried out with a collection of more than 100 tool fragments from four North American sites between 1000 and 6000 years in age, made use of two complementary techniques. First, the presence of hemoglobin on the surfaces of various tools was established by testing with "Chemstrips," plastic sticks carrying an absorbent pad impregnated with chemicals that react with the iron-containing heme compound that is present, along with globin proteins, in a functional hemoglobin molecule. Reaction with heme results in an easily recognized color change that signals the presence of blood residues. Those tools giving a positive reaction were then treated with a high salt solution, which causes the hemoglobin molecules to form crystals. The crystals were examined with the light microscope to identify their precise shapes, these shapes enabling the species of origin to be identified. Ninety tools gave positive reactions with the test strips, the blood residues deriving from species as diverse as humans, caribou, deer, moose, rabbits, sheep, grizzly bears, sea lions, and squirrels.

The publication of these results in *Science* caused intense excitement among archaeologists. Unfortunately it quickly became clear that the conclusions of the research were far from proven, because of unrecognized limitations in the methods that had been used. Chemstrips are designed for detection of blood in urine and are specific in that context, but can give false positive results when exposed to various compounds not usually found in urine but quite possibly present on the surfaces of stone tools. These include chlorophyll and oxidizing agents such as traces of bleach, or microbial enzymes such as peroxidases. Control experiments, carried out only after publication of the initial results, showed that these strips regularly give false positive reactions with soils taken from archaeological sites. Similar specificity problems have been demonstrated for the related Hemastix, which researchers initially turned to when the unsuitability of Chemstrips was first revealed.

The use of hemoglobin crystal shape as a means of identifying species of origin was also shown to be insecure. Indeed, this method has never been reproduced successfully by any other researcher and has been criticized by others on the grounds that crystal morphology is affected by the chemical environment on the tool surface and hence is not species specific, that visual examination of crystals is not sufficient to determine their precise shapes, that any degradation of the hemoglobin molecules would affect crystal shape, and that the crystals formed by hemoglobins from different species do not have significantly different shapes.

We have already described the problems that can arise when immunological tests are used to identify archaeological proteins (Section 3.3). The specificity of these methods depends on a perfect match between the antibody and the structural feature that it recognizes on the surface of its target antigen. Any diagenetic change to the antigen will reduce the specificity of this match and lead to misidentifications. These problems were recognized as early as the 1930s when blood typing was first attempted with mummified remains. Remarkably, in the next stage of research into blood on stone tools, all of these lessons were forgotten. A series of papers in the 1990s reported the use of crossover immunoelectrophoresis (CIEP) to identify hemoglobin from various species on stone tools of different ages. In parallel with these projects, other researchers were reporting equally extensive control experiments that showed that CIEP and other immunological techniques are unreliable when applied to archaeological residues. The results obtained when different immunological methods were used to examine the same protein residue were not in agreement with each other, and misidentifications occurred after artifacts were treated with blood of known origin. And the story does not end in the 1990s: in 2004 the media widely reported claims that blood residues had been identified on two million-year-old tools from Sterkfontein Cave in South Africa. The blood proteins had supposedly been preserved for this exceptional length of time because of the protective effects of absorption into the silica surface of the artifact and encasement in the clay matrix of the cave floor. This report was described as "surprising" by an expert in ancient protein degradation.

9.2.2 "Blood on stone tools" provide lessons for all of biomolecular archaeology

The controversies that have surrounded research into blood residues on stone tools illustrate, in an extreme form, one of the major technical challenges that has to be faced by biomolecular archaeologists. This is the danger that a result that is, in some respects, correct becomes discounted because the data interpretation goes beyond what is secure into the realms of the unproven, speculative, or irrelevant. Most researchers agree that the deposits seen on the surfaces of stone tools are indeed blood residues, and this in itself is an important conclusion. But many researchers are now wary of attempting to study those residues, because the research area has achieved such a high degree of notoriety that any results, however soundly based, are now treated with initial skepticism.

Data overinterpretation is a general problem for biomolecular archaeology, as it is for any area of interdisciplinary research, because it is most likely to occur when the researcher is not fully aware of the technical limitations of the methods that he or she has used, or is not fully appraised of the context in which the data should be interpreted. Successful biomolecular archaeology therefore requires a genuine collaboration between biomolecular scientists fully expert in the methods, and archaeologists equally expert in their own discipline. This will remain necessary until genuine biomolecular archaeologists with expertise in both areas emerge. If such a collaboration is not established, then it becomes increasingly likely that technical limitations will not be recognized and/or the data will not be placed in their correct context. The outcome will be either overinterpretation or misinterpretation.

The challenge of merging biomolecular science and archaeology is not difficult to solve. Unfortunately there appear to other, more complex factors, that occasionally encourage overinterpretation of the outcomes of biomolecular archaeology projects. Everybody is interested in their origins and in the human past, and it is not surprising that the public have an immense interest in our research area. The results of biomolecular archaeology projects are often reported in the media, the breadth of coverage directly proportional to how exciting the results are. There should not, of course, ever be a temptation to interpret data in such a way as to make the results more exciting than they actually are. There has also been a perception on occasions in the past that the editors of some scientific journals have used the potential extent of the media coverage as one of the factors in deciding whether or not to publish a paper reporting a biomolecular archaeology project. Scientists get more professional esteem for publishing in "high impact" journals, so once again a potential conflict of interest has to be resisted. This challenge is, in some respects, more important than that posed by contamination with modern biomolecules. Any research discipline that is looked on by other researchers as sacrificing scientific rigor for media impact quickly loses credibility. The history of science is littered with such examples.

Further Reading

Bouwman, A.S., Chilvers, E.R., Brown, K.A. & Brown, T.A. (2006) Identification of the authentic ancient DNA sequence in a human bone contaminated with modern DNA. *American Journal of Physical Anthropology*, 131, 428–31.

Cooper, A. & Poinar, H. (2000) Ancient DNA: do it right or not at all. *Science*, 289, 1139.

Downs, E.F. & Lowenstein, J.M. (1995) Identification of archaeological blood proteins: a cautionary note. *Journal of Archaeological Science*, 22, 11–16.

Gilbert, M.T.P., Bandelt, H.-J., Hofreiter, M. & Barnes, I. (2005) Assessing ancient DNA studies. *Trends in Ecology and Evolution*, 20, 541–4.

Loy, T.H. (1983) Prehistoric blood residues: identification on tool surfaces and identification of species of origin. *Science*, 220, 1269–71.

Pruvost, M., Grange, T. & Geigl, E.-M. (2005) Minimizing DNA contamination by using UNG-coupled quantitative real-time PCR on degraded DNA samples: application to ancient DNA studies. *Biotechniques*, 38, 569–75.

PART III

THE APPLICATIONS OF BIOMOLECULAR ARCHAEOLOGY

10

Identifying the Sex of Human Remains

Identification of the sex of human remains is one of the most fundamental contributions that biomolecular research can make in archaeology. Sex can sometimes be identified from the structure of a skeleton, but this requires that the skeleton be reasonably complete and that the individual concerned did not die before reaching puberty. If the osteology is inconclusive, then ancient DNA analysis is the only alternative approach to sex identification. Although the use of ancient DNA in sex identification is relatively straightforward, the process illustrates many of the issues that must be taken into account in any biomolecular archaeology project. These issues are not simply the technical ones involved in designing the project and ensuring that the results are accurate and authentic. They include an understanding of the archaeological context within which the biomolecular results will be interpreted. We must always bear in mind that the archaeological context provides the reasons for the biomolecular studies, and without an understanding of the former the latter are likely to be misdirected or even irrelevant outside of their own narrow setting.

10.1 The Archaeological Context to Human Sex Identification

If we were unfamiliar with archaeology, then we might assume that sex identification has no importance beyond the simple cataloguing of human remains found at a particular site. In reality, the archaeological context to sex identification is much more interesting than it might at first appear. The interest arises because of two issues. First, at some sites the sex ratio is significantly skewed towards males or females, despite our expectation that in the past, as today, the number of male births would have been approximately equal to the number of female ones. Second, there is a difference between sex and **gender**. We will deal first with the possibility of skewed sex ratios.

 © 2011 Terry Brown & Keri Brown

10.1.1 Various factors can result in a skewed sex ratio

In modern human societies, approximately 105 boys are born for every 100 girls, with the total number of males and females in the population evened out at about 1 : 1 because of the slightly longer life expectancy of women. Various factors can result in these figures becoming skewed, most notably the death of females during childbirth and the death of males during warfare. As recently as the 19th century, female mortality associated with childbirth could reach 30% during outbreaks of infections whose causes in the pre-microbiological era were unknown and which progressed without check in the absence of antiseptic techniques. Male mortality as a result of warfare has had an equally dramatic impact, and this is true not only of the global conflicts of the 20th century: it has been estimated that half the male population of Athens was killed during the Peloponnesian War with Sparta during the last 30 years of the 5th century BC.

Archaeological evidence can contribute to our understanding of how factors such as childbirth and warfare have influenced human demography down through the ages, but a skewed sex ratio among burials recovered from a single archaeological site often has a more subtle, and in some ways more interesting, cause. As an example, consider Grave Circle B at the Bronze Age citadel of Mycenae in Greece, the legendary home of Agamemnon and Clytemnestra. This Grave Circle spans some three to four generations during the period 1650–1550 BC, when Mycenae was establishing itself as a dominant trading and political power in the eastern Mediterranean. Very few, perhaps only 4, of the 35 occupants of Grave Circle B were female. But this should not be taken as evidence of massive mortality of pre-adult females at Mycenae. Instead, we must consider the nature of the Grave Circle, in particular the richness of the materials found alongside the skeletons, which include a face-mask of electrum (a naturally occurring gold-silver amalgam) and other goods that indicate that these individuals held high status in their society. In other words, the individuals buried in Grave Circle B were not a representative cross-section of the Mycenaean population of 1650–1550 BC, but instead were the elite members of that society. Having recognized this fact, we immediately see that the implication to be drawn from the sex ratios in Grave Circle B is that early Mycenaean society was male dominated, with relatively few women acquiring positions of status.

Most of the Grave Circle B skeletons are adults and are sufficiently intact for their sex to be identified by osteological examination. As we will see in Section 10.2, osteology cannot distinguish between male and female children, because the sex-specific differences in the skeleton do not develop until puberty. Ancient DNA therefore has an important role to play at those sites where relatively large numbers of infant skeletons have been found. The discovery of such sites has led to suggestions that some past societies practiced infanticide – the deliberate killing of infants, technically those less than one year in age. In the popular imagination, infanticide in prehistory is often equated with child sacrifice, although ancient historians realize that written reports of child sacrifice were often made by one society attempting to smear the reputation of a rival and do not always look on these accusations as the literal truth. It is perhaps more likely that infanticide in the past was carried out for much the same reasons as it is today – because of the greater value placed on males rather than females. This is illustrated by the famous letter written by a

Roman citizen around 1 BC, in which the man gives his wife instructions regarding her newborn child "if it is a boy, keep it, if a girl, discard it". But we should not assume that infanticide is always directed at female offspring, and in Section 10.4 we will examine an interesting project in which ancient DNA analysis has shown that sometimes it was the male children who were disposed of.

10.1.2 Sex is not the same as gender

The second contextual issue that we must consider is the distinction between sex and gender. Scientists are not alone in using "gender" as an interchangeable term for "sex," either through a desire to add variety to an article that would otherwise be too much about sex, or from a misplaced attempt at political correctness. Gender is not a synonym for sex: the two words have their own precise definitions and meanings. Sex is a biological characteristic determined by the functioning of various genes, many but not all located on the X and Y chromosomes. Gender, in contrast, is a cultural construction, a system of social categorization that uses the differences between the biological sexes as a way of structuring thought and practice, but is not determined by those biological differences.

The common misconception that sex and gender are the same thing is based on the equally common misconceptions that there are two biological sexes – male and female – with a clear distinction between them, and just two genders – male and female – that map precisely onto the biological sexes. None of this is correct. It is true that for 99.98% of births the distinction between the biological sexes is clear by looking at the baby, and the child can be assigned a gender that is the same as its biological sex. But about 1 in 4000 babies have ambiguous genitalia and these children are difficult to categorize. A number of syndromes exist that, while comparatively rare, occur at regular frequencies in human populations, and give rise to various genital anomalies and intersex conditions. Some of these conditions arise because the individual has an unusual complement of X and Y chromosomes, as is the case with Klinefelter's syndrome, in which a person who is anatomically male has an additional X chromosome. Others are hormonal in origin, such as androgen insensitivity syndrome. Some have complicated effects, an example being 5α-reductase deficiency, a hormonal condition where the infant at birth appears to be female, but at puberty "becomes" male. Until recently our solution to these anomalous infants was to operate on them soon after birth to "turn" them into boys or girls.

These anomalous conditions must also have existed in the past, and so past societies must have had the same problems that we have in categorizing infants. We cannot assume that in dealing with ambiguous babies these past societies felt rigidly confined to a choice of just two genders. There are various societies around the world today that have more than two genders, and there are documented examples of societies with third genders, or even more, from anthropology and from history. The earliest of these references date back to Babylon and Egypt in the 2nd millennium BC, and are found in some of the oldest written records, making it quite likely that additional genders also existed in earlier unrecorded periods of prehistory. Neither can we assume that in the past the gender set at birth was always retained throughout the lifetime of an individual, any more than it is in Western societies today.

10.1.3 Gender studies form an important part of archaeological research

The non-equivalence between sex and gender means that identification of the biological sex of a skeleton cannot be used as a direct indication of the gender role played by that individual in the society in which he or she lived. Any researcher involved in identification of the biological sex of one or more skeletons, for example by ancient DNA typing, should therefore also be aware of how gender is studied by archaeologists, so they can place their results in an appropriate context.

Gender archaeology is the name given to the discipline that investigates the cultural construction of gender in past societies, using such evidence as the material remains of past peoples and **ethnographic analogy**, the latter involving observation and analysis of diverse modern societies. Gender archaeology does not mean the study of women only – although much initial work was carried out to "discover" women in the past, especially in Paleolithic archaeology, where the emphasis at one time was largely on male activities ("Man the Hunter"). Gender archaeology should in fact mean a "gendered archaeology," where the investigation of gender roles and gender construction includes both men and women, as well as the search for "third genders." This has led to the recognition that certain roles, activities, or behaviors that we consider appropriate for a particular gender today were not necessarily the same in prehistory, and that archaeological interpretation has suffered from a tendency to project, consciously or unconsciously, our own expectations and perceptions of gender roles and activities onto the past. For example, an assumption that was once implicit in Paleolithic research was that stone tools were made only by men, but there was no firm evidence whatsoever for this; it was merely a projection of late 20th century perceptions onto the Paleolithic. We are now more aware that gender roles, activities, and behaviors are dependent on social and cultural values, which can change through time and from group to group. If we have any doubts, we need only to consider how the roles of women in western societies have changed since the 1950s.

How is gender studied in archaeology? One way is through material culture, especially depictions of identifiable men and women in cave art, figurines, carved standing stones, pottery, and metalwork. Unfortunately, most prehistoric art is ambiguous and interpretations tend to be subjective, especially as images become more abstract and the difficulty in judging the prehistoric artist's intentions increase. The temptation to stretch the limits of interpretation by assigning meaning to the more obscure images and more complex scenes runs the risk, once again, of allowing our own values to influence the conclusions we reach about past societies. Prehistoric art is fraught with these difficulties, and the same images and general features can be given different meanings by different scholars. For example, the preponderance of male depictions in cave paintings at the Neolithic cult cave of Grotta di Porto Badisco, close to Otranto in southern Italy, has been interpreted as indicative of male dominance in Neolithic Italy. On the other hand, the common occurrence of female figurines in the Neolithic in southeast Europe has not been taken as an equivalent indication of female domination, but only as evidence that in this society certain aspects of being a woman were important.

Burial evidence has also featured heavily in gender archaeology. Burials are often accompanied with **grave goods**, the nature of which might indicate the gender of the individual with whom they are buried. These artifacts can be further investigated as they occur in other contexts, such as settlements and ritual sites. Again there are interpretative problems:

is there any basis to an assumption that swords and other artifacts of war indicate a male gender, whereas jewellery and weaving combs must be associated with females? Comparison between the biological sex of a skeleton and the gender implications of the grave goods can therefore be informative, but it should also be borne in mind that grave goods may have nothing to do with gender at all, but instead indicate status, age, or achievement in a society. The tendency, still practiced by some archaeologists, to use grave goods to assign biological sex to a burial, not only confuses biological sex with gender but also fails to appreciate the complexity of grave goods.

10.2 Osteological Approaches to Sex Identification

Osteoarchaeology – the study of the structure and function of archaeological skeletons – can often be used to identify the sex of a burial. Biomolecular techniques complement rather than replace these methods, and we must therefore understand the basis to the osteological approach as well as its strengths and limitations.

10.2.1 Osteoarchaeology can identify the sex of an adult skeleton

It is possible to use osteology to identify the sex of a skeleton because adult humans, like other primates, are **sexually dimorphic**, males generally having a larger body size than females and the two sexes displaying various specific skeletal differences. The most important of these is the shape of the pelvis, females having a wider, shallower sciatic notch compared with males because of the reproductive requirements (Figure 10.1). Male and female skulls also differ in size and robustness, mainly due to the later onset of puberty in males, which allows continued increase in muscle mass. There are also differences in the detailed

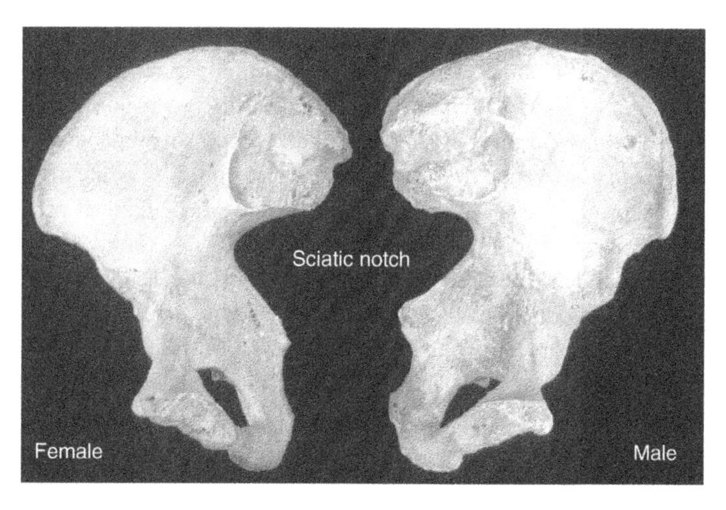

Figure 10.1 Views of the female and male innominate bones, which form one half of the human pelvis, showing the wider, shallower sciatic notch of the female skeleton. Images kindly provided by Charlotte Roberts.

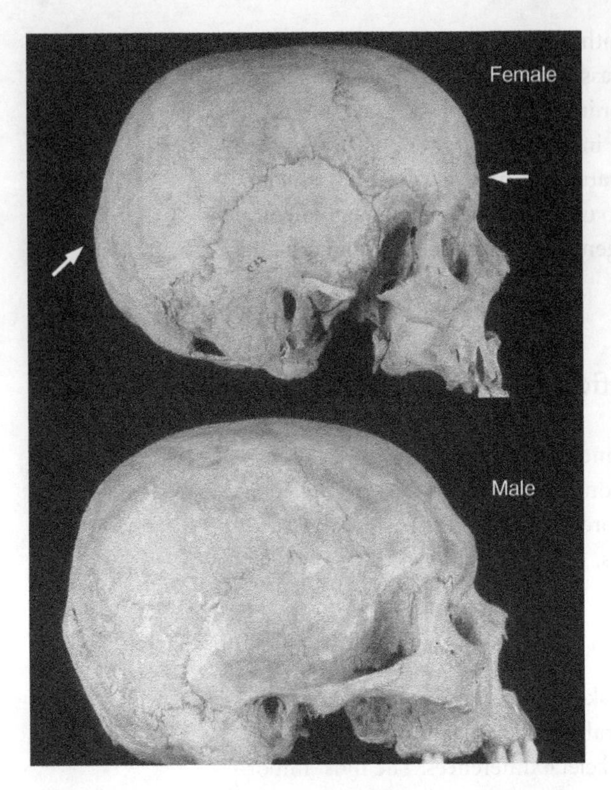

Figure 10.2 Female and male skulls, with the less distinct brow ridge and a smaller occipital protuberance of the female indicated. Images kindly provided by Charlotte Roberts.

anatomies of the skulls, including less distinct brow ridges and a smaller occipital protuberance at the rear of the head in females (Figure 10.2). Less diagnostic features include the size of the long bones and digits (more robust in males, which therefore have larger feet and hands), width of the rib cage (wider in males), and size of the teeth (larger in males).

Osteological methods of sex identification are reliable only to a certain extent, and it is important to be aware of their limitations. The main problem is that these morphological and morphometric techniques have, of necessity, been developed with modern human skeletons, which means that the standard data against which archaeological specimens are compared are representative only of modern populations. Not all modern populations display the same degree of dimorphism, and criteria that distinguish sex for one population (e.g. Caucasians) are less secure when applied to a second population (e.g. Africans). These factors must be taken into account when studying prehistoric skeletons, in order to make sure that the appropriate criteria are used for sex identification. Even then, we cannot be certain that the application of these standards to prehistoric populations is appropriate. It has been suggested that prehistoric females might have displayed greater skeletal robustness, so that in the absence of more diagnostic features such as an intact pelvis, they could appear to be males.

Osteoarchaeological identifications must therefore be based on skeletal variations that are likely to have a reasonably secure association with sex, not only in different modern populations but in the past also. If this requirement is met, then the vast majority of adult human skeletons recovered from archaeological sites can be sexed with a reasonable degree of confidence, problems arising only when the remains are poorly preserved or are of juveniles. The most reliable indicator of sex is the shape of the pelvis; and when an intact pelvis is present, archaeological remains can be identified as male or female with 95% confidence. Bear in mind, however, that 95% confidence gives an expectation that 1 in 20 skeletons will be incorrectly identified. Skulls can be used for sex identification with 85–90% confidence. Post-cranial features such as robustness or gracility of the long bones, muscle attachment markings, and size of the feet, can also be used as sex indicators, but only with a confidence of 80–90%.

10.2.2 Osteoarchaeology is less successful with children and fragmentary skeletons

The figures given above refer to adult skeletons, or at least ones from individuals who have passed through puberty, the period when the sex-specific features of the skeleton start to become clear. This means that infants and juveniles are notoriously difficult to sex. There have been attempts to devise methods of sex identification for fetuses and infants based on the structure of the pelvis, as there is measurable sexual dimorphism in the sciatic notch at an early age, but there is too much overlap between the male and female morphologies to allow this to be used as a reliable sex indicator, and the level of accuracy is only about 70%. In any case, the dimorphism is soon lost during the subsequent growth of the infant. For older children the level of accuracy when a complete skeleton is examined is about 75–80%, but without an intact pelvis, sex identification is only 50% accurate – which, with a choice of only two alternatives, is no better than a guess. The possibility of using the size of the canine teeth, which are larger in young males, to identify the sex of juveniles has also been explored, with a success rate of 75–80% on remains of known sex.

Fragmentary bones present additional problems. Intact skeletons are often found when sites are excavated, but sometimes the bones have become broken either because of the funerary practice or as a result of post-burial disturbance. Cremated bone (Section 7.1) is particularly difficult in this respect. Cremation of human remains was widespread in prehistory and during the Classical Age. Due to the high temperatures involved (up to 1000°C) the bones suffer cracking, shrinking and distortion (Figure 7.4), and the skeleton is often incomplete, either because some of the bones became completely decomposed during the cremation or because only a part of the cremated skeleton was interred, as seems to have been a common practice during the British Bronze Age. If the diagnostic bones are severely distorted or missing, then sex identification is extremely difficult, if not impossible.

10.3 Using DNA to Identify the Sex of Archaeological Skeletons

Now we can turn our attention to use of DNA to identify the biological sex of a skeleton. At the genetic level, there is very little difference between men and women. The human female has two X chromosomes, and the human male an X chromosome and a Y chromosome. It is the presence of the Y chromosome that determines maleness: when only one X chromosome is present and no Y, as in Turner's syndrome, the phenotype is female. The alternative – one Y and no X – never occurs and is presumed to be lethal. Methods that use DNA typing to identify the sex of an individual must therefore be able to distinguish between X and Y chromosomal DNA, detection of the latter indicating that the individual is male.

10.3.1 Some DNA tests simply type the presence or absence of the Y chromosome

The simplest DNA tests for sex identification use PCRs that are directed specifically at the Y chromosome. Such tests are relatively easy to design because the X and Y chromosomes are quite different from one another. The X chromosome is three times longer than Y, 153 Mb compared with 51 Mb, and only at the very ends of the two chromosomes are there regions of significant nucleotide sequence similarity (Figure 10.3). In between, the bulk of the DNA in each chromosome is unique either to X or to Y, providing considerable scope for the design of Y-specific PCRs.

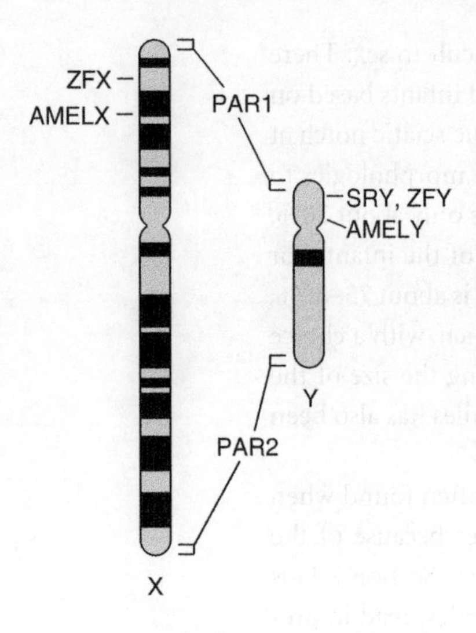

Figure 10.3 The human X and Y chromosomes, showing the locations of the two regions of sequence similarity, PAR1 and PAR2, and the positions of various genes used in DNA-based sex identification.

The first Y-specific PCR to be widely used was directed at a repetitive DNA element called DYZ1 (Table 10.1). This repeat sequence is 3.4 kb in length, has a copy number of approximately 3000, and is located only on the long arm of the Y chromosome. Because of the high copy number, PCRs for DYZ1 have great sensitivity, ideal for work with ancient DNA, and in the standard method the amplicon length is 102 bp, short enough to give a product with degraded ancient DNA templates. Y-specific sex identification tests have also been designed with the SRY (sex-determining region Y) gene, one of the central genes involved in development of male characteristics in the unborn child. This is a single copy gene so the PCR is less sensitive than that for DYZ1, but the amplicon length, at 93 bp, is even shorter. Other possible targets for Y-specific PCRs are short tandem repeats (STRs) that are present on the Y but not X chromosome. These STRs are routinely used in studies of human origins and migrations, so a number of PCR systems are available, and occasionally these are also used for sex identification.

With modern DNA, the use of a Y-specific PCR for identification of sex is usually straightforward. If the PCR gives a product, then the sample is male; if there is no product, then it is female. When used with archaeological extracts, the results can be less easy to interpret, because absence of a PCR product can have any of three meanings:

- Absence could mean that there is no Y chromosome and the specimen is female.
- Absence could mean that no ancient DNA is present in the PCR: the specimen could be female or male.
- Absence could mean that the extract contains an inhibitor that prevents the PCR from working: again, the specimen could be female or male.

The difficulty of distinguishing between a female result and the other two scenarios means that only males can be identified unambiguously if a Y-specific PCR is used with ancient DNA. Methods that give PCR products with both males and females, but enable the two to be distinguished, are therefore needed.

Table 10.1 Targets for DNA-based sex identification.

Locus	*Description*	*PCR product sizes*
DYZ1	Repeat units on Y chromosome	X no product, Y 102 bp
SRY	Gene only on Y chromosome	X no product, Y 93 bp
Y STRs	Short tandem repeats on Y chromosome	Various systems
AMELX/Y	Dimorphic gene on X and Y chromosomes	X 106 bp, Y 112 bp X 80 bp, Y 83 bp
ZFX/Y	Dimorphic gene on X and Y chromosomes	*Hae*III digestion: X 37 + 172 bp, Y 37 + 84 + 88 bp
Alphoid repeats	Dimorphic repeats on X and Y chromosomes	X 157 bp, Y 200 bp X 170 bp, Y 130 bp
DX424	Repeat units on X chromosome	X 181–199 bp, Y no product

10.3.2 Slightly different versions of the amelogenin gene are present on the X and Y chromosomes

The central regions of the human sex chromosomes are largely different from one other, and the genes that they contain are unique to X or Y; but within these areas there are also a very small number of genes that have counterparts on both chromosomes. One such gene is AMELX (on the X chromosome) and AMELY (on the Y). This is the **amelogenin** gene, which codes for a protein that is involved in deposition of enamel during tooth development. The AMELX and AMELY genes are similar, but not identical, and when the nucleotide sequences are aligned, a number of indels, positions where a segment of DNA has either been inserted into one sequence or deleted from the other sequence, are seen. If the primers for a PCR anneal either side of an indel, the products obtained from the X and Y chromosomes will have different sizes. Female DNA would give a single band when the products are examined, because females only have the X chromosome, whereas males would give two bands, one from the X chromosome and one from the Y. If the sample contains no DNA or the PCR is inhibited, no bands will be obtained. This means that there is no confusion between these technical failures and a male or female result.

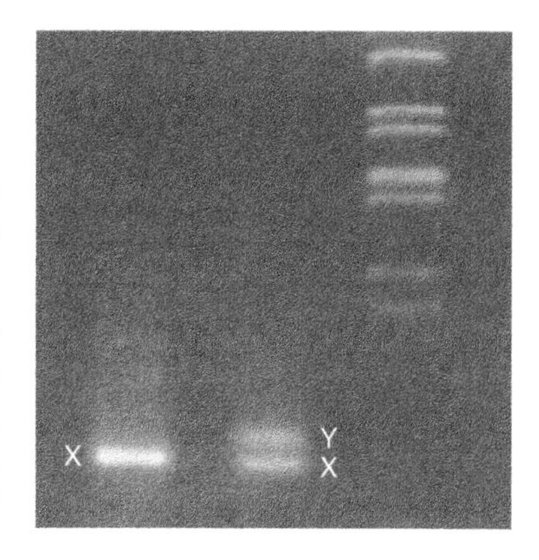

Figure 10.4 Agarose gel showing the results of PCR with amelogenin primers and female (left lane) and male DNA (center lane). The distinction between the X and Y products is clear. The righthand lane contains DNA size markers.

Of the various PCR systems that have been published for sex identification based on the amelogenin gene (see Table 10.1), the one that is now used routinely in both ancient DNA research and in forensic science gives products of 106 bp for the X chromosome and 112 bp for the Y. The difference is therefore only 6 bp, but this is sufficient for a clear distinction between the bands when the amplicons are examined by agarose gel electrophoresis (Figure 10.4).

There can still be problems though, especially with the phenomenon known as **allelic dropout**. This occurs when one of the two PCR products, almost always the larger one, fails to be synthesized in sufficient amounts to give a visible band in the gel. Allelic dropout is more common when the template DNA is short and fragmented, as is always the case with ancient DNA, because synthesis of shorter amplicons is favored, and longer ones are produced less efficiently, or possibly not at all. Allelic dropout was a significant problem with some of the earlier amelogenin PCR systems that were devised, as these gave relatively long products (e.g. 196 bp for X and 132 bp for Y in one system, and 329 bp for X and 235 bp for Y in another), but the small lengths of the 106/112 bp products are generally thought to reduce dropout to a minimum. Experience shows, however, that it is still a significant problem with archaeological specimens, and one that cannot be ignored, as the band that is likely to drop out is the longer one, for the Y chromosome. Hence the presence of just the 106 bp product is not diagnostic for a female skeleton, as this result is also obtained with male DNA if the 112 bp is lost through dropout. An even shorter set of amelogenin amplicons, 80 bp for X and 83 bp for Y, has recently been devised, but even these can still display Y allele dropout. Potential female results from either of these amelogenin tests should therefore be checked by carrying out at least four separate PCRs with each ancient DNA sample, and using two independent extracts, ideally from different parts of the skeleton.

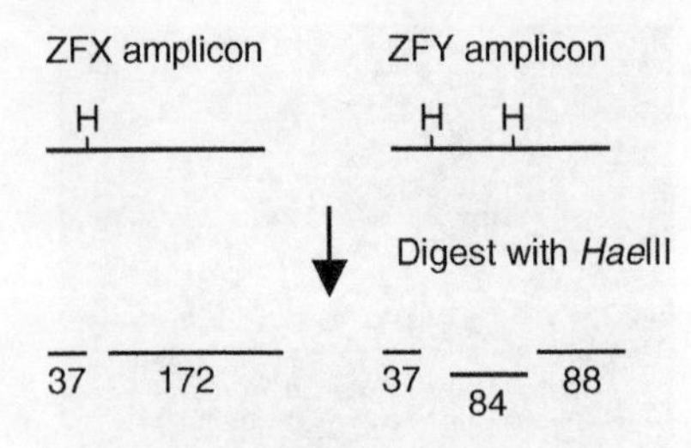

Figure 10.5 Digestion of the ZFX and ZFY amplicons with *Hae*III gives fragments of diagnostic sizes, because the ZFY amplicon contains an additional restriction site not present in ZFX. H, *Hae*III recognition sequence (5′–GGCC–3′).

10.3.3 Other sex identification PCRs can be used to check the results of amelogenin tests

Even if multiple PCRs are carried out, there will always be some uncertainty regarding any female result obtained when an amelogenin test is used with ancient DNA. As well as the possibility of Y allele dropout, there are also some rare cases where the amelogenin gene is deleted from the Y chromosome. Individuals carrying this deletion are male, because the amelogenin gene, although located on the sex chromosomes, plays no role in the physiological determination of sex, but they will be typed as female by an amelogenin PCR. The frequency of this deletion is low – less than 1 in 5000 males – but it could be much higher than this in small family groups if it was carried by one of the group's ancestors. To guard against these various problems, sex identifications using ancient DNA should be based on at least two different PCR targets, rather than depending solely on amelogenin typing.

A number of possible alternatives have been published by various researchers (see Table 10.1). These include the ZFX and ZFY genes, which code for DNA-binding proteins that regulate the expression of other genes, many of which are involved in sex determination. The PCR products from ZFX and ZFY are both the same length, 209 bp, but can be distinguished by digestion with a **restriction endonuclease.** This is an enzyme that cuts DNA at a specific sequence and hence cleaves a PCR product into two smaller fragments if the amplicon contains the recognition sequence. When treated with the restriction endonuclease called *Hae*III, the ZFX amplicon is cut into two fragments of 37 bp and 172 bp, whereas the Y amplicon is cut into three segments, of 37 bp, 84 bp, and 88 bp (Figure 10.5). PCRs have also been designed to amplify the non-coding alphoid repeat sequences present within the centromeres of the X and Y chromosomes, two different systems being used with product sizes of 157 bp and 200 bp, or 170 bp and 130 bp, for X and Y respectively. One final possibility is to combine the Y-specific DYZ1 PCR with a second PCR, carried out in the same tube, directed at the multicopy DX424 locus present in the X chromosome, which can be amplified as fragments of 181–199 bp. Although slightly complicated, as two sequences are being amplified at the same time, this system has high sensitivity because both targets are present at high copy number.

10.3.4 DNA typing cannot always give an accurate indication of biological sex

Leaving aside the technical problems caused by allelic dropout, there are still occasions when a DNA test does not reveal the correct sex of the individual being studied. This is because there are rare individuals who display **sex reversal**, and whose genetic sex (their genotype) differs from their biological sex (their phenotype). In humans, sex reversal can give rise to both male and female phenotypes (Table 10.2). DNA tests only reveal the genotype, so in these cases an amelogenin or other sex identification PCR might give misleading results.

Table 10.2 Examples of sex reversal syndromes and similar chromosomal anomalies in humans.

Syndrome	*Phenotype*	*Cause*	*Amelogenin PCR result*
Conditions resulting in an inaccurate amelogenin result			
Sex reversal	Male	XX with translocated SRY gene	Female
Sex reversal	Female	XY with deleted SRY	Male
Sex reversal	Female	XY but two copies of DAX	Male
Androgen insensitivity	Female	XY but no male androgen hormone	Male
Congenital adrenal hyperplasia	Male, female or intersex	Defect in steroid hormone synthesis	Male or female, but usually not agreeing with phenotype
Amelogenin deletion	Male	XY but deleted AMELY	Female
Conditions not affecting the amelogenin result			
Turner's	Female or infrafemale	XO, no Y chromosome	Female
Klinefelter's	Male or intersex	XXY, extra X chromosome	Male
Triple X	Female or infrafemale	XXX	Female
47,XYY	Male	XYY	Male
48,XXXX	Female	XXXX	Female
48,XXYY	Male	XXYY	Male
49,XXXXX	Female	XXXXX	Female
49,XXXXY	Male	XXXXY	Male
49,XXXYY	Male	XXXYY	Male

The female phenotype is usually determined by the absence of the Y chromosome, both for normal XX females and those with Turner's syndrome, who have just a single X chromosome and whose genotype is designated XO. But approximately one in every 20,000–25,000 individuals who are phenotypically male lack an Y chromosome and have a "female" XX genotype. In 80% of these cases, the sex reversal occurs because one of the X chromosomes carries a copy of the SRY gene, which at some stage in the person's ancestry has been transferred to it from a Y chromosome by **translocation**, a recombination event that results in a segment of one chromosome becoming attached to a different chromosome. The translocation attaches a small part of the Y chromosome, containing the SRY gene but not the amelogenin gene, on to the end of an X chromosome. Although phenotypically male, all of these XX individuals would be typed as female if sex identification was based on PCR of the amelogenin gene, though male if the SRY gene itself was assayed.

Conversely, approximately one in 20,000 females have an XY genotype. A few of these – some 10% of the total – have a version of the Y chromosome from which the SRY gene has been deleted. For these individuals, the amelogenin test will give a male result and the SRY test a female outcome. Other XY reversals occur because the X chromosome has two, rather than one, copies of the DAX gene. This gene is antagonistic to SRY but when present in a single copy, cannot prevent SRY from working; so normal XY individuals are male. But when there are two copies of DAX, SRY cannot function and the phenotype is female. This type of reversal cannot be distinguished by amelogenin and/or SRY PCRs, or indeed any of the currently available DNA tests, all of which would give a male result.

As well as these rare sex-reversed individuals, there are other, more frequent syndromes involving unusual complements of the sex chromosomes (Table 10.2). The commonest of these is Klinefelter's syndrome, in which the genotype is XXY and the phenotype usually male, though the condition has varying degrees of severity and can lead to ambiguities especially when the sex of the child is identified at birth. Ambiguities are also associated with XO and XXX genotypes, which sometimes give rise to sexually underdeveloped **infrafemales**. The presence of unusual chromosome complements has no effect on the validity of most PCR tests, as the presence of the Y chromosome always indicates a male phenotype, but these tests in their usual forms will fail to detect the extra chromosomes. These extra chromosomes can, however, be detected by quantitative versions of the amelogenin test, which make use of fluorescently tagged primers whose signals can be measured in the PCR product to determine the relative number of X and Y amplicons and hence to identify the numbers of X and Y chromosomes in the original sample. This version of the amelogenin test can be used with modern DNA, but has never been applied to archaeological material and probably would not give an accurate result, because of the fragmentary nature of ancient DNA and the concomitant possibility of partial or complete dropout of the Y and/or X alleles.

Other sex reversals and ambiguities arise because of errors in the biochemical pathways that specify development of the sexual characteristics. An example is androgen insensitivity syndrome, which results in an XY female because the male androgen hormone, although produced in rudimentary testes, is ineffective, the cell-surface receptors to which the hormone usually binds being non-functional. The gene for these receptors is located on the X chromosome and is mutated in individuals who display the syndrome. With other syndromes, the underlying genes are located on the autosomes rather than sex chromosomes. This is the case with congenital adrenal hyperplasia, a group of related syndromes in which the steroid sex hormones are not made correctly due to a mutation in one of the genes specifying the enzymes involved in their synthesis. Very specific genetic tests are needed to detect these biochemical syndromes, regular sex identification PCRs providing no indication of their presence.

10.3.5 DNA tests can also be used to identify the sex of animal remains

Sex identification of animal bones is also important in some areas of archaeology. This is particularly true in studies of animal husbandry, because the ways in which domesticated animals were used in the past can be inferred from herd structures (the ratio of male to female animals) and kill-off patterns (the age at which animals were slaughtered). For example, a preponderance of adult female cattle at an archaeological site implies that animals were being raised for milk production rather than as a source of meat or for traction.

Osteological indicators of sex are not the same in all mammals, and the sexual dimorphisms that enable male and female human skeletons to be distinguished do not hold for cattle and other farm animals. Domesticated animals are more difficult to sex by osteological means, the size and robustness of the skeleton often being the only indicator, with males bigger than females. The sizes differences are most apparent in the metapodial bones of the feet, as the feet have to bear the full weight of the animal. With cattle, the standard method is to measure the breadth of the trochlea, grooved structures at the distal ends the metacarpals, the foot bones of the forelimbs. With modern animals this test has a high degree of accuracy, but results from archaeological material are less easy to interpret, as the relative sizes of male and female cattle might not have been the same in the past as they are now.

Although the osteological indicators of sex are different in humans and domestic animals, the genetic determinants are the same. DNA sex tests similar to those used with humans have been devised for most domestic animals, their main application in modern husbandry being in identifying the sex of embryos so that herd structures can be manipulated prior to the birth of new animals. To date there have been few attempts to apply any of these tests to archaeological material. PCRs directed at the ZFX/ZFY genes have, however, been used with cattle bones from a 13th century AD site in western Sweden. These PCRs were designed to amplify a very short, 63 bp segment containing a single nucleotide polymorphism (SNP), a C in the ZFX version of the gene and a T in ZFY. Sequencing of the PCR product was therefore needed to identify which genes were present. Results were obtained for 22 animals and in every case the outcome of the DNA test was in agreement with the sex, as indicated by measurements of the metacarpals, despite the cattle from this site being relatively small and the male and female animals showing much less extreme differences in size compared with modern cattle. The DNA work therefore suggests that the osteological method is, in fact, an accurate way of identifying the sex of archaeological animal remains, at least with cattle, and DNA tests are necessary only in those cases where the metapodia are absent or when measurement of these is inconclusive.

10.4 Examples of the Application of Sex Identification in Biomolecular Archaeology

DNA-based methods for sex identification have now been applied to many archaeological assemblages, especially to fragmentary and disarticulated skeletons or the remains of juveniles, these being the types of material that are difficult, or impossible, to sex by osteological examination. Some of this work has been aimed simply at improving the cataloguing of archaeological sites by increasing the number of skeletons for which a biological sex can be assigned, but there have also be attempts to use DNA-based sex identification to tackle broader archaeological issues. To illustrate the contribution that the DNA methods can make we will consider two areas of research, the first of these concerning the practice of sex-specific infanticide in past societies.

10.4.1 Ancient DNA can improve our understanding of infanticide in past societies

We know from historical records that the Romans carried out infanticide (Section 10.1), and it is likely that this practice occurred in the outposts of the Roman Empire as well as in Italy itself. Indications of infanticide in Roman Britain have been obtained from comparisons between the ages at death of infant skeletons from this period and from medieval Britain. Romano-British skeletons show a peak of deaths at nine months, corresponding to newborn babies, whereas the age distribution for medieval infant deaths is much flatter. If the medieval distribution represents the natural pattern of deaths among infants, then the peak in the Romano-British data could indicate infanticide of newborns. That this supposed infanticide might largely have been of female babies has been inferred by the skew in the sex ratio among adult skeletons from Romano-British cemeteries, which shows an imbalance of 1.46 males to every female. Whether these two pieces of circumstantial evidence genuinely point to female infanticide could be assessed by identifying the sex ratio among the infant skeletons. So far DNA methods have not been extensively used to address this question, but those results that have been obtained provide no support for female infanticide. For example, two of the sites whose mortality patterns suggest infanticide are the Roman villas at Ancaster in Lincolnshire and Thistleton in Rutland. Amelogenin PCRs have given successful results with six infants from Ancaster, four of which are male and two female, and seven from Thistleton, five of these male and two female. These numbers are, of course, too small to make general conclusions about the situation throughout the whole of Roman Britain, but the results give no support to the hypothesis that female infanticide was carried out at these two sites.

Outside of Britain, infanticide was at one time suspected at the late Roman villa of Lugnano in Teverina, approximately 70 miles northwest of Rome in the Tiber valley. Here a cemetery containing 22 prenatal skeletons, 20 neonates, and 5 infants was discovered. Amelogenin tests enabled sex to be assigned to 9 of the skeletons, these 9 comprising 5 females and 4 males. Again, the sample size is low but the approximately equal number of males and females is more consistent with death by natural causes than infanticide. Interestingly, other DNA tests carried out with these skeletons indicated that malaria might have been endemic in the area, supporting the views of the excavators of the site, who had previously suggested that the villa had been abandoned following a malaria epidemic. This disease causes high mortality among infants and also results in miscarriages, possibly explaining the presence of the neonate and fetal skeletons.

Despite the lack of evidence for infanticide provided by the above studies, there is one ancient DNA project that has revealed the existence of this practice in Roman times. Excavations at a Roman bathhouse at Ashkelon in Israel, dating to the 4th–6th centuries AD, uncovered the skeletal remains of 100 newborn babies, most of these less than two days old, the remains buried in a sewer under the building. Amelogenin PCRs were successful with 19 of 43 skeletons that were tested, 14 of these being male and 5 female. This is the kind of skewed ratio that we might anticipate if sex-specific infanticide was being carried out, but surprisingly the bias is towards the death of male babies rather than females. The explanation is possibly that the children were the offspring of prostitutes who worked at the baths, for whom there would have been more economic sense in keeping the female infants who might subsequently take up their mothers' careers or be sold as slaves. In the ancient world, female slaves were considered to be more valuable that male slaves, especially if virgins. It was more likely to be poorer families unwilling or unable to pay expensive dowries for their daughters' marriages who resorted to female infanticide.

10.4.2 Ancient DNA enables contradictions between osteology and grave goods to be resolved

DNA-based sex identification is also helping archaeologists resolve the contradictions that sometimes arise between the biological sex of a skeleton as inferred by osteological analysis, and the gender indications of the artifacts that accompany the skeleton in the grave. As described in Section 10.1, there has always tended to be an assumption that grave goods are indicative of the gender of the individual with whom they are buried, men being buried with typically "male" objects such as weapons and shields, and females with more feminine artifacts, such as jewelry and combs. In many cases where the osteoarchaeological data are uncertain, the grave goods are used to identify the sex of a skeleton, and it has even been known for skeletons that according to their morphology are male, or female, to be assigned to the other sex because of "inconsistent" grave goods. There are a remarkable number of "robust females" recorded in site reports, in other words skeletons that look to be male but are accompanied by non-masculine artifacts.

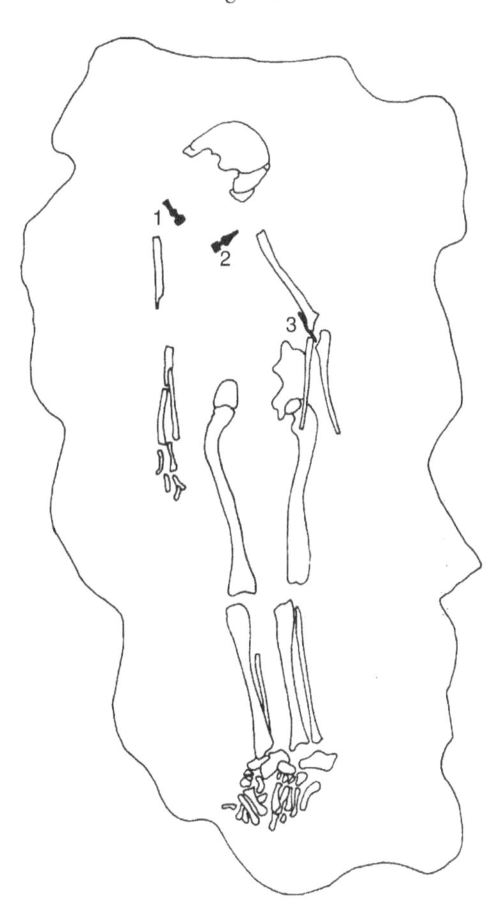

Figure 10.6 Drawing of a tall Anglo-Saxon skeleton of uncertain sex from Blacknall Field. The positions of the three dress pins found in the burial are indicated. Drawing courtesy of the Wiltshire Heritage Museum.

Inconsistent grave goods are particularly common in Viking and Anglo-Saxon cemeteries in northern Europe. Wealthy female Viking burials from the 11th to 12th centuries in Finland and the Baltic States have been found with axes, spears, and javelins, sometimes but not always alongside other more "female" artefacts, such as personal ornaments. A male Anglo-Saxon skeleton from Friesland, Holland, was accompanied by brooches, beads, and a bracelet similar to the grave goods found with female burials in the same cemetery. In Britain, inconsistent grave goods have been found in a number of Anglo-Saxon cemeteries, with multiple examples at some sites: seven males with ornaments and four females with weapons at Buckland near Dover, for example. In some cases, such as at West Heslerton in Yorkshire, ancient DNA has been used to confirm that the osteological sex identifications made for these anomalous burials are correct. In others, ancient DNA has shown that the grave goods are giving the correct indication of sex.

One of the burials in the 5th–6th century AD cemetery at Blacknall Field, near Pewsey in southern England, provides an interesting example of the conflict that can arise between osteology and grave goods. Much of the cranium and pelvis of this particular skeleton was missing, making an osteological assessment difficult (Figure 10.6), but the person's height, estimated at over 1.83 m, was exceptional for Anglo-Saxon times, when the average was 1.73 m for males and 1.52 m for females. The skeletal features therefore seem to suggest that the burial must be male, but three dress pins, typical female ornaments, were found near the shoulders of the skeleton. Amelogenin and SRY PCRs were carried out with this skeleton, as well as PCRs directed at SNPs on the X and Y chromosomes. None of these tests gave any evidence for the presence of Y chromosomal DNA, suggesting that this person, despite her height, was a female. Interestingly, a second skeleton from the same cemetery was estimated to be almost the same height, but was unambiguously female according to the structure of the pelvis. These results indicate how difficult it is to base sex identification on criteria such as height and robustness, which are looked on as good indicators in some modern populations, but clearly cannot be applied so rigorously with historic and prehistoric skeletons.

Conflicts between biological sex and grave goods also occasionally arise at southern European sites. An example was reported at the Etruscan necropolis of Monterozzi in Italy, which dates to the 4th–3rd centuries BC. One of the skeletons recovered from this site had osteological features suggesting a male sex, but was buried with ornaments identified as earrings, usually considered feminine objects. Amelogenin PCRs showed that this individual was indeed female.

Further Reading

Brown, K.A. (1998) Gender and sex: what can ancient DNA tell us? *Ancient Biomolecules*, 2, 3–15.

Cappellini, E., Ciarelli, B. & Sineo, L., *et al.* (2004) Biomolecular study of the human remains from tomb 3859 in the Etruscan metropolis of Monterozzi, Tarquinia (Viterbo, Italy). *Journal of Archaeological Science*, 31, 603–12.

Faerman, M., Bar-Gal, G.K. & Filon, D., *et al.* (1998) Determining the sex of infanticide victims from the Late Roman era through ancient DNA analysis. *Journal of Archaeological Science*, 25, 861–5.

Haas-Rochholz, H. & Weiler, G. (1997) Additional primer sets for an amelogenin gene PCR based DNA sex test. *International Journal of Legal Medicine*, 110, 312–15. [The system giving 80 and 83 bp products.]

Mays, S. (1993) Infanticide in Roman Britain. *Antiquity*, 67, 883–8.

Mays, S. & Cox, M. (2000) Sex determination in human remains. In *Human Osteology: In Archaeology and Forensic Science* (ed. Cox, M. & Mays, S.), 117–30. Cambridge University Press, Cambridge.

Mays, S. & Faerman, M. (2001) Sex identification in some putative infanticide victims from Roman Britain using ancient DNA. *Journal of Archaeological Science*, 28, 555–9.

Nakagome, Y., Nagafuchi, S. & Seki, S., *et al.* (1991) A repeating unit of the DYZ1 family on the human Y chromosome consists of segments with partial male-specificity. *Cytogenetics and Cell Research*, 56, 74–7.

Pfitzinger, H., Ludes, B. & Mangin, P. (1993) Sex determination of forensic samples: co-amplification and simultaneous detection of a Y-specific and an X-specific DNA sequence. *International Journal of Legal Medicine*, 105, 213–16. [Sex identification using a combination of DYZ1 and DX424.]

Reynolds, R. & Varlaro, J. (1996) Gender determination of forensic samples using PCR amplification of the ZFX/ZFY gene sequences. *Journal of Forensic Science*, 41, 279–86.

Sørensen, M.L.S. (2000) *Gender Archaeology*. Polity Press, Cambridge.

Sullivan, K.M., Mannucci, A., Kimpton, C.P. & Gill, P. (1993) A rapid and quantitative DNA sex test: fluorescence-based PCR analysis of X-Y homologous gene amelogenin. *Biotechniques*, 15, 636–41. [The amelogenin system with 106 and 112 bp amplicons.]

Svensson, E.M., Götherström, A. & Vretemark, M. (2008) A DNA test for sex identification in cattle confirms osteometric results. *Journal of Archaeological Science*, 35, 942–6.

11

Identifying the Kinship Relationships of Human Remains

An understanding of the kinship relationships between a set of human remains is, after sex identification, the second fundamental contribution that biomolecular research can make in archaeology. When an adult is found buried with an infant, we immediately ask if these individuals were parent and child. When a cemetery is excavated, the arrangement of the burials often suggests the presence of family groups, and if the burials are all from a similar period, then there may be broader kinship relationships within the cemetery as a whole. When probing these relationships, we are particularly interested in identifying individuals who have moved into the community from elsewhere, possibly as marriage partners for young men or women who were born and raised locally.

In contrast to sex identification, osteological examination of skeletons can give little assistance in distinguishing pairs and groups of individuals that are related to one another. Biomolecular archaeology provides the only means of carrying out this kind of analysis in a comprehensive way. DNA testing is well established as a means of identifying family and broader relationships among living people, and isotope analysis of teeth can reveal individuals who spent their early years at a location away from the place where their bodies are found. In this chapter, we will examine how these techniques are being applied to address questions of kinship that arise in archaeological research.

11.1 The Archaeological Context to Kinship Studies

As with sex identification, it is important that we understand why kinship is important in archaeology. We must continually remember that our biomolecular studies are aimed at expanding our knowledge of past people, and that these studies are meaningless unless placed in the appropriate archaeological context.

Biomolecular Archaeology: An Introduction, by Terry Brown and Keri Brown © 2011 Terry Brown & Keri Brown

11.1.1 Kinship provides a sense of identity

Kinship is important in our own lives because it provides us with a sense of who we are. Anthropologists look on an understanding of one's identity as a universal human imperative. In some cultures, kinship is set by non-biological factors such as common residence, food sharing, and participation in rituals, but in many societies family relationships and ancestry are the overriding considerations that define kinship. To appreciate the importance that we place on our own sense of identity we need only consider the success of companies who, in return for a mouth swab and credit card details, provide glossy certificates describing one's mitochondrial DNA clan mother and Y chromosome clan father.

As well as its relevance to personal identity, kinship is also an important way of organizing societies, and there is a whole wealth of anthropological literature showing how kinship is used to enable an individual to "know" their place in their society. Various lineages (lines of descent) in a society might be ranked, with individuals from an important lineage assuming positions of power, the extent of their personal status deriving from the closeness of their relationship with the leaders of the lineage. In these hierarchical societies, status and often wealth are inherited by virtue of a person's position in a network of kin, rather than being acquired during their lifetime, except perhaps by exceptional individuals who force their way into, or displace a lineage, through their personal prowess. Anyone familiar with the history of imperial Rome will understand the complexities of the relationships that can exist among a ruling elite, and how desirable it would be, in the absence of written records, to be able to infer such relationships from the skeletons of members of that elite.

Kinship analysis therefore enables us to address the critical archaeological question of whether a past society was egalitarian or hierarchical, and this in turn can throw light on the evolution of social systems in the past. In Europe, the Copper Age is looked on as an important period of transition in this respect. Also called the Chalcolithic or Eneolithic periods, the Copper Age lasted from 4500 to 2000 BC and was the time when metal working was first adopted. Immediately before the Copper Age, the late Neolithic population of Europe lived in largely agrarian communities whose burial practices suggest that their society had no hierarchical structure. But by the end of the Copper Age, these communities had adopted more complex social organizations, with greater stratification and individualization, sufficient to drive the subsequent emergence in the Early Bronze Age of the sophisticated, warrior-led societies that eventually evolved into the civilizations of classical times. The social development of the Copper Age was accompanied by increases in population density resulting from the gradual intensification of agriculture and the discovery of metallurgy, but these technological advances tell us little about the nature of the social processes occurring at that time. These processes can only be disclosed by examining the structures of the societies themselves, for example by identifying the extent to which the internal structure of a society was based on groups of biologically related people. Kinship studies of individual Copper Age communities could therefore make an important contribution to our understanding of this important period of social transition.

Table 11.1 Important post-marital residence patterns.

Name	*Description*
Patrilocal	Woman moves and the new couple live within or close to the man's family
Matrilocal	Man moves and the new couple live within or close to the woman's family
Neolocal	Man and woman live together away from either parental family
Matrifocal	Woman on own establishes new family away from either parental family
Natalocal	Man and woman stay with their own families after marriage

11.1.2 Endogamy and exogamy are important adjuncts to kinship

Endogamy and **exogamy** are aspects of social structure that are interlinked with kinship. In cultural terms, endogamy refers to marriage between two individuals who belong to the same social group, whereas exogamy is marriage between people from different groups. Endogamy, if practiced to excess, can lead to the accumulation of recessive, often harmful, genetic traits, as illustrated by the high frequency of some genetic diseases in modern endogamous groups such as the Ashkenazi Jews and the Amish. The extent to which the genetic benefits of exogamy were understood in the past is questionable, though it has been claimed that avoidance of mating with one's close relatives is an innate behavioral pattern displayed not just by humans but also by other animals. More prosaically, a major driver of exogamy among prehistoric communities might have been the need for small groups of people to forge alliances with one another in order to reduce conflict. This implies a certain amount of resource sharing, so a converse advantage of endogamy is that property can be retained and accumulated within a single social group. Endogamy also promotes bonding and is common in displaced populations that feel a need to establish a new sense of identity. The practice of endogamy and exogamy might also be stratified within a society, one possibility being that members of the ruling elite practice exogamy in order to promote political alliances, but others in the group marry among themselves.

Implicit in an exogamous marriage is the need for one or both of the partners to move away from their natal group, the group to which they were born. Anthropologists recognize a number of **post-marital residence patterns** (Table 11.1), the commonest being the **patrilocal** and **matrilocal** systems. In patrilocal residence, the woman moves and the new couple live within or close to the man's family, whereas in the matrilocal system the man moves in with woman's family. Patrilocal residence typically occurs in societies with patrilineal inheritance, a married son staying close to his parents so he can take over the homestead when his father dies. Matrilocal residence patterns are less common, most of the modern examples occurring in societies in which inheritance is passed down the female rather than male line. There are also examples of matrilocality where the man is not expected to make any commitment to raising the family and returns to his own group once his wife, who remains with her parents, has given birth.

Information on the structure of past societies as gained by kinship studies would therefore be augmented if it were possible to distinguish exogamous from endogamous marriage systems, and in the former case identify if the subsequent residence pattern was patrilocal or matrilocal. In the example used above of Copper Age Europe, it would be particularly interesting to ask if during this period of transition from an egalitarian to warrior-led society there were accompanying changes in marriage systems. We must therefore turn our attention to the methods that can be used by archaeologists to study kinship and to detect exogamy.

11.1.3 Kinship and exogamy are difficult to discern in the archaeological record

Attempts to use osteoarchaeological methods to study kinship between skeletons focus on small variations in skeletal morphology called **non-metric traits**. There are over 400 such traits in the human population, mainly involving variations in the shapes of certain skeletal structures or the presence or absence of small **super-numerary bones**, which are possessed only by some individuals. The non-metric traits of the cranium, for example, include:

- Variations in the foraminera at the base of the skull, these being the holes through which nerves enter and leave the head.
- The presence or absence of super-numerary sections of the skull called wormian bones, especially the inca bone, so named because it is relatively common in Peruvian mummies.
- Retention of infant features into adulthood, such as the metopic suture, a structure on the front of the skull comprising flexible connective tissue that enables the baby's skull to bend during birth.
- The presence or absence of bony growths at certain positions on the lower jaw.
- Various dental traits including super-numerary teeth, the number of cusps on individual teeth, the number of roots to each molar tooth, and the shapes of the incisors.

The large number of non-metric traits available for study means that detailed comparisons between different skeletons can be carried out to identify pairs or groups of individuals that share the same features. But despite much research and statistical analysis, no simple correlation has been found between non-metric traits and genetic relationships, raising questions about the ability of this type of osteological analysis to indicate family patterns. There are some instances where non-metric traits in human burial assemblages have been analyzed and the results used to suggest kinship, for example at the megalithic tomb of La Chaussee-Tirancourt, near Amiens, France, but it is now suspected that many other factors, such as environment and diet, play a significant part in the variability of these features. It is also clear that some non-metric traits, especially those associated with the joints of the arms and legs, can be influenced by repetitive activities such as squatting and so might arise from the type of work that a person does. These various complications mean that non-metric traits cannot be used with any degree of certainty to assign kinship or family relationships within a group of burials.

Some inferences can, however, be made by less direct means. In particular, the arrangements of burials in a cemetery are often thought to give indications of both status and interrelationships. It has been speculated, for example, that the linear arrangements of Bronze Age burial mounds in Wessex, in southern England, represent lineages of leaders. Similarly at Mycenae the co-occupancy of graves and the organization of the graves in Circle B into four groups has led to suggestions that the people buried in the Grave Circle, whose status is indicated by the wealth of their grave goods, were members of a single family or small number of families. The skulls of seven of these individuals were sufficiently well preserved

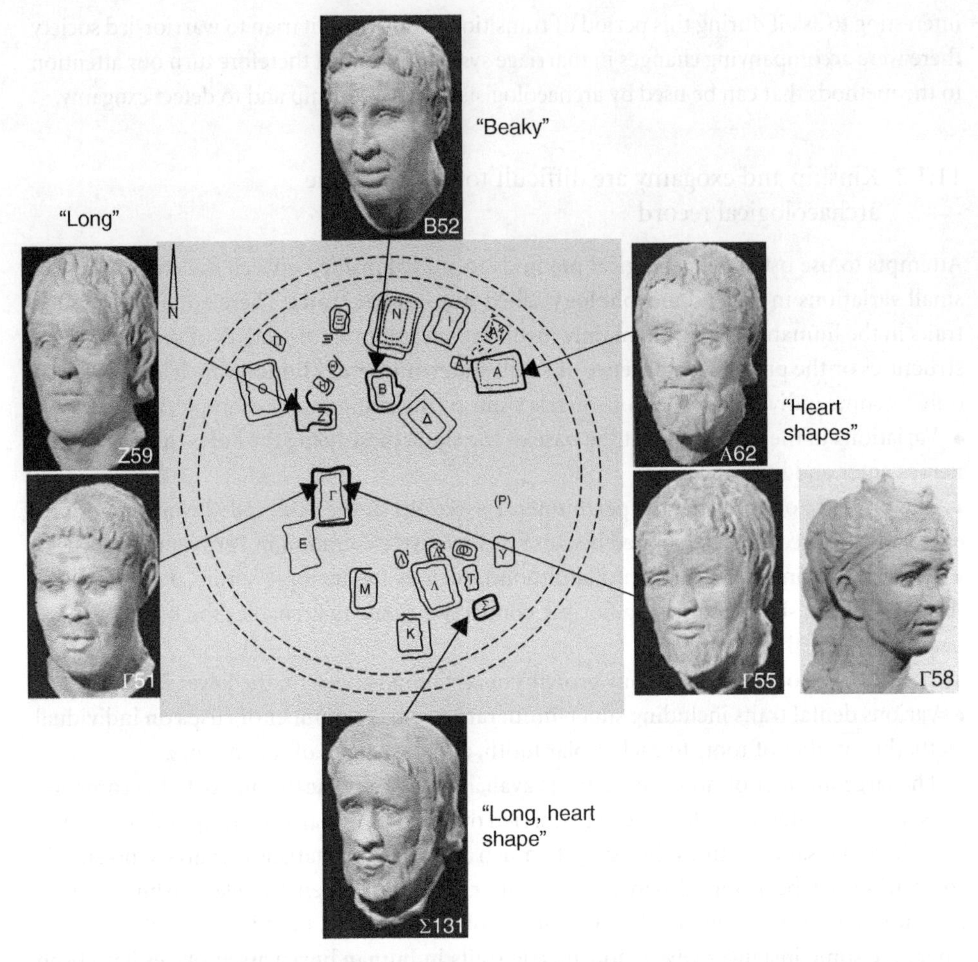

Figure 11.1 Map of Grave Circle B at Mycenae, Greece, and facial reconstructions of seven of the skulls recovered from this site, dating to 1650–1550 BC. The faces are classified as "heart-shaped," "long," and "beaky," with one individual, Σ131, intermediate between two of these types. Reprinted from Abigail Bouwman, Keri Brown, John Prag, and Terry Brown, "Kinship between burials from Grave Circle B at Mycenae revealed by ancient DNA typing," *Journal of Archaeological Science*, 35, 2580–4, copyright 2008, with permission from Elsevier and the authors.

for facial reconstructions to be attempted, the physical features thus revealed placing them in three groups, based on whether they had "heart-shaped," "long," or "beaky" faces (Figure 11.1). When the results of these facial reconstructions are viewed in three dimensions, the similarities between the members of each group are striking. In essence, this is an indirect type of non-metric trait analysis, based on variations in the structure of the skull and using the subjective, though powerful, ability that we possess to recognize and distinguish between the faces of different people. Few would argue, however, that facial reconstruction, when possible, gives anything more than an indication of possible kinship relationships.

Marriage patterns are even more difficult to study using non-biomolecular approaches. Exogamy has sometimes been implicated from careful examination of grave goods, especially when female skeletons are accompanied by artifacts associated with a different region from the one in which they are buried, suggesting a patrilocal residence pattern. With these more subtle aspects of social organization, hypotheses about the past are often drawn by ethnographic analogy, in which the behavior of historic and present-day social groups are studied in order to infer the behavior patterns of prehistoric people. This approach to the past has obvious dangers, not least the assumption that, however similar their cultures, a prehistoric social group behaved in the same way as an analogous modern population.

11.2 Using DNA to Study Kinship with Archaeological Skeletons

Now we turn our attention to the methods used by biomolecular archaeologists to assess the kinship relationships between skeletons recovered from archaeological sites. We will focus on the DNA methods that enable actual kinship relationships to be inferred, and for the time being leave to one side the isotope analyses that are used to study post-marriage residence patterns.

11.2.1 Archaeological techniques for kinship analysis are based on genetic profiling

The DNA methods were originally developed for use in forensic science as a means of identifying suspects from traces of DNA left at crime scenes or of establishing paternity in civil cases. Central to these techniques is the use of variable DNA sequences, usually short tandem repeats (STRs), to establish an individual's genetic profile (Section 2.4).

In criminal cases, the objective is to match the genetic profile of DNA from a crime scene with that of suspects or with genetic profiles that are contained in DNA databases, in order to identify possible perpetrators of the crime and to exclude innocent people from the investigation. Simple identification based on genetic profiling is less frequently used in biomolecular archaeology but has found applications, notably in determining if different parts of one or more disarticulated skeletons belong to the same individual. The more common use of genetic profiling in biomolecular archaeology is equivalent to paternity testing, where the analysis involves comparisons between profiles to establish if two or more individuals could be related, using the principles of inheritance as described in Section 2.4.

A key objective for forensic scientists has been to develop genetic profiling to the stage where the statistical likelihood of two individuals on the planet sharing the same profile is so low as to be considered implausible when DNA evidence is presented in a court of law. The current methodology, called CODIS (Combined DNA Index System), makes use of 13 STRs (Table 11.2) with sufficient variability to give only a 1 in 10^{15} chance that two individuals, other than identical twins, have the same profile. As the world population is around 6×10^{9}, a match between a profile obtained from a crime scene and that of a suspect is highly suggestive. Each STR is typed by PCRs with primers that are fluorescently labelled and which anneal either side of the variable repeat region (Figure 11.2). The alleles present

Table 11.2 The CODIS set of STRs.

Locus	*Chromosome*	*Repeat type*
TPOX	2	AATG
D3S1358	3	AGAT
FGA	4	Complex tetranucleotide
D5S818	5	AGAT
CSF1PO	5	AGAT
D7S820	7	GATA
D8S1179	8	TATC
TH01	11	AATG
VWA	12	AGAT
D13S317	13	GATA
D16S539	16	GATA
D18S51	18	GAAA
D21S11	21	TCTA

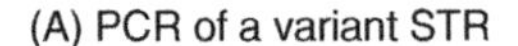

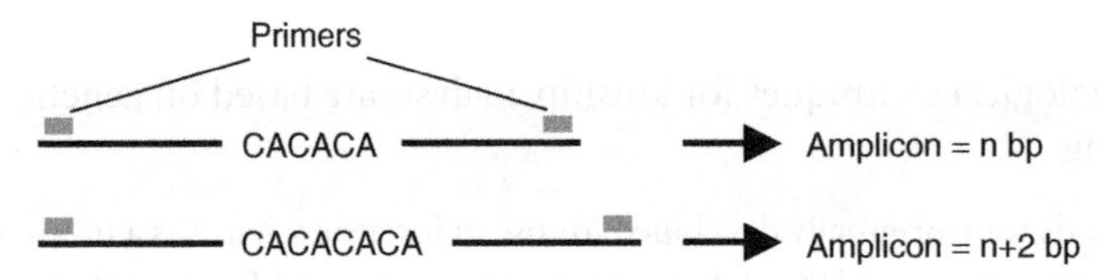

(B) Sizing of PCR products by capillary gel electrophoresis

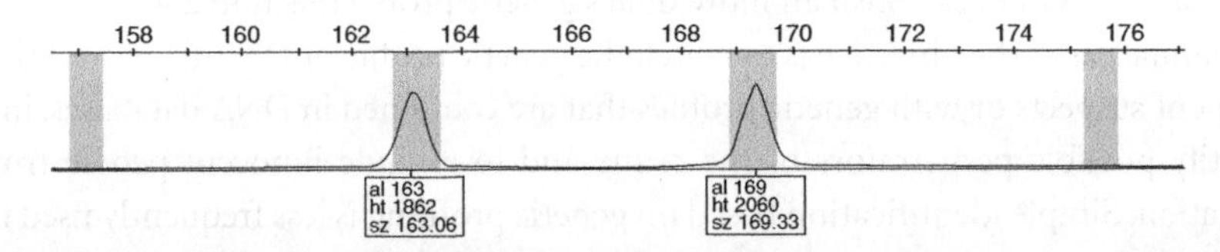

Figure 11.2 STR typing. (A) Two alleles of an STR differ by one CA repeat unit. The PCR products will therefore differ in length by 2 bp. (B) A capillary gel electrophoretogram obtained after PCR of an STR in maize. Two peaks are seen, showing that this STR is heterozygous in the plant that was tested. The gray areas indicate the expected locations of the peaks corresponding to the different alleles for this STR. Image in (B) kindly provided by Claudia Grimaldo.

at the STR are then typed by measuring the sizes of the amplicons by capillary gel electrophoresis. Two or more STRs can be typed together in a **multiplex PCR** if their amplicon sizes do not overlap, or if the individual primer pairs are labelled with different fluorescent markers, enabling the amplicons to be distinguished in the capillary gel.

The same principles are used in attempts to obtain genetic profiles from archaeological remains, but the details are different. Several of the PCRs in the CODIS set give relatively long amplicons, unsuitable for use with ancient DNA, and so the ones chosen are usually those whose longest products do not exceed 175 bp. These PCRs are usually accompanied with additional ones not always carried out in forensics, including ones that type STRs on the Y chromosome, which can give information specifically on paternal relationships. With ancient DNA, the expectation is always that only a few of the attempted PCRs carried out with any skeleton will yield results. It therefore is sensible to try typing as many STRs as possible in the hope that the information obtained from those PCRs that are successful will enable the kinship relationships to be pieced together.

11.2.2 Various complications can arise when STRs are typed in archaeological material

Even if the PCRs are successful, there might be problems in identifying the STR alleles that are present in an ancient DNA sample. Allelic dropout, which was described in Section 10.3 as a complicating factor in identification of sex by the amelogenin test, is a more substantial problem when STRs are typed because of the relatively large size of some of the expected amplicons. Alleles with larger repeat numbers are therefore quite likely to drop out if the ancient DNA is as fragmentary as is usually the case. The problem is compounded by the difficulty in knowing if dropout has occurred, as it is quite possible for an individual to be homozygous for a particular STR, meaning that both copies of the STR are the same length and the PCR product comprises just a single-sized amplicon. Failure to detect two alleles of different lengths does not therefore mean that one allele has failed to amplify. Homozygosity should, however, be distinguishable from allelic dropout by the height of the peak in the electrophoretogram: if both alleles are the same length, then twice as much PCR product should be synthesized, giving a double-height peak.

Stuttering is a second substantial problem that often occurs when STRs are amplified from ancient DNA. Rather than each allele being revealed in the electrophoretogram as a simple peak representing a single amplicon whose length identifies the allele, a series of linked peaks of varying heights are seen, indicating that amplicons of different lengths have been synthesized (Figure 11.3). Stuttering is thought to be caused by the polymerase enzyme slipping as it copies the repeated DNA units within the STR, so that amplicons that contain repeat numbers slightly different to the correct value are synthesized, slippage of this kind being promoted in some unknown way by the damaged nature of ancient DNA templates. Stuttering can also be caused by short deletions that routinely occur when ancient DNA is being amplified, these deletions not necessarily being within the repeat region and possibly giving rise to amplicons whose lengths are intermediate between the lengths expected for true alleles. Stuttering due to slippage is particularly acute with dinucleotide repeats but less so with longer repeat motifs such as tetranucleotides and pentanucleotides, although with all STRs slippage is greater for alleles

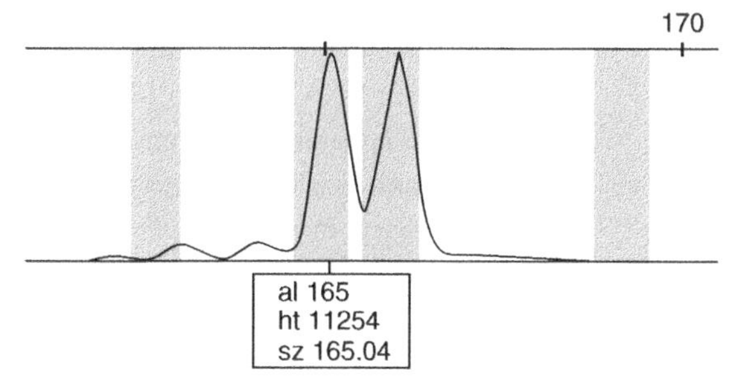

Figure 11.3 Stuttering. In this electrophoretogram, the peak on the left is preceded by a series of small stutter peaks. This example is not serious and the correct peak can easily be identified. Difficulties arise when the stutter peaks are similar in height to the correct peak, which might therefore be misidentified. Image kindly provided by Claudia Grimaldo.

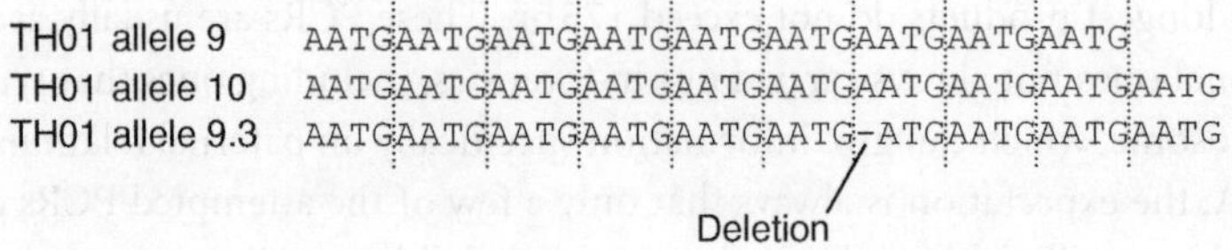

TH01 allele 9 AATGAATGAATGAATGAATGAATGAATGAATGAATG
TH01 allele 10 AATGAATGAATGAATGAATGAATGAATGAATGAATGAATG
TH01 allele 9.3 AATGAATGAATGAATGAATGAATG–ATGAATGAATGAATG
Deletion

Figure 11.4 Sequences of TH01 alleles 9, 10, and the microvariant 9.3.

with a relatively high repeat number. It is generally assumed that the largest peak in the stutter product indicates the correct allele length, but this might not always be the case, especially with tetranucleotide STRs, which often give rise to a stutter product that is one repeat shorter than the correct one.

When STRs are being typed, it is also necessary to beware of additional technical difficulties that apply also to modern DNA, but that might be more troublesome when ancient DNA is being studied. **Non-template addition** refers to the tendency of some types of DNA polymerase to add an additional A to the 3′-end of a polynucleotide it has just synthesized. If both ends of an amplicon have a non-template addition, then its effective length increases by 1 bp, resulting in a shift in the position of the amplicon in the electrophoretogram. This can occur during any PCR, not just ones directed at STRs, but such a small size change is immaterial in most other applications where the length of the amplicon provides the required information – in the amelogenin sex test, for example. It is, however, very material if a mononucleotide STR is being typed, as a single base-pair change would lead to an incorrect allele identification. Consistency is important: either the PCR should be carried out so that all amplicons become adenylated at both ends (usually achieved by increasing the length of the final synthesis step during the PCR) and this length change taken into account when mononucleotide alleles are typed, or a polymerase that has negligible non-template activity should be used so that none of the amplicons become modified in this way. An intermediate situation – in which some amplicons are adenylated and some are not – will lead to a stutter product that might be easy to identify and discount when STRs are being amplified from modern DNA, but that will simply increase the confusion with ancient DNA, when stuttering is also occurring because of the damaged templates.

An apparent length anomaly will also occur if one of the alleles being typed is a **microvariant**, in which the repeat motif is not entirely regular. For example, within the CODIS set, the repeat motif in TH01 is AATG, except in the microvariant allele 9.3 in which the seventh repeat is ATG (Figure 11.4): in other words, this is a single base-pair deleted version of allele 10, which has ten perfect AATG units. Microvariant alleles exist for most STRs, especially for the larger and more variable ones. Those that are known are included in the databases used by the software that interprets STR electrophoretograms and reads out the allele identities for the operator, but there is at least one instance of a novel microvariant not known in modern populations being found in an archaeological specimen. Unthinking reliance on automated identification would have mistyped this allele. It is also possible for rare variations in the regions flanking one or more alleles of an STR to prevent primer annealing so that no PCR product is obtained, resulting in a **null allele**. If no product at all is synthesized, then the null allele is easily recognized, but, as with allelic dropout, a null at one of the pair of alleles is difficult to detect. Null alleles have been rigorously checked for in the CODIS set, so this problem is only likely to arise with less well-characterized PCRs.

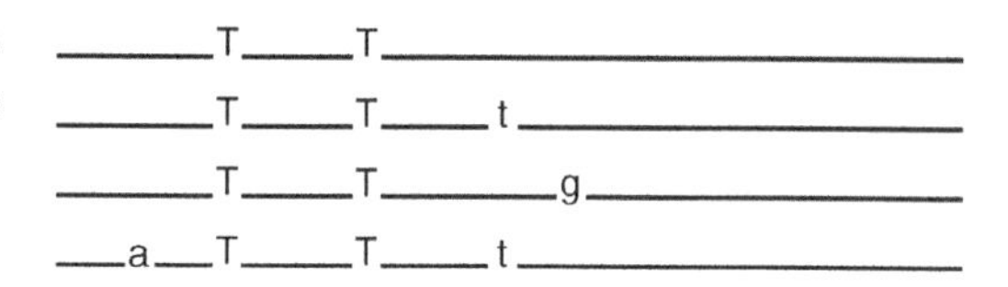

Figure 11.5 The relationship between haplotypes and a haplogroup. This haplogroup differs from the Cambridge Reference Sequence (CRS)—the first human mitochondrial sequence to be obtained—by two T's shown as capitals in the top drawing. The individual haplotypes within this haplogroup all have these two T's, but some also have their own variations, shown as small letters in the lower three drawings.

11.2.3 Mitochondrial DNA gives additional information on kinship

STRs provide very precise information on kinship relationships but are often difficult to type in ancient DNA. This is not just because of allelic dropout, stuttering, and the various other complications described above. The main problem is that STRs are located in the nuclear genome, so there are only two copies of each STR per cell. This low copy number means that if only a small amount of ancient DNA is present in a specimen, or the ancient DNA is highly degraded, then there simply might not be any intact copies of the STR available for amplification. In all ancient DNA projects, it is preferable to use multicopy markers, as there is a greater chance that one or more amplifiable copies of these markers will be present in the DNA extract. For these reasons, mitochondrial DNA is often used when information on kinship is being sought. There are about 8000 identical copies of the mitochondrial genome per human cell, so markers located on mitochondrial DNA are four thousand times more abundant than nuclear markers. Frequently, it is possible to obtain mitochondrial PCR products from ancient extracts that lack amplifiable nuclear DNA.

Mitochondrial DNA can be used to study kinship because it exists as a series of sequence variants, called **haplotypes**, almost 7000 of which are known in the modern human population. These haplotypes fall into approximately 90 **haplogroups**. Each member of a haplogroup possesses a small set of shared-sequence polymorphisms, which define that haplogroup, along with additional polymorphisms unique to the haplotype (Figure 11.5). Most of these polymorphisms are located in the two hypervariable regions of the mitochondrial DNA (Figure 2.10), though some positions elsewhere in the genome must also be typed in order to distinguish a few of the variants. The two hypervariable regions span 1150 bp of DNA and so cannot be amplified as a single PCR when ancient DNA is being studied. Instead 10 or more overlapping PCRs must be carried out to build up the complete sequence. This is a daunting task and in most cases only a few of these PCRs are attempted, or are successful, and the aim is not to discriminate every haplotype but simply to identify the haplogroup to which an individual belongs.

Mitochondrial DNA is inherited solely through the maternal line, so siblings and their mother share the same haplogroup, but the father's might be different (Figure 11.6). Identification of the haplogroups possessed by a group of skeletons that are buried together can therefore give an indication of possible kinship relationships. It must be kept in mind that the demonstration that two individuals have the same haplogroup is by no means a proof that they are closely related, as there are many more people on the planet than there are haplogroups. Two people could therefore belong to the same haplogroup by chance, or more precisely because of a deep ancestral relationship that might go back

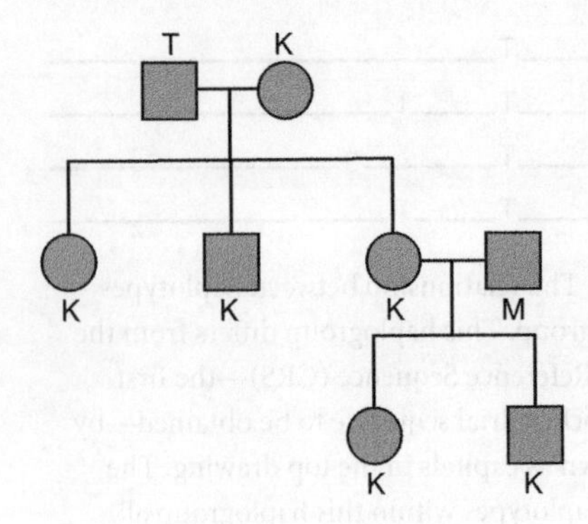

Figure 11.6 Maternal inheritance of mitochondrial DNA haplogroups. Circles are females, and squares males. The original parents are a female with haplogroup K and male with haplogroup T. All three of their children inherit their mother's haplogroup and so are K. One marries a man with haplogroup M. Their children are haplogroup K.

hundreds if not thousands of years. Inferences of kinship from mitochondrial DNA must therefore take into account the frequency of each haplogroup in the population as a whole. There are extensive data on haplogroup frequencies in modern populations, but these are not ideal for judging the likelihood that two archaeological skeletons with the same haplogroup are in fact closely related. This is because of the added complication that in an inbred population, certain haplogroups will become very frequent and the same one might be possessed by the majority of the members of that population. The control data that are needed is therefore not the frequency of haplogroups in modern populations, but the frequency in the particular population to which the skeletons belonged. This means that a substantial number of skeletons have to be typed before kinship relationships can be identified with any degree of confidence, and such large-scale typing is rarely possible with ancient material. Note, however, that if two individuals have different haplogroups, then they cannot be full siblings or mother and child. In these situations, PCR of a single short segment of the hypervariable regions containing a polymorphism possessed by one but not the other skeleton enables a secure, albeit negative, conclusion on kinship to be drawn.

11.3 Examples of the Application of Kinship Analysis in Biomolecular Archaeology

So far there have been only a handful of projects in which kinship within burial groups has been understood as a result of ancient DNA typing, the relative lack of success in this area emphasizing the difficulty in amplifying STRs consistently from archaeological specimens. Not surprisingly, this is easier with relatively recent specimens, and several of the successful projects have concerned people who lived and died during historic rather than prehistoric times.

11.3.1 Genetic profiling of ancient DNA was first used to identify the Romanov skeletons

Although more in the realms of forensic science rather than biomolecular archaeology, the work carried out in the early 1990s to identify the remains of the Romanovs provides the clearest illustration of the way in which STR analysis can be used to assess kinship relationships between human skeletons. The Romanovs and their descendents ruled Russia from the early 17th century until the time of the Russian Revolution when Tsar Nicholas II was deposed and he and his wife, the Tsarina Alexandra, and their five children imprisoned. On July 17, 1918, all seven, along with their doctor and three servants, were murdered and their bodies disposed of in a shallow roadside grave near Yekaterinburg in the Urals. In 1991, after the fall of communism, the remains were recovered with the intention that they should be given a more fitting burial.

Table 11.3 STR genotypes obtained from the skeletons thought to include the Romanovs.

Skeleton	*VWA*	*TH01*	*F13A1*	*FES/FPS*	*ACTBP2*
Male adult 1	14, 20	9, 10	6, 16	10, 11	not done
Male adult 2	17, 17	6, 10	5, 7	10, 11	11, 30
Male adult 3	15, 16	7, 10	7, 7	12, 12	11, 32
Male adult 4	15, 17	6, 9	5, 7	8, 10	not done
Female adult 1	15, 16	8, 8	3, 5	12, 13	32, 36
Female adult 2	16, 17	6, 6	6, 7	11, 12	not done
Child 1	15, 16	8, 10	5, 7	12, 13	11, 32
Child 2	15, 16	7, 8	3, 7	12, 13	11, 36
Child 3	15, 16	8, 10	3, 7	12, 13	32, 36

Although it was suspected that the bones recovered were indeed those of the Romanovs, the possibility that they belonged to some other unfortunate group of people could not be discounted. Nine skeletons had been found in the grave, six adults and three subadults, the latter being the term used for a person, typically in his or her teens, who has passed through the juvenile period but does not yet display the full set of adult skeletal features. Osteological examination suggested that four of the adults were male and two female, and the three subadults were all female. If these were the remains of the Romanovs, then their son Alexei and one their daughters were, for some, reason, absent. The bodies showed signs of violence, consistent with reports of their treatment during and after death, and at least some of the remains were clearly aristocratic, as their teeth were filled with porcelain, silver and gold, dentistry well beyond the means of the average Russian of the early 20th century. Facial reconstructions also showed likenesses to photographic images of Tsar Nicholas and his family. But DNA tests were thought to be the only way of proving the identity of the bodies and distinguishing the adult Romanovs from their doctor and servants.

The first set of tests used amelogenin PCRs to establish if ancient DNA was preserved in the bones and, if possible, to obtain independent evidence of the sex of each skeleton. PCR products were obtained for each skeleton, the sex typings agreeing with the osteological identifications, indicating that a more detailed ancient DNA analysis might be feasible. STR analysis was therefore carried out to test the hypothesis that the three subadults were siblings and that two of the adults were their parents, as would be the case if indeed these were the Romanovs. Results were obtained for five tetranucleotide STRs, these immediately showing that the three subadults could be siblings, as they had identical VWA and FES/FPS genotypes and shared alleles at each of the three other loci (Table 11.3). The TH01 data show that female adult 2 cannot be the mother of the children because she only possesses allele 6, which none of the children has. Female adult 1, however, has allele 8, which all three children have. Examination of the other STRs confirms that she could be the mother of each of the children. The TH01 data exclude male adult 4 as a possible father of the children, and the VWA results exclude male adults 1 and 2. When all the STRs are taken into

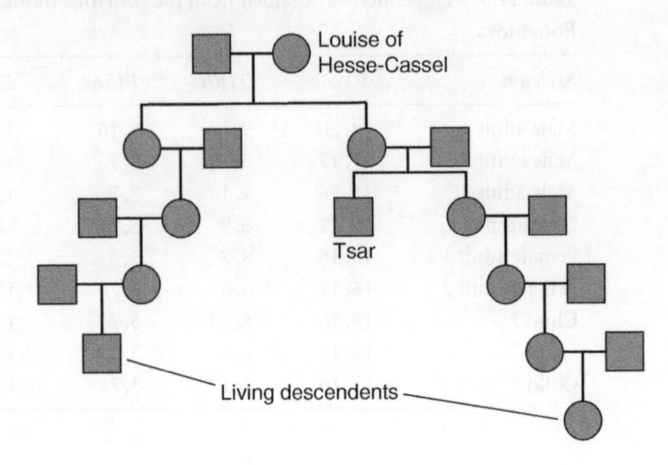

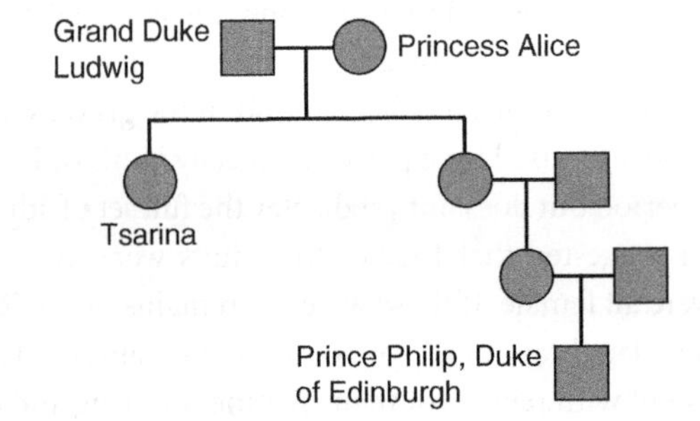

Figure 11.7 Relationships of living individuals with the Tsar and Tsarina.

account, male adult 3 could be the father of the children. Note that all these conclusions can be drawn simply from the TH01 and VWA results: the other STR data simply provide corroborating evidence.

11.3.2 Mitochondrial DNA was used to link the Romanov skeletons with living relatives

In a genuine archaeological project, where there would be no attempt to link the remains with named individuals, the analysis would not have been taken beyond the STR results described above, but in this case further proof of the identity of the skeletons was required. This was achieved by amplifying parts of the mitochondrial hypervariable regions from each of the skeletons and comparing the sequences with those of two living matrilineal descendants of Tsar Nicholas's grandmother, Louise of Hesse-Cassel, and with Prince Philip, the Duke of Edinburgh, whose maternal grandmother was Princess Victoria of Hesse, the Tsarina Alexandra's sister (Figure 11.7). The mitochondrial DNA sequences from four of the female skeletons – the three subadults and the adult female identified as their mother – were exactly the same as that of Prince Philip, strong evidence that the four females were members of the same lineage.

With the adult male thought to be the Tsar the analysis was more complicated, as two sequences were present among the clones obtained from his PCR product. These sequences differed at a single position which was either a C or a T, the former four times more frequent than the latter. This could indicate contamination, but instead was interpreted as showing that the Tsar's mitochondrial DNA was **heteroplasmic**, an infrequent situation where two different haplotypes coexist within the same cells. The two living matrilineal descendents of the Tsar's grandmother both had the haplotype with a T at this position, suggesting that the mutation producing the C variant had occurred very recently in the Tsar's lineage. Support for this hypothesis was subsequently provided by analysis of ancient DNA from the Tsar's brother, Grand Duke George Alexandrovich, who died in 1899, which showed that he also displayed heteroplasmy at the same position in his mitochondrial DNA. On balance, the evidence suggested that the Tsar's remains had been correctly identified.

The DNA results were considered sufficiently convincing for the skeletons identified as the Tsar, Tsarina, and their three daughters to be given a state funeral in St Petersburg in 1998. Recently, however, doubts have been raised about the DNA studies. These are partly on technical grounds, some of the amplicons obtained from the bone extracts being relatively long (over 1.2 kb) which, according to the criteria now used to test the authenticity of ancient DNA results (Section 9.1), would indicate that contaminating rather than ancient DNA might have been amplified. Mitochondrial DNA sequences have also been obtained from a finger supposedly taken from the body of Grand Duchess Elisabeth, another sister of the Tsarina Alexandra, who was murdered in 1918, on the same day as the royal family. This sequence does not match that of the skeleton identified as the Tsarina, but the identity of the finger cannot be assured and it has been suggested that it might have come from an unrelated person.

The final mystery concerns the whereabouts of the remains of the two other children. During the middle decades of the 20th century several women claimed to be a Romanov princess, because even before the bones were recovered there had been rumors that one of the girls, Anastasia, had escaped the clutches of the Bolsheviks and fled to the West. One of the most famous of these claimants was Anna Anderson, whose case was first widely publicized in the 1920s. Anna Anderson died in 1984, but she left an archived tissue sample whose mitochondrial DNA does not match the Tsarina's and suggests a woman of Polish origin. There have also been various people claiming to be descended from Tsarevich Alexei. But these stories are almost certainly romances, as the partially cremated bodies of two other subadults, found near Yekaterinburg in 2007, have now been examined, and their mitochondrial haplotypes suggest that they are the missing Romanov children.

11.3.3 A detailed STR analysis has been carried out with the remains of the Earls of Königsfeld

The Romanovs are not the only historic group who have been studied by STR analysis. Eight skeletons were discovered during excavations in 1993 at St Margaretha's Church in Reichersdorf, Lower Bavaria. Memorial stones suggested that these were the remains of seven generations of the Earls of Königsfeld, who had been buried in the church during the 16th–18th centuries. The memorial stones were placed by the wall of the church rather than on the graves themselves, but the family tree was known from historic records and each

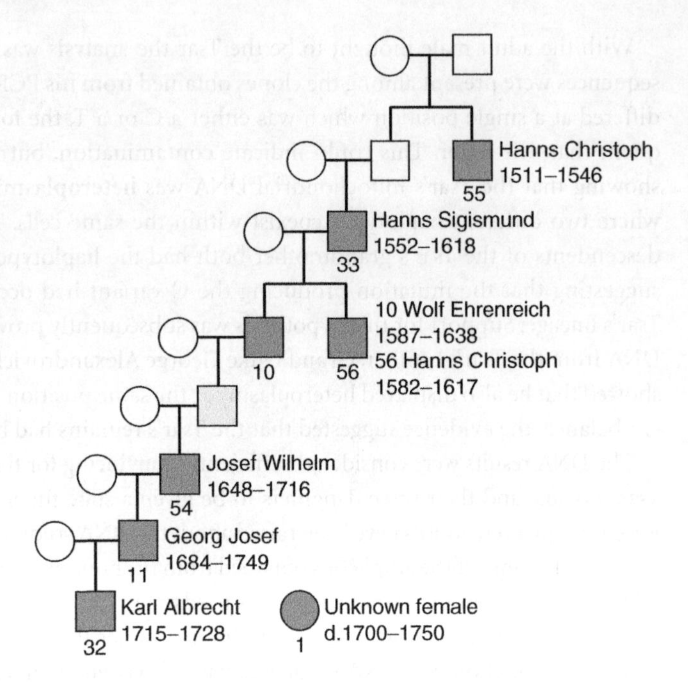

Figure 11.8 Family tree of the Earls of Königsfeld. Solid squares and circles indicate individuals supposed to be buried in St Margaretha's Church, with names, dates, and the grave numbers. The unidentified female burial is also indicated. The light gray square is Franz Nikolaus (1619–88), whose grave was destroyed and could not be excavated.

skeleton could be tentatively identified based on its age at death and the nature of its grave goods, which gave an approximate indication of the date of burial (Figure 11.8). These archaeological studies provided the first surprise as they revealed that the burial in grave Ma1 was female, as the skeleton had typically female morphological features and the grave contained the remnants of female clothes. None of the female members of the Königsfeld family was recorded as having been buried in the church, raising questions about this woman's identity. The problem became more intriguing when a second skeleton, Ma32, which had been typed as male from its osteology, was repeatedly shown to be another female when a series of amelogenin tests was carried out.

Initially the DNA analysis focussed on the six male skeletons in order to see if these followed a single paternal line of descent, as expected from the genealogy of the of Earls of Königsfeld. As the male line was being examined, four STRs on the Y chromosome were typed. One skeleton, Ma56, yielded no PCR products and had to be excluded from any further analysis. Positive results were obtained for the other five. Four had identical alleles at each of the STRs, entirely as expected for a series of fathers and sons, but the fifth skeleton, Ma11, had different alleles at three of the STRs, making it impossible for him to be a member of this lineage. Autosomal STRs were also typed for these five male skeletons and the two females. The results were consistent with the linear descent of four of the males as indicated by the Y chromosome results, but again showed that Ma11 was not a member of this male line. Ma11 could, however, have been the father of Ma32, one of the females, with the other female, Ma1, as the mother.

How do we untangle the complicated relationships revealed by the ancient DNA typing? The first question is the identity of Ma32, the skeleton typed as female by the ancient DNA results. The archaeological evidence had suggested that Ma32 was the skeleton of Karl

Albrecht, the last male member of the lineage buried at St Margaretha's, who had died in 1728 at the age of only 13. Instead, it seems likely that the skeleton that was recovered belonged to one of his four sisters, all of whom died young and who, according to historical records, might have been buried together with Karl Albrecht in the same grave. Their mother, identified by the DNA evidence as skeleton Ma1, was Maria Anna, who died in 1722. Maria Anna's husband was Georg Josef, whose burial was, according to the archaeological evidence, Ma11, the anomalous male who does not belong in the Königsfeld lineage. Either he was the result of an adulterous affair or skeleton Ma11 (and possibly his wife and daughter also) is not a member of the Königsfelds after all. An answer could be found by testing living paternal descendants of the Earls of Königsfeld, but unfortunately their lineage became extinct in 1815. Comparisons between Y chromosome haplotypes and the surnames of living people suggest that many male lineages have been broken in a manner similar to that seen with the Earls of Königsfeld. In societies where surnames are inherited through the male line, there should be a correspondence between surname and Y chromosome haplotype, but this correspondence is often only approximate, reflecting the frequency of adultery in past and recent societies, and making even more nebulous the information provided by so-called "ancestry" companies.

11.3.4 Kinship analysis at a Canadian pioneer cemetery

Now we move on to a more typical archaeological study, one where the names of the individuals are not known, and where the amount of information that could be obtained was severely restricted by the problems inherent in working with ancient DNA. This project studied 27 adults, 6 subadults and 5 infants (less than 1 year old) from part of a pioneer cemetery in Durham, Upper Ontario, excavated in 1992 in order to rebury these individuals away from an area close to a roadside. From the coffin fittings found in the graves the burials were dated from 1827 to after 1900. The historical records suggested that the 117 people buried in the cemetery were members of a small number of families, with only 37 surnames present in total and 77 of the burials belonging to just 11 of these surnames. The organization of the cemetery, with some graves surrounded by a low wall, and others clustered into groups, suggested that the burials were not randomly distributed but might be arranged to represent "family plots," each containing a related set of individuals (Figure 11.9). To test this hypothesis, STR, mitochondrial DNA, and amelogenin PCRs were attempted.

The results illustrate many of the problems that limit our ability to carry out meaningful kinship studies with archaeological material (Table 11.4). Four of the 27 skeletons could not be typed either because no PCR products were obtained or those products contained mixtures of amplicons preventing the correct ancient DNA sequences from being identified. Each of the remaining 23 skeletons yielded PCR products when part of the multicopy mitochondrial genome was amplified, but the autosomal STR TH01 could be typed in only 18 of these, with some uncertainty in 4 cases, and only 13 skeletons gave results with an amelogenin PCR. Attempts to amplify other STRs gave inconsistent and generally unreliable results.

The amelogenin PCR used in this study was not the standard one described in Section 10.3, but a modified version on which the X chromosome sequence, at 155 bp, is longer

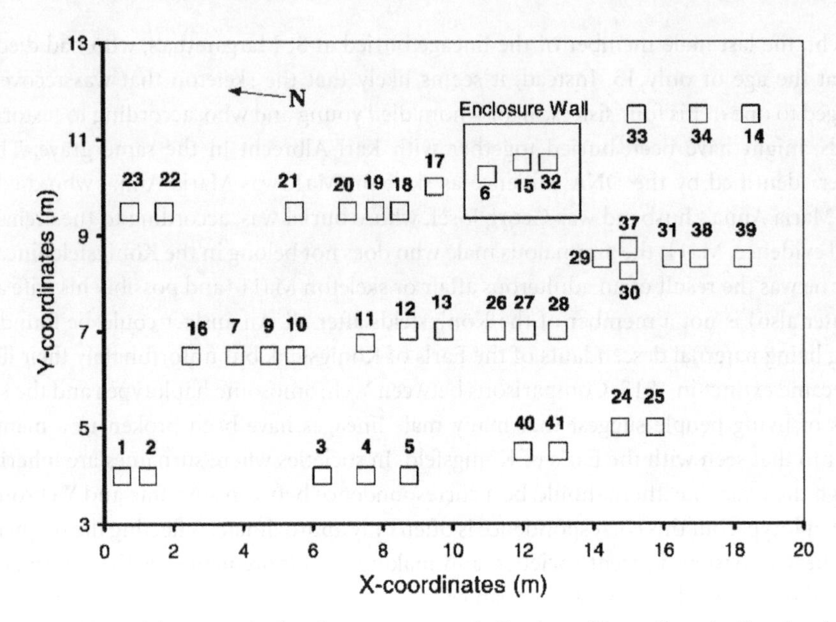

Figure 11.9 Map of the graves in the pioneer cemetery in Durham, Upper Ontario. Reprinted from J.C. Dudar, J.S. Waye, and S.R. Saunders (2003), "Determination of a kinship system using ancient DNA, mortuary practice, and historic records in an Upper Canadian pioneer cemetery," *International Journal of Osteoarchaeology*, 13, 232–46, with permission from John Wiley & Sons, and the authors.

than the Y sequence, which is 120 bp. The idea is to favor amplification of the Y product so that if allelic dropout occurs, then it is the X product that is lost, and hence males can be identified with greater certainty than with the standard test, in which it is the Y product that is more likely to be lost. Using this clever modification, a sex could be assigned to 13 skeletons, in each case agreeing with the sex identified by osteological examination, a concordance that gives confidence that genuine ancient DNA was being typed.

The STR results were too limited to allow any conclusions. Although TH01 genotypes could be determined for 18 skeletons, one STR is insufficient for carrying out a kinship analysis. Reliance therefore had to be placed on the mitochondrial DNA results and the kinship analysis restricted to maternal relationships. The mitochondrial PCR amplified a 261 bp region of hypervariable region II, the 23 skeletons giving a total of 18 different haplotypes, which were designated A to S. When these haplotypes were mapped onto a plan of the burials, distinct clusters could be seen, supporting the hypothesis that the cemetery was organized as a set of family plots (Figure 11.10). Further interesting information emerged when the mitochondrial DNA haplotypes were compared with the dates of the burials. Each burial could be assigned to one of four phases (1827–50, 1850–78, 1878–1900 and post-1900) based on the precise details of the coffin fittings. Note that each of these phases is approximately equivalent to a single human generation. Each haplotype seemed to persist for two burial phases, corresponding to two generations, before being replaced by a new sequence.

One interpretation of this pattern is that it indicates a patrilocal residence system, with a woman moving into her husband's family when she marries and then, when her life is done,

Table 11.4 Results obtained with skeletons from the Canadian pioneer cemetery.

Burial	*Mitochondrial DNA type*	*TH01 genotype*	*Amelogenin PCR products*
6	J	7, 9.3	Y
7	A	9.3, 9.3	–
9	I	9, 9.3	–
12	D	9.3, 9.3	XY
13	D	9, 9.3	X
14	D	–	–
15	J	7, 9.3	X
16	F	9, 10	X
17	G	6, 7	Y
18	K	7, 10?	–
20	J	7, 9.3	–
22	F	7, 9.3	XY
23	F	7, 9.3?	Y
24	D	–	–
25	D	5, 9	Y
26	H	–	–
28	A	–	–
29	K	–	
31	C	6, 6	Y
34	D	8, 9.3?	Y
38	C	6, 9.3	XY
39	I	6, 9.3	–
41	J	6, 8	Y

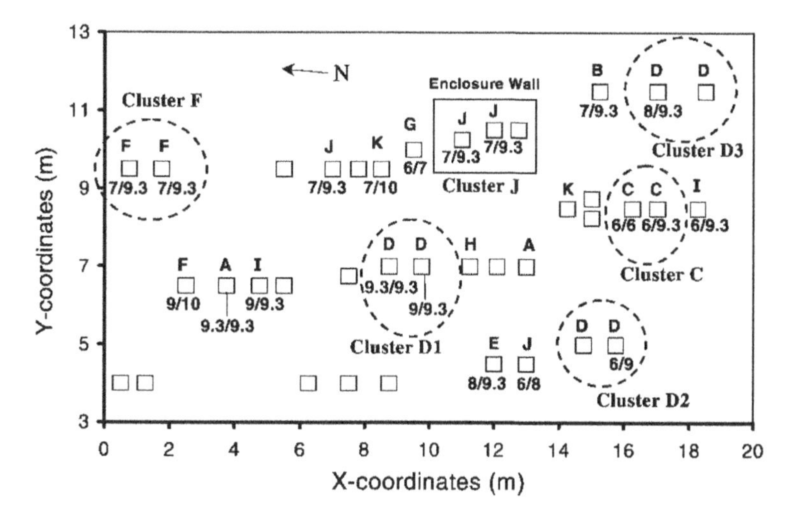

Figure 11.10 DNA results at the pioneer cemetery. Letters are mitochondrial DNA haplotypes, and numbers are the genotypes for TH01. Clusters with similar DNA results are circled. Reprinted from J.C. Dudar, J.S. Waye, and S.R. Saunders (2003), "Determination of a kinship system using ancient DNA, mortuary practice, and historic records in an Upper Canadian pioneer cemetery," *International Journal of Osteoarchaeology*, 13, 232–46, with permission from John Wiley & Sons, and the authors.

becoming the earliest possessor of a particular mitochondrial haplotype to appear in the family plot in the cemetery. Her sons inherit her mitochondrial DNA, stay in the family group and are themselves buried alongside her when they die. But her daughters leave to marry into other patrilocal groups, and are buried with their new families. Because the sons do not pass their mitochondrial DNA on to their offspring, a particular haplotype occupies just two interment phases within a single family plot, before being replaced by the haplotypes belonging to the next generation of brides to marry into the family. So although in this project the limitations of ancient DNA analysis prevented multiple loci from being typed, those data that could be obtained provide an interesting insight into social organization, one entirely consistent with our historical knowledge of life in pioneer farmstead communities,.

11.3.5 With older archaeological specimens, kinship studies increasingly depend on mitochondrial DNA

With older archaeological material, greater dependence must be placed on mitochondrial DNA as a source of information from which to infer kinship, nuclear DNA being increasingly less likely to survive the further back in time we go. The difficulty in carrying out kinship analysis with older material is illustrated by the attempts that have been made to determine whether the wealthy individuals buried together in Grave Circle B at Mycenae were members of a single family or small number of families. Twenty-two of these skeletons have been tested with a more comprehensive suite of PCRs than have been applied to any other archaeological group, these PCRs directed at 7 different segments of the mitochondrial hypervariable region, 17 autosomal STRs, 5Y chromosome STRs, 2Y chromosome SNPs, and the amelogenin locus. The success rate was very low. Nuclear amplification products were occasionally obtained, but too sporadically for any of these results to be authenticated. With mitochondrial PCRs, 18 of the 22 skeletons never gave an amplification product of the correct size, or if they did, then that product was considered to be a contaminant because it was accompanied by contaminated negative controls, was entirely made up of sequences known to be possessed by somebody who had handled the bones, or was not human mitochondrial DNA. This left just four skeletons for which authentic mitochondrial DNA sequences of varying lengths could be obtained. When the sequences were analyzed, it was found that the skeletons designated Γ55 and Γ58 belonged to mitochondrial haplogroup UK, Z59 belonged to haplogroup U5a1a, and A62, whose DNA was the least well preserved of these four, was one of H, HV1, J, U, U3, U4, and U7 (but not UK or U5a1a).

Although the success rate was low, the DNA results yielded interesting new information about Mycenae. It was probably not a coincidence that the four skeletons that yielded ancient DNA sequences were four of the ones for which facial reconstructions had been performed, as reconstructions could only be done with the best preserved material. The facial reconstructions had suggested that Γ55 and Γ58, an adult male and female, respectively, might be related, as both had "heart-shaped" faces, and both were buried together in the same grave (Figure 11.1). This hypothesis was supported by their shared mitochondrial haplogroup. Z59 is an adult male from a different grave, a member of the "long face" group and hence possibly not a member of the same family as Γ55 and Γ58. Again, this hypothesis was supported by the mitochondrial DNA data, as Z59 had a different haplogroup to Γ55 and Γ58, and hence belongs to a different maternal lineage. A62 was another adult male, the third member of the group with "heart-shaped" faces, and so possibly a member of the same family as Γ55 and Γ58. Although A62 gave the least complete DNA results, the partial sequence that was obtained ruled out the possibility that he had the UK haplogroup, and hence it could be concluded that he did not share maternal descent with Γ55 and Γ58. A non-maternal relationship (e.g. the same father but different mothers) could not be discounted in the absence of data from the Y chromosome.

The DNA data therefore gave some support to the notion that the burials in Grave Circle B are arranged into family groups, as Γ55 and Γ58 have the same haplogroup. Γ55 and Γ58 are also noteworthy in that Γ55, the male, is the only person in Grave Circle B to be buried with a facemask, and Γ58 is one of only four skeletons – of 35 in all – that is definitely female, an underrepresentation that suggests that those females who are buried in Grave Circle B had some special reason for being there. The pair were of about the same age when they died, and were probably contemporaneous, as the remains of Γ58 were still articulated at the time they were moved aside to allow the interment of Γ55, suggesting that they were buried within a few months of each other. Their maternal relationship could therefore indicate that they were brother and sister. Initially it was thought that Γ55 and Γ58 were husband and wife, Γ58 being present in the Grave Circle only because she had married an important man. The demonstration that Γ55 and Γ58 are more likely to be brother and sister changes our perceptions regarding the reasons why Γ58 was buried in this high status and male-dominated Grave Circle. It would appear that she is there because she held a position of status due to her own birthright or personal authority. This particular kinship analysis has therefore given us a novel insight into the role of women in early Mycenaean society.

11.3.6 Strontium isotope analysis can contribute to kinship studies by revealing examples of exogamy

The four case studies we have considered so far have taken us from relatively straightforward projects in which the sole objective has been to piece together a family lineage, to ones where an understanding of kinship has been supplemented by an insight into social organization, the demonstration of a patrilocal residence system in the pioneer community, and the inference that some Mycenaean women held high status as their birthright rather than as a consequence of their marriage. To conclude this chapter we will examine a final case study in which aspects of social organization are revealed by combining DNA typing with analysis of strontium isotope ratios in the teeth of the skeletons being studied.

We return to the period at the end of the European Neolithic, described earlier in this chapter, when early farming groups were adopting a more structured social organization. Near Eulau in central Germany, excavation of a 4600-year-old cemetery revealed four groups of burials, comprising 13 individuals in total, a mixture of adult males and females as well as children. The arrangement of the skeletons in these burials suggested that each was a family group comprising one or both parents and their children (Figure 11.11). Close cooperation between the archaeologists working on the cemetery and scientific experts enabled ancient DNA extractions to be carry out very soon after the skeletons were

Figure 11.11 Family grave 99 at Eulau, Germany. There are four skeletons. The upper two are adults, with a male aged 40–60 years on the left, and a female aged 35–50 years on the right. Below are two male children aged 4–5 and 8–9 years. The DNA analysis showed that the adult male had mitochondrial haplogroup U and Y haplogroup R1a. The adult female was mitochondrial haplogroup K. The two children both were haplogroups K and R1a, indicating that they could be offspring of the two adults. Image kindly provided by Landesamt für Denkmalpflege und Archaeologie Sachsen-Anhalt, Robert Ganslmeier (State Office for Heritage Management and Archaeology Saxony-Anhalt). Thanks also to Bettina Stoll-Tucker and Wolfgang Haak.

discovered, minimizing the opportunity for post-excavation DNA decay (Section 7.1). Sequences of the mitochondrial hypervariable I could be obtained from 9 of the 12 individuals that were examined, the haplotypes that were discovered supported the hypothesis that each burial was a family group. Importantly, different haplotypes were identified in different family groups, showing that the population as a whole was not hugely inbred, and giving confidence that the relationships inferred from the data were real and not due to unrelated individuals sharing the same haplotype by chance.

As well as the DNA analysis, strontium isotope ratios were also measured in the teeth of these individuals. In Section 6.2, we learnt that the ratio of the strontium isotopes ^{87}Sr and ^{86}Sr in geological formations varies from place to place, and that the $^{87}Sr/^{86}Sr$ ratio in a person's teeth reflects the geological ratio of the locality where they spent their childhood, this being the period when the adult teeth are developing and the strontium isotope ratios in those teeth are established. When the $^{87}Sr/^{86}Sr$ ratios in the teeth of the Eulau skeletons were examined, it was found that all of the adult males and all of the children had strontium signatures similar to those in the sediments surrounding the graves. In other words, these individuals probably grew up in the area in which their skeletons were buried. The three adult females in the graves, on the other hand, had quite different $^{87}Sr/^{86}Sr$ ratios. These values were more similar to those in deposits from the Harz mountains, some 60 km away from Eulau. The implication is that in this society marriages were exogamous and patrilocal, and in the case of these particular burials, the women moved from the Harz mountains to join their husbands' community.

Patrilocal exogamy has also been suggested by studies of sites from about 500 years earlier at Flomborn, Schwetzingen, and Dillingen in the lowlands of southwestern Germany. In this work, the analysis was less comprehensive, as no DNA typing was attempted and only strontium isotope ratios were studied. As at Eulau some of the skeletons gave $^{87}Sr/^{86}Sr$ values significantly different from the value measured from the local geology, these values being more characteristic of upland areas some 50–100 km away. The majority, but not all, of these incomers were female, suggesting that they might have been exogamous marriage partners for the local men. Many of those local males were buried with a shoe-last adze, a type of stone implement thought to be used as a hoe. The implication is that these local people grew crops, and the absence of shoe-last adzes from the burials of the incoming people could, similarly, suggest that they were not cultivators, but perhaps foragers or animal herders. There is a danger of allowing speculation to run away from the solid data, but we begin to see how a powerful combination can be made by linking the molecular data with other forms of archaeological evidence, such as grave goods, to infer aspects of prehistoric social organization.

Further Reading

Allen, W.L. & Richardson, J.B. (1971) The reconstruction of kinship from archaeological data: the concepts, the methods, and the feasibility. *American Antiquity*, 36, 41–53. [Describes the difficulties in using archaeological information to assess kinship.]

Bentley, R.A., Price, T.D. & Stephan, E. (2004) Determining the "local" $^{87}Sr/^{86}Sr$ range for archaeological skeletons: a case study from Neolithic Europe. *Journal of Archaeological Science*, 31, 365–75. [Strontium analyses in the lowlands of southwestern Germany.]

Bouwman, A.S., Brown, K.A., Prag, A.J.N. & Brown, T.A. (2008) Kinship between burials from Grave Circle B at Mycenae revealed by ancient DNA typing. *Journal of Archaeological Science*, 35, 2580–4.

Butler, J. (2006) Genetics and genomics of core STR loci used in human identity testing. *Journal of Forensic Science*, 51, 253–65. [The CODIS set.]

Coble, M.D., Loreille, O.M. & Wadhams, M.J., *et al.* (2009) Mystery solved: the Identification of the two missing Romanov children using DNA analysis. *PLoS ONE* 4(3), e4838.

Dudar, J.C., Waye, J.S. & Saunders, S.R. (2003) Determination of a kinship system using ancient DNA, mortuary practice, and historic records in an Upper Canadian pioneer cemetery. *International Journal of Osteoarchaeology*, 13, 232–46.

Gerstenberger, J., Hummel, S., Schultes, T., Häck, B. & Herrmann, B. (1999) Reconstruction of a historical genealogy by means of STR analysis and Y-haplotyping of ancient DNA. *European Journal of Human Genetics*, 7, 469–77. [The Earls of Königsfeld.]

Gill, P., Ivanov, P.L. & Kimpton, C., *et al.* (1994) Identification of the remains of the Romanov family by DNA analysis. *Nature Genetics*, 6, 130–5.

Haak, W., Brandt, G. & de Jong, H.N., *et al.* (2008) Ancient DNA, strontium isotopes, and osteological analyses shed light on social and kinship organization of the Later Stone Age. *Proceedings of the National Academy of Sciences USA*, 105, 18226–31. [The project at Eulau.]

Scarre, C.J. (1984) Kin-groups in megalithic burials. *Nature*, 311, 512–13. [Use of non-metric traits to assess kinship.]

Tyrrell, A. (2000) Skeletal non-metric traits and the assessment of inter- and intra-population diversity: past problems and future potential. In *Human Osteology: In Archaeology and Forensic Science* (ed. Cox, M. & Mays, S.), 289–306. Cambridge University Press, Cambridge.

12

Studying the Diets of Past People

In the previous two chapters, we examined how biomolecular archaeology, principally using DNA, is used to obtain information on the identities of past peoples and to throw some light on their social organizations. Now we will turn our attention to their diets. These **paleodietary studies** encompass two linked but distinct questions, which we will deal with separately, one in this chapter and one in the next.

The first question concerns the types of food that were eaten by the people living at a particular place at a particular time. To address this issue, biomolecular archaeologists make use of a range of approaches, including chemical analysis of the residues found in cooking vessels, the study of stable carbon and nitrogen isotope ratios in skeletons, and DNA analysis of genetic markers associated with particular types of diet such as ones including consumption of milk. These areas of biomolecular archaeology are described in this chapter.

The second question concerns not the diets themselves but the way in which past people obtained their food, their mode of **subsistence**. Central to this issue is the transition from a hunting-gathering lifestyle to one based on agriculture, and much of the research in this area is aimed at understanding when and where particular animals and plants were domesticated, and then how agriculture spread and developed in different parts of the world. We will examine all of those topics in Chapter 13.

12.1 The Archaeological Approach to Diet

Anthropologists consider the procurement, preparation and consumption of food to be important activities that not only bond societies and family groups together, but also contribute to one's sense of identity, the nature of the food consumed, and the effort put into acquiring it, often reflecting an individual's gender, status, and rank in society. The important

 © 2011 Terry Brown & Keri Brown

questions concern not only what people eat, but why they eat it. To understand the latter point we need only think of the social role of communal eating in our own lives, whether a family meal, the first date at a restaurant, or the formal dinner at a science conference. Archaeologists are equally interested in these issues but often their focus is, of necessity, on more specific aspects of the diets of past peoples, those aspects being the ones that can actually be inferred from the archaeological record. These studies make use not only of biomolecular methods, but also of non-molecular approaches to identifying the types of food that were consumed at particular sites. Before looking at the biomolecular techniques, we will briefly consider how these broader archaeological approaches are used to study past diets.

12.1.1 Diet can be reconstructed from animal and plant remains

The most direct way of identifying the types of food consumed at an archaeological site is to look for the remains of that food, in the form of animal bones and plant remnants such as preserved seeds. This approach is equally valid for hunter-gatherer and agricultural sites but the interpretations placed on the data have to take account of the mode of subsistence.

The minimum number of individual animals present at a site at a particular time can be estimated from the bone assemblages found at different stratigraphies, and these figures can be used to calculate the amount of meat that was available and the energy that could be obtained by its consumption. But at agricultural sites care has to be taken to distinguish animals that were used for meat production from those used for other purposes. An understanding of herd structures (Section 10.3) is therefore an important corollary to the use of bone assemblages in dietary reconstruction, the presence of young adult animals implying meat production and a preponderance of adult females implying that animals were kept as a source of milk rather than meat. The possibility that some animals were used for traction rather than meat production must also be considered, as indicated by the presence of artifacts such as yokes and bridles, and by osteological examination, which can reveal bone changes that result from the stresses and strains caused by haulage of heavy loads. Adding to the complications is the likelihood that animals that were primarily used for milk production or traction might also have been eaten. A marine or freshwater component to the diet, whether of fish or invertebrates, can also be identified and to a certain extent quantified from the skeletal remains found at a site.

Plant remains can also be used in paleodietary reconstruction, but here the inferences that can be made are much less precise. Often the remains are in the form of burnt seeds from plants that were gathered by nomadic groups or grown by agriculturalists. Seeds can be recovered by careful sieving of archaeological soils, but they can only be used for qualitative purposes, to identify the types of plants that were being eaten, and not for quantitative analyses of the contribution of individual plants to the diet. This is because seeds such as barley and wheat grains are the actual foodstuff, not the leftovers (as is the case with animal bones), so the number of seeds of a particular species that are recovered is not a direct measure of the amount of that plant that was consumed.

Macroscopic and microscopic examination of coprolites, where these are available, can also give indications of diet. Traces of bone or hair are usually taken as evidence of the presence of meat in the diet, and identification of seeds, pollen, and phytoliths will show which types of plant were consumed. Stomach contents can be equally informative on those rare occasions when they are preserved in mummified bodies. Microscopic examination of the stomach contents of the Tyrolean iceman (Section 7.2) has revealed that his last meal was bread made from einkorn wheat. Similarly, Lindow Man, a bog body from Cheshire, England, ate a cereal bread shortly before his death.

In carrying out dietary reconstructions, it is important not to place a personal interpretation on the data that are acquired. The types of food that are eaten are strongly influenced by cultural preferences, and our preferences today might not be the same as those of people living in the past. Many insects, cockroaches in particular, are nutritious but not widely consumed in modern societies. Dietary reconstructions must therefore take account of the full range of possible food remains found at a site, not just the ones that we might consider to be palatable.

12.1.2 Examination of tooth microwear can give an indirect indication of diet

Dietary reconstruction from preserved food remains is only possible when those remains are present at the site that is being studied. This is not always the case, especially when we move further back into the past and try to understand the diets of hominids such as Neanderthals and *Homo erectus*. The skeletons of these early humans are only rarely accompanied by extensive archaeological remains from which their diets can be reconstructed.

In the absence of direct evidence in the form of food remains, it is possible to obtain some information on diet by **dental microwear** analysis. In this type of research, the surfaces of teeth are examined with the scanning electron microscope to identify scratches and other types of damage that build up over an individual's lifetime and reflect the types of food that have been eaten. With humans and other primates, it is usually the molars that are examined most closely as these are the teeth used for chewing. If the diet consists mainly of hard nuts and seeds, which are cracked open in the mouth, the surfaces of the molars become pitted with small indentations. If, on the other hand, the diet is rich in leaves or meat, which have to be chewed before swallowing, then the molar surfaces become scratched with fine lines. In a more sophisticated form, microwear analysis can also indicate the way in which the teeth as a whole are used. Modern humans do not extensively use their incisors for chewing, so these teeth show relatively little scratching. Some primates such as macaques, on the other hand, prepare food for swallowing mainly by chewing with their front teeth, so with these animals the incisors are much more damaged. The way in which early hominids dealt with their food can therefore be deduced from the microwear patterns of different parts of their dentition.

In archaeology, microwear studies are also made on the preserved teeth of farm animals, in order to make inferences about husbandry. Studies with sheep, for example, have shown that grazing gives rise to distinctive microwear patterns characterized by striations on the molars, caused by chewing the fibrous material in grass. Sheep that are allowed to browse in woodland pastures, or raised on fodder collected from woodlands, generally have fewer striations on their teeth because the vegetation they are eating is less abrasive (Figure 12.1). The microwear patterns on archaeological remains have to be interpreted carefully, but in one study it has been suggested that at a Norse settlement in Greenland the pastureland was overgrazed, because the teeth of sheep found at the site displayed not only the striations associated with grazing, but additional damage thought to be caused by soil in the diet. The implication was that the sheep were eating grass that had been cropped close to the ground and so were ingesting small amounts of soil at the same time. In another study, it was shown that sheep allowed to graze on seaweed, a form of husbandry that is common today in Scotland and the islands of the north Atlantic, develop characteristic microwear patterns.

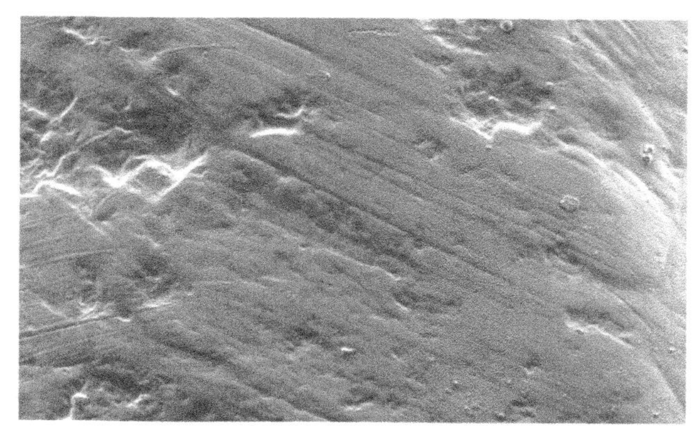

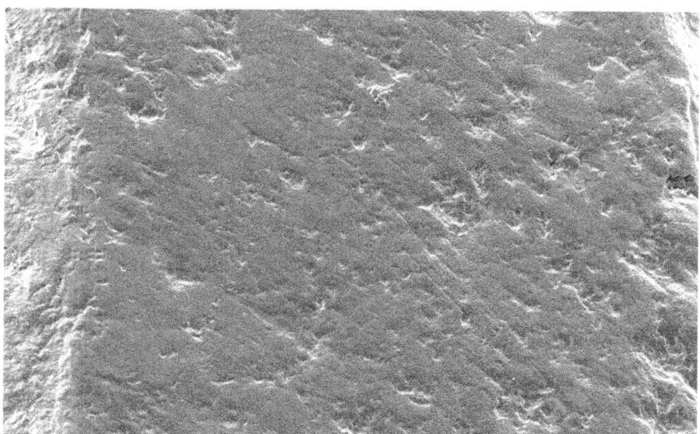

Figure 12.1 Differences in the dental microwear patterns for grazing (upper panel) and browsing (lower panel) sheep from Gotland, Sweden. The view in each panel is approximately 250 μm across. Reprinted from Ingrid Mainland, "Dental microwear in grazing and browsing Gotland sheep (*Ovis aries*) and its implications for dietary reconstruction," *Journal of Archaeological Science*, 30, 1513–27, copyright 2003, with permission from Elsevier and the authors.

12.2 Studying Diet by Organic Residue Analysis and Stable Isotope Measurements

Examination of bones and plant remains at archaeological sites, or of microwear patterns on the teeth of humans and their animals, can provide some information on past diets, but these approaches are limited in what they can achieve. Animal and plant remains may not be present, and if they are, then they may not be entirely representative of the diet as a whole. Microwear studies can only give a very broad indication of the types of food that were eaten. Studies of past diets are becoming increasingly dependent on biomolecular analysis, with two types of approach particularly important. The first of these is organic residue analysis, in which the remains of foodstuffs are examined on potsherds from cooking vessels and storage containers in order to identify the types of food that were being used at an archaeological site. The second approach is stable isotope analysis, which can be carried out either on skeletal remains, providing information on the diets of individual people, or on compounds purified from food remains, broadening the scope of organic residue analysis. We will explore each of these areas of research in turn.

12.2.1 Food residues can be recovered from the remains of pottery vessels

Organic residue analysis is important in several areas of biomolecular archaeology, and we will encounter different examples when we study prehistoric technology in Chapter 14. In the dietary context, the types of residue that are analyzed are derived from foodstuffs that were cooked or stored in ceramic vessels – pots made from fired clay. Pottery is one of the earliest human inventions, the oldest ceramic vessels being made in Japan over 12,000 years ago, the technology then spreading to China around 10,000 years ago and being present throughout Eurasia by 5000 BP, by which time pottery had also been independently invented in the Americas. Pottery vessels are occasionally recovered intact, but more usually are found as broken fragments called **potsherds**, which are very durable and are often found in large quantities, sufficient to reconstruct the appearance of the original vessels. The shapes and decorative styles of these pots are used as cultural indicators and in some cases as a means of dating a site, which is possible because particular pottery styles are associated with different periods.

In the vast majority of cases, the food residues that are studied are recovered from within the pottery matrix of a potsherd, having become absorbed into the wall of the original vessel during the cooking process. These residues will therefore be fats or other lipids, and occasionally proteins, which became solubilized when the contents of the vessel were heated. Residues of liquids such as wines and cooking oils can also become absorbed in the walls of vessels in which they are stored. Absorbed residues are not visible to the naked eye, but are very common, probably being present in 80% or more of all archaeological pottery assemblages. They are recovered by grinding a small piece of a potsherd into a powder and then soaking for a short period in an organic solvent, such as a mixture of chloroform and methanol, possibly with sonication to aid removal of the absorbed molecules from the pottery matrix. A subsequent aqueous extraction in a mild alkali solution can also be used if it is suspected that the potsherd might contain compounds that are insoluble in organic solvents. The compounds that are present in the extracts are then identified by techniques such as gas chromatography-mass spectrometry (Section 4.2). Absorbed residues are so common in potsherds that organic analysis has become a powerful and broadly applicable means of studying past diet, with a broad range of foodstuffs identified from different periods and geographical regions.

As well as absorbed food residues, the remains of foodstuffs are also sometimes recovered as visible deposits on the surfaces of potsherds. These are usually sooty and in some cases may derive from the fuel used to cook the food, rather than the food itself, so ideally a distinction needs to be made between deposits on the inner and outer surfaces of vessels. Those on the inner surface are likely to have resulted from foodstuffs that became burnt while being cooked. As with absorbed residues, their analysis can reveal the type of food that was being prepared, even if in these cases that food was probably spoiled and did not itself get eaten. Less common, and spectacular when found, are intact vessels in which the unburnt remnants of their original contents are still preserved.

12.2.2 Various technical challenges must be met if residue analysis is to be successful

Although organic residue analysis has been very successful in identifying foodstuffs and other materials that were processed or stored in pottery vessels, it is by no means a routine area of research. A number of challenges must be met in order for the analysis to be successful. The first of these is the essential requirement that the residues being studied derive from the vessel's contents and are not more recent contaminants. The most likely source of contamination is organic compounds from the burial environment, such as plant residues present in the soil, but contamination could also result from transfer of residues between potsherds that were buried close together, particularly in a damp environment, or from handling during or after excavation. Often, contaminants will be recognized as non-endogenous because their identities will not match the residues expected from genuine foodstuffs, but this is not always the case. For example, soil compounds from decaying vegetation could be mistaken for residues of plant products that had been cooked or stored in the vessel, and human lipids deposited on a potsherd by handling might be typed as "animal fat" if the residue analysis is not carried out carefully enough. Precautions therefore need to be taken to avoid confusing contaminants and real food residues. Cleaning the surface of a potsherd to remove absorbed contaminants runs the risk of also removing genuine residues, which might be present in only the outer 2 mm or so of the ceramic matrix. One suggestion for those samples where contamination is suspected is to compare the absorbed residues on the inner and outer surfaces of the potsherd, because those from material cooked or stored within the vessel should be present only on the inside.

A whole new set of "contamination" issues arise if a vessel has been used to cook or store different items at different times. Detection of multiple residues in unusual combinations should alert the researcher to the possibility that a vessel was reused in this way, rather than indicating the existence of some complex foodstuff or exotic beverage.

An additional requirement is the ability to extrapolate from the identities of the compounds that are detected in the archaeological residue to the identities of the compounds that were present in the foodstuff. The problem is that those compounds will have undergone chemical transformations, first because of the heating and other physical factors involved in the cooking process, and second because of further changes occurring during diagenesis. It is therefore essential that the decay pathways of the relevant compounds are understood. It is equally essential that compounds, or decay products, that act as biomarkers for different types of foodstuff be known. An example of the type of information that is needed was described in Section 8.4, where we learnt that long-chain ketones of the general formula $CH_3(CH_2)_mCO(CH_2)_nCH_3$ have been shown to be specific decay products of, and hence biomarkers for, the saturated fatty acids found in the triglycerides of particular types of animal fat.

In many cases, a foodstuff can be identified with confidence only if the archaeological context of the residue is taken into account. Without some knowledge of the likely foods that were stored or processed at a particular site, the chemical information obtained by residue analysis is unlikely to be sufficient to make an unequivocal identification. There are just too many types of food, and individual biomarkers do not have sufficient specificity unless the range of the analysis can be constrained in some way. The problems have been highlighted by a test project in which 11 research groups expert in organic residue analysis were given potsherds that contained absorbed residues from camel milk. The researchers did not know the identity of this residue and had to carry out their analyses "blind." None of the groups correctly identified the residue, most not getting close, though one did suggest that it might be milk of a herbivore (Table 12.1). Their success would have been much greater had they known from which part of the world and which archaeological period the potsherd

Table 12.1 Results of a "blind" test for detection of camel milk absorbed into a potsherd.

*Method used**	*Result*	*Interpretation*
CIEP	No proteins detected	Proteins degraded by cooking
GC/MS	16:0, 18:0 and 18:1 fatty acids detected, 16:0 most abundant, 18:1 least	Veal, egg, or goat milk
GC	Medium chain fatty acids detected, 18:0 and 18:1 least abundant	Decomposed roots, tubers, or berries
GC/MS	14:0, 16:0, 18:0, and 18:1 fatty acids and cholesterol detected	Veal or egg
GC	No results	No interpretation
IRMS	$\delta^{13}C$ and $\delta^{15}N$ values obtained	Milk of a herbivore
HTGC, GC/MS	Triglycerides detected, 16:0 fatty acids more abundant than 18:0	Animal product

* Abbreviations: CIEP, crossover immunoelectrophoresis; GC, gas chromatography; HTGC, high temperature gas chromatography; IRMS, isotope ratio mass spectrometry; MS, mass spectrometry.

had come, as this would have enabled them to plan the most appropriate analytical approach, and to compare the compound identifications that they made with the types of food that realistically might have been expected at that place and time.

12.2.3 Information on diet can be obtained by stable isotope analysis of skeletal components

In Chapter 6, we explored the various types of dietary information that can be obtained by analysis of the stable carbon isotopes ^{12}C and ^{13}C, and those of nitrogen, ^{13}N and ^{14}N. We learnt that when applied to skeletal remains, these isotope studies can address three aspects of diet:

- The presence of a high proportion of maize in the diet can be identified because this gives rise to skeletal $\delta^{13}C$ values higher than those of people who only eat the much commoner C_3 plants.
- A diet rich in marine animals – fish and invertebrates – can be distinguished from one based on terrestrial animals because of differences in the $\delta^{13}C$ and $\delta^{15}N$ values. Marine and freshwater inputs can also be separated by $\delta^{13}C$ and $\delta^{15}N$ measurements.
- A diet rich in terrestrial meat can be distinguished from a more omnivorous or vegetarian one because of differences in the resulting $\delta^{13}C$ and $\delta^{15}N$ values. A weaning infant, who essentially feeds off its mother, appears to have a very carnivorous diet.

When using stable isotope analysis to study these aspects of diet, it is important to have a precise understanding of the relationship between the isotope content of the diet and the isotope content of different components of the skeleton, so that informed choices can be made about the sampling strategy. Bone material undergoes constant replacement, so isotope ratios in bones represent the dietary intake during the last 10 years of an individual's life. Teeth, on the other hand, are formed during adolescence (between 6 and 13 years in modern western societies) and not subsequently replenished, so the isotope ratios in teeth indicate the dietary isotope intake only during this period. In fact, teeth can be used for quite detailed analysis of isotope intake during adolescence, because a child's teeth do not all form at the same time, so measurements taken from different teeth in the same jaw can give a series of isotope values for different periods. The surface enamel on an individual tooth is laid down as a series of layers over a period of two to three years, so microsampling within these layers can indicate seasonal variations in isotope intake. Finally, hair contains a record of carbon and nitrogen isotopes in the diet in the months immediately prior to sampling, or prior to death if the hair is from an archaeological specimen, with a time series of isotope measurements obtainable by analyzing segments along the length of a hair (Figure 7.6).

When bone is being used, a choice must be made between examining the bulk collagen fraction, which can be obtained by demineralizing the bone by mild acid treatment, or the inorganic bioapatite fraction, obtained by dissolving the organic component in sodium hypochlorite solution. Collagen, being a protein, contains both carbon and nitrogen, and can therefore be used as a source of both $\delta^{13}C$ and $\delta^{15}N$ values. Bioapatite, on the other hand, is primarily calcium hydroxyphosphate with absorbed carbonates, and so contains carbon but not nitrogen. Bone collagen and bioapatite do not contain identical records of the human diet. Experimental feeding studies with animals have shown that collagen is formed mainly from the amino acids contained in dietary protein, whereas the carbon in bioapatite has contributions from ingested protein, carbohydrates, and fats. The $\delta^{13}C$ values obtained from collagen and bioapatite are therefore indicative of different components of the diet.

With teeth, similar considerations apply. The dentine within a tooth contains a mixture of proteins and so can provide $\delta^{13}C$ and $\delta^{15}N$ measurements, whereas the enamel on a tooth's surface is made of bioapatite and so yields only $\delta^{13}C$ values. Again, the $\delta^{13}C$ values in the protein and mineral fractions reflect different aspects of the diet. Hair is made of keratin, a protein, and so is used in both $\delta^{13}C$ and $\delta^{15}N$ studies, and has no mineral component.

12.2.4 Compound specific isotope studies extend the range of organic residue analysis

Stable isotope analysis becomes even more important as a means of studying past diet when we progress from the bulk studies outlined above to more detailed analyses of the isotope ratios in individual biomolecules. These **compound specific isotope studies** are carried out by linking a gas chromatograph (GC) to the isotope ratio mass spectrometer (IRMS) used for stable isotope measurements. Now, rather than analyzing bulk extracts that are loaded directly into the IRMS, the compounds within a complex mixture can first be separated by the GC and the isotope ratios within these individual molecules measured.

Compound specific isotope studies have been particularly informative when applied to organic residues absorbed into potsherds. Although the lipids and other chemicals derived from food undergo decay during the cooking process, and break down even further during diagenesis, their stable isotope ratios remain unchanged. This has been shown by laboratory experiments in which triglycerides from animal fats have been heated in the presence of powdered ceramics to mimic the decay process occurring in archaeological potsherds. The resulting decay products retain the same $\delta^{13}C$ values as their parent molecules, even after extensive amounts of chemical breakdown. These $\delta^{13}C$ values can therefore be used as an aid to identification of biomarkers for specific foodstuffs among the decay products detected in a residue.

This approach to residue analysis is being applied to the detection and mapping of the use of dairy products in the past. The triglycerides present in animal fats are primarily made up of saturated fatty acids with 16 or 18 carbons – the ones designated $C_{16:0}$ and $C_{18:0}$ – with the latter more predominant in ruminants such as cows and sheep compared with monogastric animals such as pigs. Typing the relative amounts of the different fatty acids in a residue can therefore give an indication of the animal that was being cooked in the original vessel. This can be achieved with reasonable accuracy by conventional residue analysis, but simply typing the amounts of the two types of fatty acid cannot make the important distinction between triglycerides from adipose tissue – flesh – and those from milk. It is only possible to say that the residue derives from animal fats, which is much less useful than identifying whether a vessel was being used to cook meat or to prepare a dairy product. This question is important not only with regard to diet, as we will see in Section 12.3, but it also underlies our understanding of the prehistoric development of dairying as an agricultural technology.

Compound specific stable isotope analysis allows adipose and milk fats to be distinguished. This is because of a subtle difference in the way in which the $C_{16:0}$ and $C_{18:0}$ acids in the two types of fat are synthesized. Adipose tissues are able to make both $C_{16:0}$ and $C_{18:0}$ fatty acids, using acetate and other substrates derived from carbohydrates contained in the diet. This process involves the isotopic shift that results in the $\delta^{13}C$ value of the synthesized biomolecule being 1–2‰ higher than that of the dietary carbon. The mammary gland, on the other hand, can synthesize its own $C_{16:0}$ fatty acids, but not the $C_{18:0}$ versions. These it obtains from fatty acids contained in the plants ingested by the animal, most of which are unsaturated but can be converted to saturated $C_{18:0}$ fatty acid in the rumen. This conversion simply involves removal of hydrogens and does not result in an isotopic shift. The $\delta^{13}C$ value of the $C_{18:0}$ fatty acid present in milk triglycerides, or in any dairy product, is therefore slightly lower than that in triglycerides from adipose tissue (Figure 12.2). The critical distinction between meat and milk is therefore possible by compound specific analysis of the $\delta^{13}C$ ratios in the $C_{18:0}$ fatty acids present in a food residue.

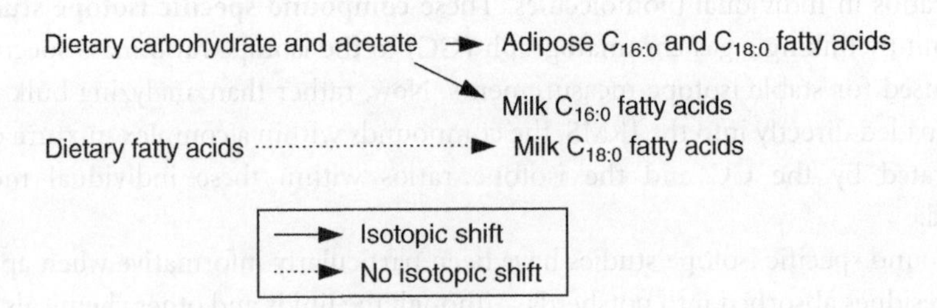

Figure 12.2 The distinction between the synthesis of $C_{16:0}$ and $C_{18:0}$ fatty acids in the adipose and mammary tissues of a ruminant.

12.3 Examples of the Use of Stable Isotope and Residue Analysis in Studies of Past Diet

There are many examples of the use of stable isotope studies, both bulk and compound specific, and residue analysis in the study of past diet. To illustrate the breadth of this area of biomolecular archaeology we will examine four topics, starting with the use of bulk stable isotope analysis to understand the diets of our more remote ancestors.

12.3.1 Diets before agriculture

Before the emergence of agriculture some 11,000 years ago, all humans were hunter-gatherers, obtaining their meat from wild game and their plants from the natural vegetation. Neanderthals, who lived throughout Europe and the adjacent parts of Asia until some 30,000 years ago, acquired their food exclusively in this way. But what exactly did they eat?

Collagen is a relatively stable biomolecule and can survive in bone for many thousands of years (Section 8.3). Measurements can therefore be made of both the $\delta^{13}C$ and $\delta^{15}N$ content of Neanderthal skeletons to make inferences about their diet. This was first done with a 40,000-year-old skeleton from France, and subsequently with other Neanderthals from France, Belgium, and Croatia, the last of these including the specimens that have provided the ancient DNA being used to sequence the Neanderthal genome (Section 16.4). The European vegetation is made up exclusively of plants that follow the C_3 photosynthetic pathway, and no marine input is expected in skeletons from inland sites. The $\delta^{13}C$ and $\delta^{15}N$ values obtained from these skeletons should therefore reflect the trophic level that Neanderthals occupied in their ecosystem. To establish this level, it is necessary to have control isotope data from animals that lived at about the same time and whose trophic levels we know. Measurements have therefore been taken from the fossil bones of typical herbivores including reindeer and bison, and carnivores such as wolves and hyenas. Comparisons with these control data show that the $\delta^{15}N$ values of the Neanderthals skeletons are on average 5‰ higher than those of contemporary herbivore skeletons, and are similar to the values for the carnivores (Figure 12.3). The implication is that the Neanderthal diet was rich in meat and included relatively little plant material, and also that the animals they ate were herbivores whose $\delta^{15}N$ values were relatively high for their trophic level. The latter include mammoths, traditionally looked on as an important prey for Neanderthals due the presence of mammoth bones at some Neanderthal sites, as well as bears.

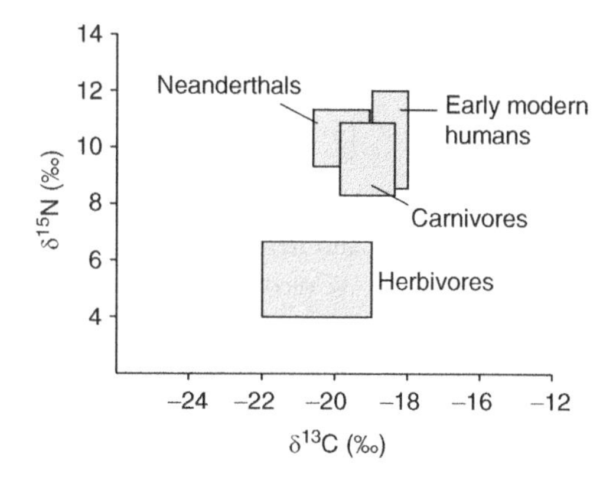

Figure 12.3 Stable isotope measurements from collagen for Neanderthals and early modern humans, compared with those for carnivores and herbivores, illustrating the high meat contents of the two types of human diet. All measurements are from European skeletons, so the $\delta^{13}C$ values do not vary greatly.

Anatomically modern humans – members of our own species *Homo sapiens* – arrived in Europe some 45,000–50,000 years ago and so shared the continent with Neanderthals for approximately 20 millennia. Studies of early modern human skeletons reveal $\delta^{15}N$ values that appear to be even higher than those of Neanderthals. This has been interpreted as indicating that modern humans, as well as eating the same game as Neanderthals, broadened their diet by including freshwater fish or birds. This is because fish and birds feeding on freshwater vegetation and algae have slightly higher $\delta^{15}N$ values than animals that feed on terrestrial vegetation, because of the different ways in which nitrogen is cycled in the two environments (Section 6.2). Inclusion of river fish or birds in the human diet therefore has a positive impact on the $\delta^{15}N$ content of the skeleton.

The notion that early modern humans in Europe had a broader diet than the local Neanderthals is attractive, but closer examination of these data, and of the entire use of stable isotope measurements to assess the trophic levels of prehistoric humans, forces a note of caution. It has been argued that, due in part to the small sample sizes that are inevitable in this type of work, there are no statistically significant differences between the $\delta^{15}N$ values of Neanderthal and early modern human skeletons. To achieve the degree of precision needed to identify fine differences between their diets we would need to sample many more skeletons than are available. Furthermore, identification of the trophic level occupied by a past human is based on the assumption that the shift in $\delta^{15}N$ values that occurs from herbivores to carnivores, which is approximately +3 ‰ in animals, is the same in humans. This has never been tested because the necessary experiments – stable isotope measurements of collagen extracted from the bones of people fed different diets for lengthy periods – have never, of course, been carried out. It is also possible that the extent of this isotopic shift depends on the amount of food that is consumed, experiments carried out with animals suggesting that overfeeding may increase the difference between the $\delta^{15}N$ values of the diet and of bone collagen, at least in herbivores. We therefore need to develop a better understanding of the factors that influence the stable isotope content of the human skeleton before placing too much emphasis on small differences in the values obtained from archaeological material.

12.3.2 Studying the relationship between diet and status in past societies

Stable isotopes can also be used to study the diets of people from a single past community, and hence to distinguish differences that might be linked to status or other social factors. We recognize such distinctions in modern societies, and most of the studies that have been carried out with archaeological material suggest that they also existed in the past. One example comes from Cahokia in Illinois, which was first occupied during the Late Woodland period at about AD 650. Some 400 years later, during the Mississippian period, the population of Cahokia began to increase dramatically and by AD 1250 it is thought that several thousand people lived there in a highly complex society. In the two centuries leading up to AD 1250, large burial mounds were erected, each one containing multiple inhumations. The research that we will examine concerns Mound 72, from which 272 skeletons were recovered, dating to AD 1050–1150.

The burials in Mound 72 can be separated into two distinct groups interpreted as "high status" and "low status." The high status burials were accompanied by elaborate grave goods, including one male laid on a platform of over 10,000 shell beads arranged in the shape of a bird of prey, and others placed on wooden structures. Paleopathological examination of these skeletons suggested that most of the high status people had enjoyed good health and high levels of nutrition while they were alive. In contrast, the more numerous low status burials had less elaborate grave goods, or none at all, were buried in mass graves, and many of the skeletons had pathological features indicative of poor health and a low quality diet. The majority of these low status individuals were young females, possibly sacrificial victims. Examination of non-metric traits on the teeth of the skeletons, although an imprecise way of establishing kinship relationships, suggested that the high and low status people formed two social groups with little interbreeding between them.

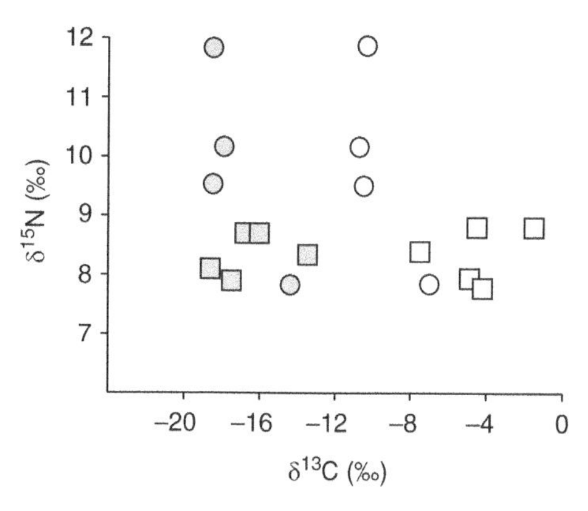

Figure 12.4 Stable isotope data obtained from the Cahokia skeletons. Circles are high status individuals and squares are low status. Gray circles and squares are values obtained when the $\delta^{13}C$ and $\delta^{15}N$ obtained from collagen are compared. Open circles and squares are values from comparison between $\delta^{13}C$ in bioapatite and $\delta^{15}N$ in collagen.

Twenty-five skeletons from Mound 72 were tested, nine of which contained sufficient bone collagen for measurement of $\delta^{13}C$ and $\delta^{15}N$ ratios. The results displayed some overlaps between the two social groups, but generally the high status skeletons gave higher $\delta^{15}N$ values than the low status ones (Figure 12.4). This could indicate that the higher status people ate more meat, and hence occupied a slightly higher trophic level than the members of the low status group. A corollary to this could be that the low status people had a greater proportion of maize in their diets, which should be revealed by an elevated value for $\delta^{13}C$ because of the unusual features of the photosynthetic pathway of this C_4 plant. In fact, when collagen $\delta^{13}C$ values were measured, only small differences were seen between the two groups, but the bioapatite fraction of their bones gave a different story. The $\delta^{13}C$ values for bioapatite in the high status skeletons fell between −10.0 ‰ and −6.7 ‰, whereas, with one exception, those for the low status people were significantly higher, at −4.6 ‰ to −1.6 ‰. The only C_4 plant available to the people living at Cahokia at that time was maize, as all the indigenous vegetation was made up of the much commoner C_3 plants. It can therefore be assumed that the difference between the $\delta^{13}C$ signals for the two groups of people is entirely due to the amount of maize that each ate. According to this model, the diets of the high status individuals contained an average of 45% maize, compared with 80% maize for the low status ones. The bioapatite measurements therefore support the deduction from the collagen $\delta^{15}N$ values that the diet of the low status people was less rich in meat.

Could the differences in the estimated amounts of maize in the diets of these two groups of people be an artifact of the experimental system? We must address the divergence between the two sets of $\delta^{13}C$ data – those from collagen and those from bioapatite – and the problem that only the latter revealed the higher maize intake of the low status individuals. It is possible for the $\delta^{13}C$ ratios in bioapatite to increase during diagenesis due to exchange of carbon with environmental carbonates and carbon dioxide, but this effect is likely to be small and insufficient to give the elevated $\delta^{13}C$ values measured in the bioapatite of these skeletons. In any case, environmental exchange should increase the $\delta^{13}C$ values of all skeletons in a similar fashion, and not lead to a difference between those of high and low status. The possibility that the bioapatite results are artifactual can therefore be discounted, at least as far as environmental exchange is concerned. The differences between the collagen and bioapatite results are more likely to reflect the differences in the dietary inputs into these two components of bone, collagen reflecting only the protein part of the diet, but bioapatite being formed from carbon taken also from dietary carbohydrate and fat. Maize has a high carbohydrate content and is a relatively poor source of protein, so its presence in the diet is likely to be much more apparent from measurements of $\delta^{13}C$ in bioapatite rather than collagen.

12.3.3 The origins of dairying in prehistoric Europe

We learnt in Section 12.2 how compound specific stable isotope analysis of $C_{18:0}$ fatty acids enables milk or milk products to be identified in residues absorbed in potsherds. Now we will look at how this technique has been used to study the origins of dairying.

Farming began in the Fertile Crescent of southwest Asia some 12,000 years ago and gradually spread into neighboring parts of Europe and Asia. Initially, domestic animals such as cattle, sheep, and goats were probably used solely as a source of meat, but at some stage the early farmers began to utilize these animals for **secondary products**, ones that can be acquired without killing the animal. Milk, wool, and traction – the use of an animal to pull a plough or cart – are examples of secondary products. Milk was probably the most important, as it provides an additional source of nutrients to complement those obtained from the carcass of an animal. The adoption of dairying is therefore looked on as an important part of the secondary products "revolution," and a key stage in the development of an increasingly sophisticated agricultural technology by prehistoric farmers.

Tracing the advent of dairying in the archaeological record is quite difficult. Only a few types of artifact, such as ceramic milk strainers, can specifically be associated with dairying, and evidence from artistic depictions of farming activities is ambiguous. Animal bone assemblages can give some clues, as a preponderance of adult females in these assemblages suggests that animals were not being kept just for meat production. Taken together, these different types of evidence appear to suggest that dairying did not begin until 2000–4000 years after cattle and other animals were originally domesticated.

Studies of potsherd residues have now shown that the dates assigned to the secondary products revolution were too recent, and that milk was being used as early as 9000 BP, only 1000 years after farming began. This work involved analysis of a large number of potsherds, 2225 in all, from 23 archaeological sites in the Levant, Anatolia, and southeast Europe, dating from 9000–5500 BP. Each potsherd was ground into a powder, the lipid residue extracted, and the compounds present in the extract separated by gas chromatography. One surprise was that only 255 of these potsherds contained absorbed residues, a substantially smaller proportion of positives than anticipated from past experience with pottery assemblages from northern Europe and elsewhere. In these earlier studies, it had not been unusual for as many as 80% of the potsherds to contain residues. The relatively low success rate might have been due to increased diagenesis in the warmer climates of southeast Europe, Anatolia, and the Levant, or could possibly reflect differences in the absorptive powers of the types of clay used in different parts of the world.

Most of the 255 potsherds that gave positive results contained animal fats, as shown by the presence of peaks for $C_{16:0}$ and $C_{18:0}$ fatty acids in the gas chromatographs. As we discussed

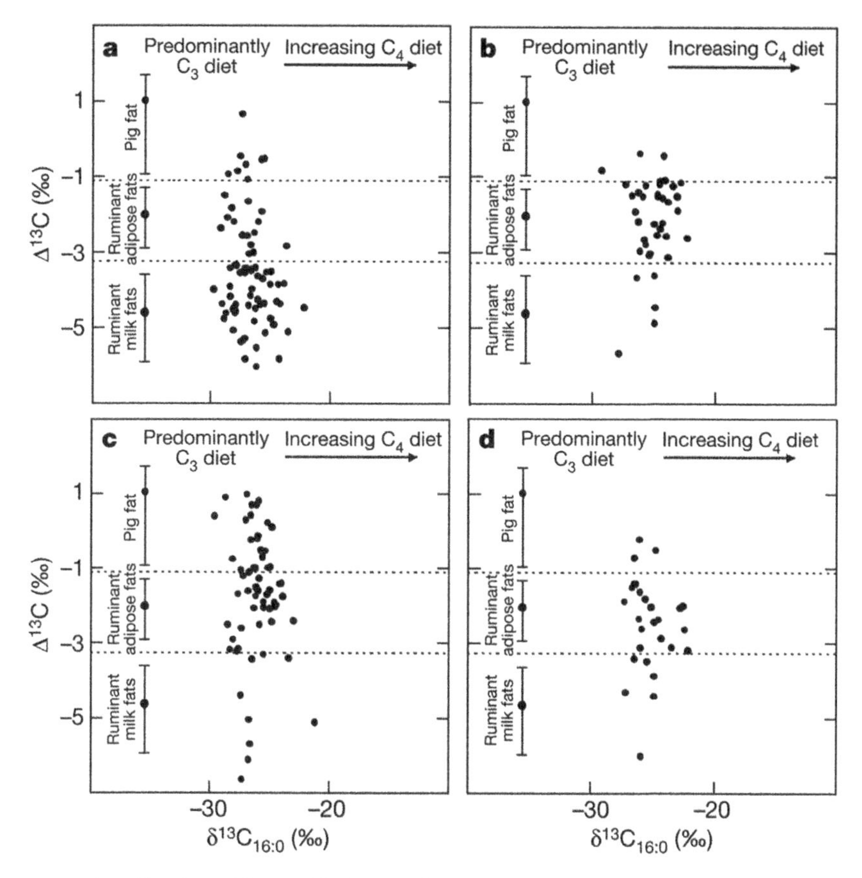

Figure 12.5 Identification of fatty acids from milk in potsherd extracts. In these graphs, the $\delta^{13}C$ value for the $C_{16:0}$ fatty acid component of a potsherd extract is plotted against the $\Delta^{13}C$ value, which is defined as $\delta^{13}C_{18:0}–\delta^{13}C_{16:0}$. This plot enables ruminant milk and adipose fats to be distinguished, as well as fats from pigs, which are non-ruminants. Pottery samples are from (a) northwestern Anatolia, (b) central Anatolia, (c) southwestern Europe/northern Greece, and (d) eastern Anatolia and the Levant. Reprinted by permission from Macmillan Publishers Ltd: Richard Evershed, Sebastian Payne, Andrew Sherratt, Mark Copley, Jennifer Coolidge *et al.*, "Earliest date for milk use in the Near East and southwestern Europe linked to cattle herding," *Nature*, 455, 528–31, copyright 2010, with permission also of the authors.

in Section 12.2, these fatty acids are characteristic of animal fats, but their presence alone does not allow a distinction to be made between adipose and milk fats. The $\delta^{13}C$ values were therefore measured for the fatty acids as a whole and for the $C_{18:0}$ fraction on its own, to identify which, if any, of the residues contained $C_{18:0}$ fatty acids with the lower $\delta^{13}C$ values expected if these were derived from milk triglycerides (Figure 12.5). The results showed that milk was being used in all the regions from which potsherds were obtained, including the earliest sites, and that dairying was particularly widespread in northern Anatolia. In this area, 70% of the potsherds that contained residues gave strong signals for the presence of milk. In contrast, at sites from eastern Turkey and the Levant, closer to the areas where farming began, the proportion of potsherds showing signs of milk usage were much smaller. This distribution almost certainly relates to the differences in the vegetation patterns of the various regions, parts of northern Anatolia providing much better grazing conditions than eastern Turkey and the Levant and hence being more suitable for rearing herbivores such as cattle.

There is an interesting corollary to this work, which we will return to later when we examine the role of genetics in studies of past diets. Humans are naturally intolerant of milk, being unable to digest it after having been weaned. Many modern groups are still lactose intolerant, and those populations that can drink milk possess a relatively recent mutation that inactivates the intolerant response. The first farmers were probably lactose intolerant and could not have used unprocessed milk without experiencing digestive problems. Laboratory experiments have shown that residues of fresh milk do not become tightly absorbed into ceramic surfaces and might not survive well in the archaeological record. Those potsherds that contained milk residues are therefore likely to come from pots that were used to process milk into products such as cheese, or to store those products. Individuals who are lactose intolerant are able to digest processed milk products better than fresh milk. It would therefore appear that the earliest dairy farmers not only recognized the nutritive value of milk, but also understood how to convert milk into what would have been, for them, more palatable dairy products.

12.3.4 Detecting proteins in food residues

So far our discussion of organic residue analysis has focussed entirely on detection of lipids. This is because lipids are relatively stable biomolecules and when they do break down, during cooking or diagenesis, they often give rise to characteristic decay products that act as biomarkers for the type of food from which they are derived. But lipids are not the only type of ancient biomolecule likely to be present in food residues absorbed in potsherds. Proteins from the food that was being cooked or stored are also likely to be present. Could these be used to extend the range of organic residue analysis in the study of past diet?

The problem with proteins is that traditional detection methods based on immunological techniques are likely to give false results when applied to partially degraded proteins from archaeological materials (Section 3.3). Some progress has been made in using immunoassays in analysis of food residues, but it is recognized that any method that relies on the immunological examination of ancient protein will be subject to errors and uncertainties. The best of this work has been directed at identification of casein in potsherds from European sites, in which the antibodies used to screen the potsherds were raised against heat-treated casein. These antibodies might therefore be more specific for the partially degraded versions of casein expected to be present on archaeological materials than ones prepared with modern casein. Casein is a protein that is only found in milk, so its detection could provide a complement to the lipid studies of dairying described above.

The development of proteomic techniques suitable for use with archaeological materials raises the possibility that reliable identifications of proteins in food residues might one day be possible. The potential of this approach has been demonstrated by examination of a single potsherd from an Inuit site in Alaska, dating to AD 1200–1400. At this coastal site the local population was likely to have utilized marine resources in their diet, in particular muscle and blubber from harbor, gray, and ringed seals and from the beluga whale.

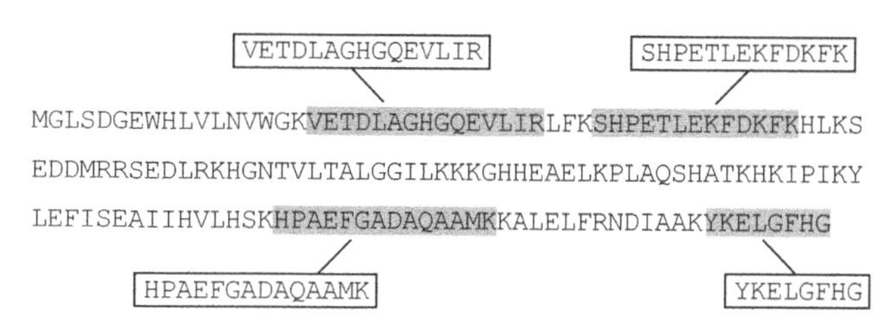

Figure 12.6 Seal myoglobin peptides identified in an Inuit potsherd. The complete amino acid sequence of seal myoglobin is shown, using the one-letter abbreviations for the amino acids. The four peptides identified in the potsherd extract are boxed and their positions in the intact protein indicated.

Before the potsherd could be studied, it was necessary to obtain some control data. Modern tissue samples from the seal and whale species were therefore heated on the surfaces of ceramic fragments, to mimic the cooking process occurring in the past, and the fragments then crushed and any proteins that had become absorbed within the pottery matrix extracted by soaking in an aqueous buffer. The extract was treated with trypsin, which cuts protein chains at positions adjacent to lysine or arginine amino acids and hence produces a defined set of peptides. The resulting mixture was examined by MALDI-TOF mass spectrometry, to give a peptide mass fingerprint, and the amino acid sequences of these peptides deduced, when possible, by comparison with the expected mass-to-charge ratios of peptides predicted to result from trypsin digestion of known seal and whale proteins. A number of such proteins could be detected, including ones from the muscle protein myoglobin, a major component of the meat of these animals.

The experiments described so far established that proteins from the meat and blubber of the animals being cooked do become absorbed into ceramics, and their peptide products are detectable by proteomic techniques. The next question was whether peptides could also be detected in the archaeological potsherd and, if so, whether these could be assigned to either of the species from which control data had been obtained. Residues extracted from the potsherd did indeed contain protein remnants which, after treatment with trypsin, yielded peptides whose mass fingerprints could be measured and amino acid sequences deduced. These included four peptides whose sequences indicated they were derived from seal myoglobin, two of these peptides being the same as ones detected in the control experiments (Figure 12.6).

This work demonstrates the potential of paleoproteomics in the study of archaeological food residues, but there is still some way to go before the approach becomes as important as lipid detection. Only a small number of peptides were detected in either the control or the archaeological residues, and these equated to only very partial coverage of the entire myoglobin amino acid sequence. With such limited coverage it was not possible to distinguish which type of seal the archaeological peptides came from, and with each of these peptides there were possible matches with other marine species and, in three cases, non-marine species as well. In fact, the degree of specificity was little better than that expected from an immunological test. The key point, however, is that the lack of specificity of immunological methods is inherent in that type of detection system, but with paleoproteomics the specificity can be improved by increasing the sensitivity of the method so that a greater number of peptides are detectable. We can anticipate that future technical developments will lead to increased sensitivity, and it is for this reason that paleoproteomics is likely in the future to make an important contribution to the study of past diets.

12.4 Using Genetics to Study Past Diets

Ancient DNA has so far been only of limited usefulness in studying past diets, but more conventional types of genetics are making a contribution. In these studies, DNA sequences are obtained from living people and techniques borrowed from evolutionary genetics used to examine those sequences in a way that tells us something about the diets of prehistoric human populations. The information that is obtained is very broad, relating to populations as a whole rather than the diets of individual groups of people. Genetics can, however, throw light on areas that are impenetrable to the more conventional biomolecular approaches to diet based on residue analysis and stable isotope studies. There are two areas in particular in which genetics has had an important impact, the first of these concerning the ability of humans to consume milk.

12.4.1 The ability of humans to digest milk evolved after the beginning of agriculture

Milk and milk products form such an important part of the modern western diet that it can be surprising to learn that our ancestors were unable to digest milk. Milk is rich in the disaccharide carbohydrate called lactose, but lactose cannot itself be used as a nutrient; first it must be broken into its constituent monosaccharides, glucose and galactose. This ability is conferred by the lactase enzyme, which is present in the epithelial cells lining the insides of the small intestine. Mammals synthesize lactase only during the early weeks of their lives, when they are dependent on the mother's milk for most of their nutrients. After weaning, the lactase enzyme is no longer synthesized and the ability to hydrolyze lactose, and hence to digest milk, gradually disappears. Adult mammals are therefore lactose intolerant because of the non-persistence of their lactase enzymes.

Some modern humans are lactose tolerant because synthesis of their lactase enzyme is not switched off after weaning. This phenotype, also called **lactase persistence**, is displayed by many people from northern Europe, or from populations that trace ancestry back to northern Europe. Lactase persistence is also present in a small number of farming groups in Africa, the Middle East, and southern Asia. In contrast, most people from Asia and Africa retain the ancestral, intolerant phenotype and cannot drink milk or consume milk products such as yoghurt, cheese, or ice cream.

Examination of the region of the human genome containing the gene for the lactase enzyme has shown that several different mutations, each of which results in lactase persistence, are present in the population as a whole. This indicates that lactose tolerance arose on multiple occasions and is a typical example of convergent evolution. Among Europeans, the commonest mutation involves replacement of a C with a T in the *MCM6* gene. This gene does not itself code for lactase, but is located close to the lactase gene and is thought to influence its expression. It is therefore quite logical that a mutation in the *MCM6* gene should result in continued synthesis of lactase into adulthood.

What can genetics tell us about the origin of lactase persistence? Under some circumstances it is possible to use the degree of divergence between a set of DNA sequences to estimate the period of time that has elapsed since those sequences began to evolve from their common ancestor. A set of sequences that have a relatively recent common ancestor will have had less time to acquire new polymorphisms than a set that began to diverge in the more distant past. If an estimate can be made of the rate at which new polymorphisms accumulate, then the differences between the sequences can be used to deduce approximately when in the past they began to diverge from their common ancestor.

If we wish to estimate when lactase persistence first appeared in Europeans, then we need to know the overall sequence diversity of the region of the genome surrounding the C to T mutation in modern Europeans. This analysis has been carried out in two ways. The first of these involved examining 101 single nucleotide polymorphisms (SNPs) in the 1.6 Mb of DNA on either side of the lactase persistence mutation. These SNPs are known to be variable in the human population as a whole, but do not themselves affect the lactase status of the individual, either because they fall in the regions between genes, or, if they are located in a gene, then the sequence alteration has no effect on the synthesis or activity of the lactase enzyme. It was discovered that these SNPs display remarkably little variability among lactose tolerant Europeans, suggesting that the persistence mutation arose recently in human evolutionary terms, probably within the last 2000–20,000 years. In a similar study, the degree of variability displayed by short tandem repeats (STRs) located near to the *MCM6* gene has been measured. Again, the degree of variability in lactose tolerant individuals was found to be relatively low, the age estimates in this case being 7450–12,300 years.

The genetic studies therefore suggest that, at least in Europeans, lactase persistence might have evolved soon after the advent of dairying, which according to the organic residue studies described above was just over 9000 years ago. A second implication of the genetic studies is that, since it first appeared in the prehistoric European population, the persistence allele has been under strong positive selection. This is consistent with our expectation that, once dairying had become widespread, lactose tolerance would have become highly advantageous. Possible links between dairying and lactase persistence have been explored by computer simulation modelling, which has suggested that selection for the *MCM6* mutation began approximately 7500 years ago in the Balkans and adjacent regions of central Europe.

12.4.2 Were early humans cannibals?

Cannibalism is an emotive subject, and until recently many archaeologists denied that it took place in human prehistory. But cannibalism has been observed in our closest relative, the chimpanzee, and in many other animal species, and has occurred in a number of modern human societies such as the Fore of Papua New Guinea. There is also evidence that it occurred in the past, from butchery cutmarks on archaeological human bones, from long bones that have been broken presumably to access the marrow, and from skulls whose bases have been removed, possibly so the brain can be taken out. These indications of cannibalism have been found at Atapuerca, in Spain, which was occupied by *Homo antecessor* around 780,000 BP, Moula Guercy, France, where

Neanderthals lived some 100,000 years ago, and at Gough's Cave, England, an important modern human site occupied from 13,500 to 11,500 BP. The broad range of dates for these sites suggests that cannibalism has been part of human prehistory for a very long time indeed.

Cannibalism was practiced by the Fore tribe of Papua New Guinea from 1900 to 1950, when the custom was outlawed by the Australian government, who had jurisdiction over the country at that time. The bodies of the deceased belonged to the mother's family, and were consumed in feasts by relatives, mainly women and children. Internal organs, including the brain and spinal cord, were eaten. From the 1920s, the Fore developed a high incidence of kuru, the laughing disease, which affected mainly women and children, with 1% of the population as a whole dying annually at its peak. The Fore ascribed the disease to witchcraft, but its true cause was found to be more interesting. Injection into chimpanzees of brain tissue taken from autopsies of individuals who had died of kuru suggested that a slow-growing virus was responsible for the disease, this virus being spread among the Fore by cannibalism. We now know that kuru is caused by the unusual proteins called prions, and that kuru is therefore a transmissible spongiform encephalopathy similar to Creutzfeldt-Jacob disease. In these rare disorders, the infecting prion protein causes similar endogenous proteins in the patient's cells to undergo a structural change and to accumulate in the brain, giving rise to a neurodegenerative disease.

Why is the discovery of kuru in the Fore relevant to prehistoric diet? The genes for human prion proteins exist as several different alleles. Among the Fore, there is currently a high incidence of older women who are heterozygous for the prion gene, with the allele called M129V always being one of the pair that they possess. The notation indicates that a methionine (M) amino acid at position 129 in the prion protein is changed to a valine (V). In younger Fore women, the incidence of heterozygotes is significantly lower. As the younger women were born after the abolition of cannibalism, the implication is that M129V heterozygosity confers some degree of protection against kuru, and so became selected among the women who were alive when cannibalism was being practiced. It is thought that heterozygosity does not actually confer immunity to kuru, but delays its onset by some 10 years or so.

The M129V allele is common in the modern human population as a whole. Could its high frequency be a genetic relic of cannibalism among early humans? It has been estimated that the allele first appeared some 500,000 years ago, or even earlier, which would concur with the dates for human cannibalism in the archaeological record. It is an interesting hypothesis, and one that finds favor among those archaeologists who are willing to look beyond the taboos of modern society. It is a controversial hypothesis, however, with much debate focussing on the veracity of the date assigned for the first appearance of the M129V allele. A more recent analysis, using a larger dataset of prion allele frequencies in the modern population, has suggested that M129V did not evolve until much more recently, not until 100,000–60,000 BP. It has been pointed out that the date of 500,000 BP was originally estimated through use of a model that did not appear to take account of the massive increase in the size of the human population which has taken place over the last half million years. It has also been reported that a second prion allele, G127V, is involved in resistance to kuru, as this allele is present in many of the elderly Fore women who do not possess the

M129V allele. However, G127V is not present in the wider modern population. On balance, it seems likely that the selection pressures that led to the current high frequency of the M129V allele were much more complex than the simple, cannibalism-driven scenario originally proposed.

Further Reading

Ambrose, S.H., Buikstra, J. & Krueger, H.W. (2003) Status and gender differences in diet at Mound 72, Cahokia, revealed by isotopic analysis of bone. *Journal of Anthropological Archaeology*, 22, 217–226.

Barnard, H., Ambrose, S.H. & Beehr, D.E., *et al.* (2007) Mixed results of seven methods for organic residue analysis applied to one vessel with the residue of a known foodstuff. *Journal of Archaeological Science*, 34, 28–37.

Bersaglieri, T., Sabeti, P.C. & Patterson, N., *et al.* (2004) Genetic signatures of strong recent positive selection at the lactase gene. *American Journal of Human Genetics*, 74, 1111–20.

Buckley, M., Kansa, S.W. & Howard, S., *et al.* (2010) Distinguishing between archaeological sheep and goat bones using a single collagen peptide. *Journal of Archaeological Science*, 37, 13–20. [Protein profiling as an aid to identification of animal bone assemblages.]

Coelho, M., Luiselli, D. & Bertorelle, G., *et al.* (2005) Microsatellite variation and evolution of human lactase persistence. *Human Genetics*, 117, 329–39.

Copley, M.S., Berstan, R. & Dudd, S.N., *et al.* (2003) Direct chemical evidence for widespread dairying in prehistoric Britain. *Proceedings of the National Academy of Sciences USA*, 100, 1524–9.

Evershed, R.P. (2008) Organic residue analysis in archaeology: the archaeological biomarker revolution. *Archaeometry*, 50, 895–924.

Evershed, R.P., Payne, S. & Sherratt, A.G., *et al.* (2008) Earliest date for milk use in the Near East and southeastern Europe linked to cattle herding. *Nature*, 455, 528–31.

Itan, Y., Powell, A., Beaumont, M.A., Burger, J. & Thomas, M.G. (2009) The origins of lactase persistence in Europe. *PLoS Computational Biology*, 5(8), e1000491. [Computer modelling of the spread of lactase persistence.]

Lee-Thorp, J.A. (2008) On isotopes and old bones. *Archaeometry*, 50, 925–50. [Covers all aspects of stable isotope studies of paleodiet.]

Mainland, I.L. (2003) Dental microwear in grazing and browsing Gotland sheep (*Ovis aries*) and its implications for dietary reconstruction. *Journal of Archaeological Science* 30, 1513–27.

Mead, S., Stumpf, M.P.H. & Whitfield, J., *et al.* (2003) Balancing selection at the prion protein gene consistent with prehistoric kurulike epidemics. *Science* 300, 640–3.

O'Connor, T. (2000) *The Archaeology of Animal Bones.* Sutton Publishing, Stroud.

Richards, M.P., Pettitt, P.B., Stiner, M.C. & Trinkaus, E. (2001) Stable isotope evidence for increasing dietary breadth in the European mid-Upper Paleolithic. *Proceedings of the National Academy of Sciences USA*, 98, 6528–32. [Early modern human diet.]

Richards, M.P., Pettitt, P.B. & Trinkaus, E., *et al.* (2000) Neanderthal diet at Vindija and Neanderthal predation: the evidence from stable isotopes. *Proceedings of the National Academy of Sciences USA*, 97, 7663–6.

Schoeninger, M.J. & Moore, K. (1992) Bone stable isotope studies in archaeology. *Journal of World Prehistory*, 6, 247–96.

Solazzo, C., Fitzhugh, W.W., Rolando, C. & Tokarski, C. (2008) Identification of protein remains in archaeological potsherds by proteomics. *Analytical Chemistry*, 80, 4590–7. [The Inuit potsherd study.]

Soldevila, M., Andrés, A.M. & Ramirez-Soriano, A., *et al.* (2005) The prion protein gene in humans revisited: lessons from a worldwide resequencing study. *Genome Research*, 16, 231–9.

Stern, B., Heron, C., Serpico, M. & Bourriau, J. (2000) A comparison of methods for establishing fatty acid concentration gradients across potsherds: a case study using late Bronze Age Canaanite amphorae. *Archaeometry*, 42, 399–414.

13

Studying the Origins and Spread of Agriculture

Until about 11,000 years ago, all humans were hunter-gatherers, obtaining their food by hunting wild game and gathering wild plants. Today the vast majority of the people on the planet depend on food from farmed animals and cultivated crops, using wild resources to provide only a minor part of their diet, or none at all for many individuals in the more "developed" societies. The dependence of our modern world on agriculture needs no emphasis: without it only a fraction of the current human population could be supported.

Biomolecular archaeology is contributing to our understanding of the origins and spread of agriculture in a number of ways. Most importantly, it is enabling the evolutionary origins of domesticated animals and plants to be located, providing novel insights into the events involved in the transition from hunting-gathering to agriculture. These studies have relied almost exclusively on DNA, being based on analysis of sequences and genotypes obtained from archaeological specimens and from living animals and plants. Biomolecular archaeology is also playing an increasingly important role in studies of the spread of agriculture from its centers of origin to new parts of the world. Again, this research is largely dependent on DNA, but it also makes use of stable isotope analysis to identify changes in diet that signal the arrival of agriculture at a particular place, and identification of starch grains and other microfossils diagnostic of specific crop plants. Finally, biomolecular archaeology is being used to follow the development of agriculture after its initial establishment. The work that is being carried out on dairying, which we examined in the context of past diet (Section 12.3), illustrates this application of biomolecular archaeology, and there are other examples that we will study at the end of this chapter.

Biomolecular Archaeology: An Introduction, by Terry Brown and Keri Brown © 2011 Terry Brown & Keri Brown

Table 13.1 Distinctive phenotypes of domesticated plants and animals.

Phenotype	*Description*
Plants, especially cereals	
Loss of seed dispersal	Due to the tough rachis mutation, which results in the grain remaining attached to the mature ear
Loss of seed dispersal aids	Such as hairs, hooks, and awns, which facilitate wind and animal dispersive processes
Increase in seed size	Can arise by human selection for larger seeds, or by tillage, larger seeds surviving deeper burial
Loss of sensitivity to environment	The seeds of most crops germinate soon after planting, whereas wild versions often germinate only in response to environmental cues such as day length and temperature
Synchronous flowering and ripening	Selected by cultivation practices
Compact growth habit	Selected by harvesting methods that preferentially sample plants of similar size and shape
Enhanced culinary chemistry	Such as improved breadmaking quality of wheat and changes to the sugar–starch balance in maize
Reduction of defensive armour	Such as loss of spines from *Dioscorea* yams
Reduction in defensive toxins	In yams and various legumes
Animals	
Smaller stature	Many domesticated animals are smaller and fatter than their wild counterparts
Less developed dentition	As part of the "tameness" of domesticated animals

13.1 Archaeological Studies of Prehistoric Agriculture

The adoption of agriculture is such an important event in the human past that it has been and continues to be studied from a number of perspectives by researchers in a variety of disciplines, including anthropologists, ecologists, plant and animal geneticists, social scientists, and evolutionary biologists, to mention just a few. Even within archaeology, distinct approaches to understanding prehistoric agriculture are being taken by specialists in different subdisciplines such as theoretical archaeology, gender studies, and bioarchaeology. Here we will summarize the current knowledge and thinking regarding the origins and global spread of agriculture.

13.1.1 Agriculture began independently in different parts of the world

The time at which agriculture was adopted in different parts of the world can be identified by searching for the remains of domesticated animals and plants, and for artifacts specifically associated with agricultural activity, such as implements used in ploughing, and grindstones used for processing grain. The domesticated versions of most plants, and to a lesser extent animals, have distinctive phenotypes compared with their wild ancestors (Table 13.1), and so these can be used as markers for agriculture, even in parts of the world where their

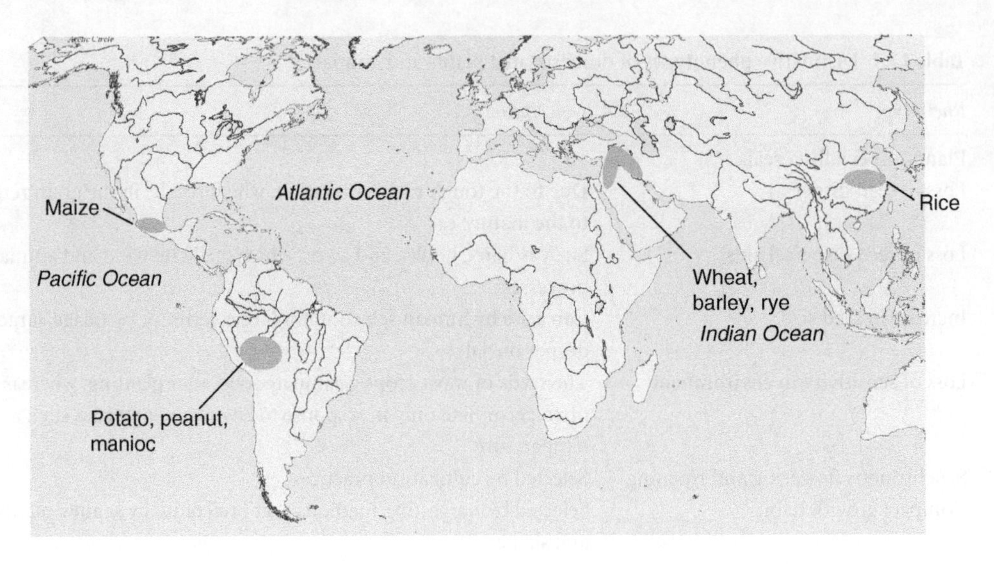

Figure 13.1 Locations of four of the primary centers for the origins of agriculture.

wild relatives are also found. The presence of the remains of domesticated animals or plants in well-dated contexts at archaeological sites can therefore chart not just the spread of agriculture into new regions, but also the very beginnings of agriculture in areas from which the first animals and plants were domesticated.

These studies have shown that there are at least four primary centers where agriculture originated (Figure 13.1). The centers are most clearly defined by the types of plants that were taken into cultivation:

- Mesoamerica, where maize, the domesticated version of the wild grass teosinte, first appears.
- The Yangtze region of southeast Asia, where rice was first cultivated.
- The "Fertile Crescent," a region of southwest Asia comprising the valleys of the Tigris, Euphrates, and Jordan rivers, and their adjacent hilly flanks, where wheat, barley, rye, and various other crops originate.
- The lowland and highland regions of central South America, which are the sources of potato, peanut, and manioc.

Because of their geographic separation, it is clear that agriculture must have arisen independently in each of these four areas. In contrast, there are various secondary centers where important crops were domesticated, but probably not until after agriculture had already reached those regions. Distinguishing primary and secondary centers is important in order to understand the social, economic, and/or environmental factors that accompanied and possibly drove the initial adoption of agriculture, as opposed to the different factors responsible for the increased sophistication of agriculture after it had spread to secondary areas.

There has been considerable debate about why prehistoric people first became farmers. At one time, the origin of agriculture in the Fertile Crescent was looked on as a "revolution" that encompassed not just farming but also as a broader package of cultural changes that included an increased use of pottery, new burial customs, and the development of large settled communities. Agriculture is still looked on as one of an important series of changes that occurred at the beginning of the Neolithic period, but the notion that these events were part of an inexorable human urge for social progress is now considered naive. In fact, it is difficult to identify those advantages that led to agriculture becoming the dominant lifestyle, and easier to find reasons in favor of hunting and gathering. It has been argued that agriculture resulted in a poorer diet because it made humans dependent on a smaller variety of plants and animals, and led to increased population densities which in turn enabled diseases to spread more quickly through a community. We are dependent on agriculture today because the world's population is so huge that without it many people would starve, so we might imagine that in prehistoric times a driver to the adoption of agriculture was the need to obtain surplus food supplies to support an increasing population. However, this is possible, at least to a certain extent, by an intensification of hunting and gathering, as happened in Australia and other regions where agriculture was never adopted until the arrival of Europeans. It therefore seems unlikely that an increased demand for food was the sole reason that humans adopted agriculture in regions such as the Fertile Crescent.

One key fact about the origin of agriculture is that it occurred at similar times in different parts of the world. This has prompted suggestions that agriculture somehow became favored by environmental changes occurring 11,000 years ago. The Last Glacial Maximum, when the planet was deep in the most recent ice age, was 20,000 years ago, and the current interglacial period, the Holocene, began about 11,000 years ago. Possibly the stabilization of the climate at the start of the Holocene led to changes in vegetation patterns or to the migration of animals that stimulated the move to farming in certain parts of the world. Attempts have also been made to explain the transition in terms of co-evolution between humans and their plant resources. The debates continue and a consensus between archaeologists, anthropologists, social scientists, and evolutionary biologists still seems a long way off.

13.1.2 The transition from hunting-gathering to agriculture was gradual rather than rapid

The debate over *why* agriculture began is matched by an equivalent debate about *how* it began. When the Neolithic Revolution was the favored concept for the origin of agriculture, it was thought that farming arose rapidly, with perhaps just a few human generations elapsing between a society that was entirely dependent on hunting and gathering as its mode of subsistence to one that was wholly agricultural. This rapid transition was supported by experiments that showed that the phenotypic traits associated with domestication of wheat and barley (see Table 13.1) can become fixed in a population that has been cultivated for as little as 20 years. Early genetic studies suggested that most crops could be traced back to a small geographical region in which they were originally domesticated, examples being southeast Turkey for wheat and the central Levant for barley. All of these lines of evidence pointed towards a rapid transition and the popular image of agriculture as a key "invention" of the human past.

In the last few years, archaeologists have accumulated increasing evidence that, at least in the Old World, the changeover from hunting-gathering to agriculture was a much longer and more drawn-out process than previously recognized. Some of the most compelling evidence has come from archaeobotanical studies of the charred remains of wheat and barley found at some of the earliest farming sites in the Fertile Crescent, near to the upper reaches of the Euphrates River. Careful examination of the chaff found in these grain assemblages enables wild plants to be distinguished from domesticated ones. The key difference lies in the structure of the ear (the seed head), which does not shatter when domesticated wheat or barley reaches maturity, so that the spikelets (the dispersal units containing the grains) remain attached to the ear after ripening. This is an important domestication trait that aids collection of the grain from the mature plants. The ears of wild wheat and barley are different because they shatter at maturity, releasing the spikelets. The non-shattering phenotype is caused by a mutation that changes the structure of the rachis node, the point at which the spikelet attaches to the ear. The "tough" rachis of a domesticated plant can be broken only by threshing, leaving a jagged break at the base of each spikelet, which is easily distinguished from the smooth abscission scar seen on a wild spikelet, even when charred archaeological remains are examined.

If agriculture arose rapidly in the Fertile Crescent, then we would expect that all of the grain assemblages at the earliest farming sites would be from plants with the domesticated phenotype. All of the chaff in these assemblages should therefore have the tough version of the rachis. But this is not the case with assemblages from six upper Euphrates sites dating from 10,200 to 6500 BP. The domesticated version of the rachis is first seen around 9250 BP, but in assemblages of this age it makes up less that 10% of the total, the bulk of the grains coming from plants with wild-type ears. Gradually the proportion of domesticated grains increases, but even at 6500 BP, well after the traditional start point for agriculture, a small amount of wild-type rachis fragments are still present in grain assemblages. Fixation of the tough rachis was therefore very slow, despite this phenotype being looked on as the central component of the "domestication syndrome," the suite of characters that distinguishes the cultivated version of a plant from its wild progenitor.

The implication of the recent archaeobotanical studies is that it took a few millennia for prehistoric people in the Fertile Crescent to make the change from a mode of subsistence fully based in hunting and gathering to one entirely dependent on agriculture. The same is thought to be true for the beginning of rice cultivation in southeast Asia, as here also the incidence of non-shattering ears increases only gradually in assemblages of different ages. Current thinking emphasizes the possible intermediate stages between hunting-gathering and agriculture, which for plants include an increased specialization in particular species, improvement of the soil on which wild strands are growing, and removal of weeds. These activities could gradually lead to tended areas of wild plants whose management might over time become more intense, leading to gradual emergence of plants displaying one or more domestication traits, and eventually resulting in emergence of a fully domesticated crop. Equivalent scenarios can be imagined for a gradual transition from hunting to herding of animals, starting with attempts to increase the numbers of a particular species by management of their natural habitat, and culminating in human control over the entire lifecycle of a herd of animals.

13.2 Biomolecular Studies of the Origins of Domesticated Animals and Plants

Biomolecular archaeology can contribute to the debate regarding the transition from hunting-gathering to agriculture by identifying the evolutionary relationships between domesticated animals and plants and their wild progenitors. The resulting information can help to establish whether a species was domesticated just once or on multiple occasions, and may also be able to indicate the geographical locations of these domestications.

Evolutionary studies involve comparisons between DNA sequences or genotypes from multiple individuals, in this case animals or plants from the relevant wild and domesticated populations. With animals, both living and archaeological specimens can be used, because the preserved bones and teeth of livestock are sufficiently widespread for ancient DNA to be obtained from multiple samples from different geographical regions. In contrast, plant remains containing ancient DNA are less common in the archaeological record (Section 7.3), so studies of crops have to rely almost entirely on living plants.

The easiest way to understand how evolutionary comparisons between wild and domesticated populations are used to study the origins of a domesticate is by examining case studies that illustrate the key features of the methodology. We will do this by considering the research that is being carried out with rice and cattle.

13.2.1 Five subpopulations of domesticated rice have been identified by typing short tandem repeats

Rice was the most important crop to be taken into cultivation in southeast Asia, today accounting for approximately 20% of the calorific intake of the human population. According to textual evidence, as early as AD 100 the Chinese recognized two types of rice, which today we call *indica* and *japonica*. These two varieties have different colored leaves and the grains are different lengths, *indica* being long-grained rice and *japonica* having shorter grains. They also have different environmental preferences, with *indica* being cultivated submerged in paddy fields in the hot areas of tropical Asia, and *japonica* grown in dryer and cooler regions of East Asia and in the highlands of the tropics. In the 1950s, a third morphological variety, originally called *javanica*, but now described as *tropical japonica*, was recognized, with medium grains and grown in the mountains of tropical southeast Asia in places such as Indonesia and the Philippines.

Before we can address any issues concerning the domestication of rice, we need to understand the evolutionary relationships between the *indica*, *japonica*, and *tropical japonica* varieties. Are these as distinct in genetic terms as they appear to be from their different morphologies, and how far back in time must we go to find their common ancestor? One way to answer these questions is by typing polymorphic DNA sequences in members of these different groups. By "polymorphic" we mean sequences that show variation within a particular species and which can be used to study relationships between different members of that species. We have already studied one area of biomolecular archaeology—kinship analysis—in which polymorphic DNA sequences are used to identify relationships between individuals (Section 11.2). Short

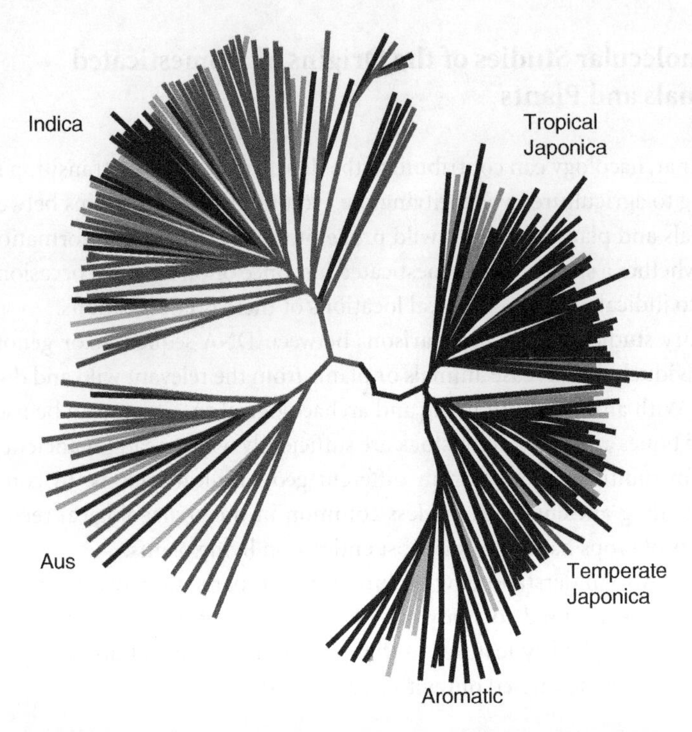

Figure 13.2 Phylogenetic tree of domesticated rice accessions constructed from STR genotype data. Image kindly provided by Susan McCouch.

tandem repeats (STRs) are the type of polymorphic DNA sequence most frequently used in kinship analysis, and these can also be used to investigate the evolutionary relationships between crop plants. The principle is exactly the same. A set of STRs is typed by PCR, and the alleles identified from the lengths of the amplicons. The alleles present in different plants are then compared to identify which plants are most closely related and which less so. With a crop, the plants that are typed are called **accessions**; and when the origins of agriculture are being studied, it is important that these are carefully chosen to represent all of the morphological variation in the crop, and that they are taken from all the geographical regions in which the crop is now grown. They can be acquired directly from indigenous farmers or, in many cases, obtained from seed collections maintained by plant breeding institutes.

To study the relationships between the different varieties of cultivated rice, 169 STRs were typed in 234 accessions, these coming from all parts of the world in which Asian rice is now grown and including multiple examples of the *indica*, *japonica*, and *tropical japonica* varieties. When a human kinship analysis is carried out, the alleles possessed by the different people who are typed are compared by eye, but this is not possible with the vast amount of information obtained when STRs are used in a broader evolutionary study such as this one. Instead the data have to be analyzed by computer. Usually, the data are converted into a phylogenetic tree whose branching pattern indicates the relationships between the various accessions. When this was done with the rice STR results, a clear distinction could be seen between *indica* rice and the temperate and tropical *japonica* groups (Figure 13.2). All of the *indica* accessions were located on one of side of the tree, and the two types of *japonica* on the other. The distinction between *indica* and *japonica* that was initially made on morphological grounds was therefore supported by the genetic analysis. The same was true for the temperate and tropical varieties of *japonica*, because although accessions of both types were clustered together, there was a clear division between the two.

Inspection of the phylogenetic tree shown in Figure 13.2 reveals additional subvarieties of rice, distinct from the three types we have already described. The clearest of these is on the *indica* side of the tree, where the accessions form two clusters, one of which is made up entirely of plants belonging to the *aus* group. This type of rice has been looked on as a specialized subvariety of *indica* that flowers early in the growing season and is able to grow in relatively dry conditions. It is found mainly in the highland regions of Bangladesh. If *aus* is indeed a subvariety of *indica* rice, then we would expect the *aus* accessions to root within the *indica* group, not alongside it as seen in the tree. The STR data therefore suggest that *aus* is more distinct from *indica* than previously thought, a discovery that, as we will see later, has important implications for the number of times that rice was domesticated.

The phylogenetic tree also provides interesting information about another subvariety of rice called *aromatic*. These include the basmati rices of India, Pakistan, and Nepal, which, as the name suggests, have a pleasant aroma. *Aromatic* rices have long grains and have therefore been looked on as another subvariety of *indica*, but the phylogenetic tree suggests that this interpretation is incorrect, and that *aromatic* is in fact more closely related to *japonica*.

13.2.2 Rice was domesticated on multiple occasions

So far the genetic analysis we have described has provided novel information on the evolutionary relationships of different varieties of cultivated rice, but has revealed nothing about the origins of the crop. Now we will take the analysis slightly further to see whether the genetic data can tell us whether rice was domesticated once or on multiple occasions.

It is difficult to infer the number of domestications simply by looking at a phylogenetic tree. It is quite possible that the five groups of cultivated rice revealed by the tree shown in Figure 13.2 evolved after the first plants were taken into cultivation, their evolution into separate populations promoted by their distribution into different geographical regions. On the other hand, the presence of these distinct groups might indicate that there were two or more domestications, each one giving rise to one or more of the different groups of rice that we see today. To answer this question we would need to assign a date to each of the splits at the base of the tree leading to the different groups. If, for example, the split between the *indica* and *japonica* sides of the tree occurred more than 10,000 years ago, then we could infer that these two major types of rice arose from separate domestications.

Unfortunately, there are no accurate methods for assigning dates directly to the splits in a phylogenetic tree derived from STR genotypes. Instead less direct approaches must be used. One possibility is to measure the amount of genetic difference between pairs such as *indica* and *japonica*, and *indica* and *aus*. There are several ways of doing this, the most robust being to calculate the statistic referred to as the **fixation index** or **F_{ST} value**, which specifically indicates the degree of difference between the allele frequencies present in two populations. An F_{ST} value of less than 0.05 indicates that two populations are not significantly

Table 13.2 F_{ST} values for rice populations.

	Aus	*Indica*	Temperate *japonica*	Tropical *japonica*
Aromatic	0.28	0.34	0.29	0.23
Aus		0.25	0.42	0.35
Indica			0.43	0.36
Temperate *japonica*				0.20

different from one another. When F_{ST} values were calculated for all the pairs of rice populations, the figures were found to range between 0.20 for the tropical and temperate versions of *japonica*, and 0.43 for *indica* and temperate *japonica* (Table 13.2). What do these figures tell us? The most important observation is that, in general, they are higher than the F_{ST} values obtained when comparisons are made between different types of cultivated maize, a plant thought to have been domesticated just once. For pairs of maize populations, most F_{ST} values are less than 0.20, and few are higher than 0.30. The implication is that the origin of at least some of the rice groups pre-dates that of the maize populations. Both species were domesticated at about the same time, so possibly this is evidence that some of the rice groups were domesticated separately, the split between the groups occurring not after domestication, but in the ancestral wild rice population.

The possibility that the *indica* and *japonica* groups of rice arose from independent domestications has also been suggested by a second genetic study in which single nucleotide polymorphisms (SNPs) were typed in 203 accessions of cultivated rice. The most convincing evidence, however, has come from a comparison of the complete genome sequences of the 9311 and Nipponbare cultivars of rice, which are representatives of the *indica* and *japonica* types, respectively. With DNA sequences it is possible to use the **molecular clock** (Section 16.2) to assign a date to the time of the most recent common ancestor of two or more genomes. When applied to the 9311 and Nipponbare genomes, the molecular clock shows that the *indica* and *japonica* groups split from one another between 200,000 and 400,000 years ago, well before rice was domesticated. It is now thought that the *aus* and *aromatic* groups also pre-date domestication, because both of these groups possess alleles that are not present in *indica* or the *japonica* groups. The presence of "private" alleles is not proof of independent domestication, but is highly suggestive. Only the temperate and tropical *japonica* groups are now thought to have diverged since domestication.

13.2.3 Genetics can also reveal where rice was domesticated

From an archaeological standpoint, the most useful information that genetics can provide about the origins of agriculture is the identities of places at which different animals and plants were domesticated. This knowledge can be combined with archaeological examination of early farming sites in those areas to gain a better understanding of the specific events that led to the adoption of agriculture in different parts of the world. How might genetic analysis enable us to locate the places where rice was domesticated?

The key to understanding where an animal or plant was domesticated is identification of the wild population that has the closest genetic relationship with the domesticated version. The assumption that this wild population is the progenitor of the domesticate is often justified, but care must be taken to distinguish, if possible, between a wild population that is the direct ancestor, in other words the one from which the original captive animals or cultivated plants were taken, and other populations that have contributed alleles to the domesticate by cross-hybridization – cross-breeding in the case of animals and cross-pollination with plants (Figure 13.3). If we really have identified the directly ancestral wild population, then its location might indicate where the animal or plant was domesticated. Again we make an assumption, in this case that the wild population has not moved since it was used as the source of the domesticates. This is clearly a possibility with wild animals, which might move into new territories, but is equally possible with plants, as vegetation patterns change in response to the environment and, importantly, to human intervention.

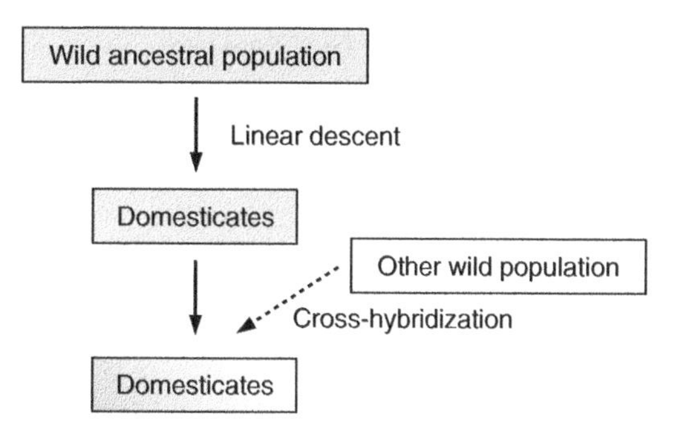

Figure 13.3 The distinction between linear descent and cross-hybridization in evolution of a domesticated population.

The only type of wild rice that grows in southwest Asia is *Oryza rubipogon*, and it has long been recognized that this species is the ancestor of cultivated rice, *Oryza sativa*. But *O. rubipogon* grows throughout Asia and simply knowing that it is the ancestral species does not help us understand where rice was domesticated. We need to make genetic comparisons between different populations of *O. rubipogon* and the various groups of cultivated rice that we think might have been domesticated separately. It would be possible to use STRs to carry out this analysis but geneticists now believe that more accurate indications of evolutionary relationships can be obtained by typing SNPs. This is mainly because of uncertainties about the way in which STRs evolve, which make it difficult to design methods for using them to infer the detailed relationships between plant accessions.

To compare wild populations of *O. rubipogon* with cultivated rice, SNPs were typed by sequencing three short segments of the rice genome in various accessions of *O. rubipogon* and *O. sativa*. We will consider the results for one of these segments, a 1300 bp sequence from the region containing a pseudogene for one of the subunits of an ATPase enzyme. This 1300 bp segment contained 66 SNPs in the 94 *O. rubipogon* and 194 *O. sativa* accessions that were typed. The combination of SNPs in a particular sequence is called a haplotype, a term that we met in Section 2.4 when we first discussed haplotypes of human mitochondrial DNA. Altogether, there were 40 haplotypes of the ATPase region, 32 of which were present in wild rice and 12 in the cultivated accessions, which means that just 4 were present in both the wild and cultivated species.

When comparing haplotypes in different populations of the same species, or in two closely related species such as *O. rubipogon* and *O. sativa*, it is more accurate to depict their relationships as a network rather than as a tree. This is because, as described in Section 2.5, a tree cannot be constructed if the dataset includes ancestral haplotypes from which others are descended, which is usually the case when closely related populations or species are examined. This is exactly what is seen with the ATPase haplotypes, the network containing several central nodes, representing

(A) Haplotype network for the ATPase gene

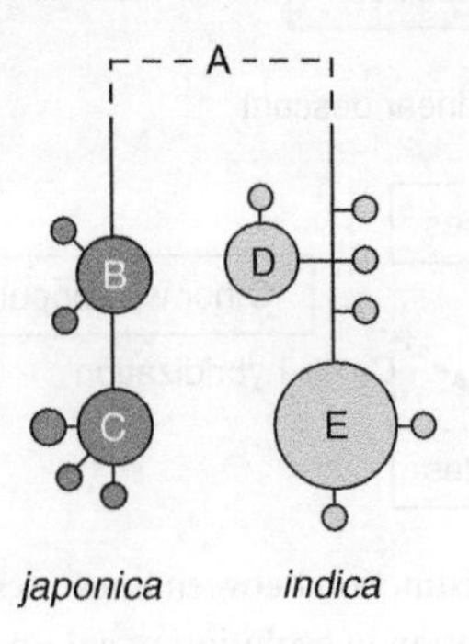

(B) Distribution of haplogroups in wild rice

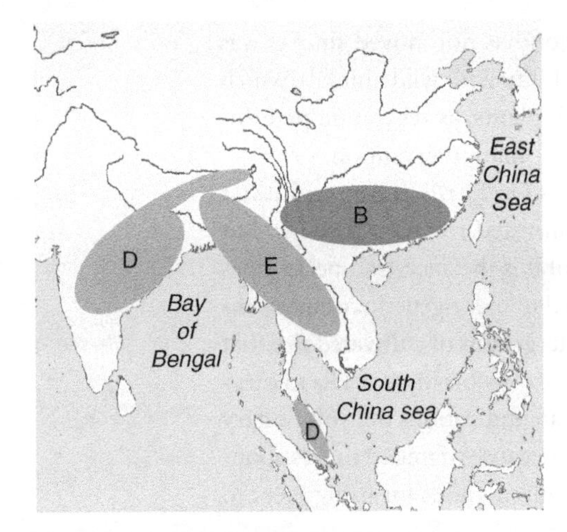

Figure 13.4 Identifying the wild origins of *indica* and *japonica* rice. (A) Simplified haplotype network for the ATPase gene. Each circle represents a haplotype, the sizes of the circles indicating the number of accessions that possess that haplotype. Haplogroups B and C include domesticated *japonica* accessions, and haplogroups D and E include *indica* varieties. Haplogroup A, found only in wild rice, has a complex structure and is not shown in the network, though its position is indicated. (B) Geographical locations of wild accessions with B, D, and E haplotypes.

ancestral haplotypes, from which families of derived haplotypes project (Figure 13.4). One of these groups of haplotypes surrounds node A and is only found in wild rice. The other four, B–E, are found in both wild and cultivated accessions. The important point, however, is that haplotypes falling into these four groups are not evenly distributed among cultivated rice. Haplogroups B and C are most frequent in the *japonica* population, and D and E are commoner in *indica*. This suggests that *japonica* rice was domesticated from a population of wild *O. rubipogon* that contained B and C haplotypes, and *indica* from an *O. rubipogon* population including D and E haplotypes.

If we now examine the geographical distributions of haplogroups B–E in wild rice, we will begin to identify the possible locations of the domestications that gave rise to the *japonica* and *indica* types of cultivated rice. Haplogroup C, as it turns out, cannot help us much in this regard, as it is present in just a single wild accession, but haplogroup B is commoner in the wild, though restricted to south China and Taiwan. The clear implication is that this region is the source of the plants that, after domestication, gave rise to the modern *japonica* types. A similar analysis of haplogroups D and E shows that these are found mainly in India and parts of Indochina, placing the domestication of *indica* rice in this area.

The genetic analysis of the ATPase region therefore suggests that *japonica* rice originates east of the Himalayas, and *indica* to the south. The two types were therefore domesticated separately, each within the region where cultivation of that type of rice subsequently developed. Examination of the haplotypes for the other two genomic segments that were sequenced gives further support to this conclusion, and additionally suggests that *aus*, originally thought to be a subtype of *indica*, was independently domesticated in India, again close to its present-day center of cultivation in Bangladesh.

13.2.4 Cattle domestication has been studied by typing mitochondrial DNA

Now we move on to the equivalent work that is being carried out on animal domestication. The principles of this research are the same as those governing the studies of rice domestication that we have just examined, but some of the details are different. The main difference is that animal studies tend not to be based on nuclear markers such as STRs. Instead, they depend largely on mitochondrial DNA typing.

We have already seen how in humans the mitochondrial genome exists as a series of haplotypes (Section 2.4). The same is true for all mammals and, as with humans, individual haplotypes can be identified by sequencing parts of the mitochondrial DNA, especially the two hypervariable regions. When this is done with domesticated cattle, the haplotypes that are discovered fall into two separate groups (Figure 13.5A). The first of these is the T "super-haplogroup," which is found in the taurine type of domesticated cattle, which do not have humps and are given the species name *Bos taurus*. Sometimes these are referred to as "European" cattle but they are found also in Asia and Africa, so the name is not particularly useful. T is designated a super-haplogroup because it comprises five distinct haplogroups, called T, T1, T2, T3, and T4. These are not equally common through Eurasia and Africa. Most genuinely European cattle have the T3 haplogroup, and most taurine breeds from Africa are T1. In Asia, both T3 and T4 are found, T3 being distributed throughout the continent and T4 confined to small localized areas of eastern Siberia and Japan. All of the haplogroups except T4 are common in cattle from the Near East, suggesting that this is where *Bos taurus* was domesticated. This is in agreement with the archaeological evidence, the earliest domesticated cattle bones being found at sites in the Indus valley. The different frequencies of haplogroups that we now see in Europe, Africa, and Asia are assumed to reflect the frequencies in the animals first transported out of the Near East towards these new areas as agriculture spread, with T4 evolving from T3 at some period after taurine cattle had reached eastern Asia.

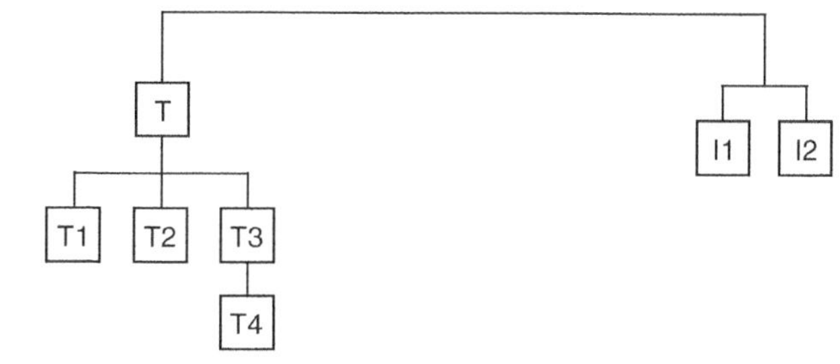

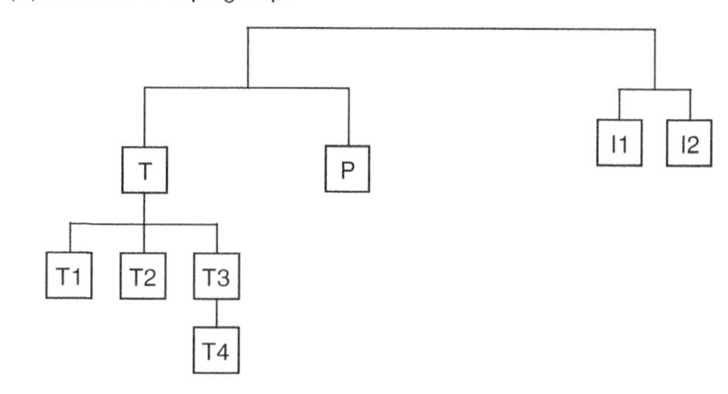

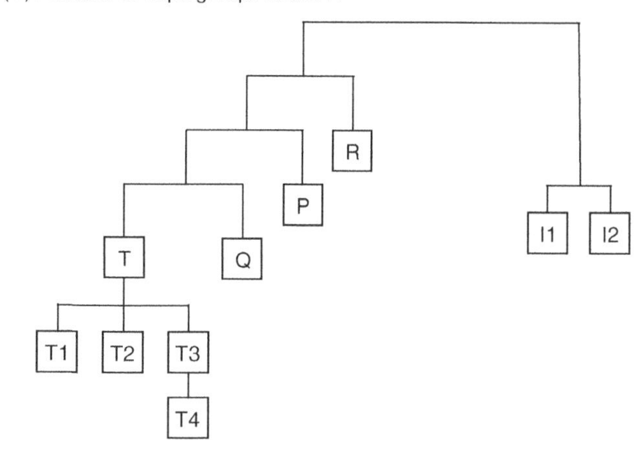

Figure 13.5 Relationships between the haplogroups of domestic cattle and wild aurochsen.

The second mitochondrial super-haplogroup found in domesticated cattle is I, which is less variable with just two haplogroups recognized, I1 and I2. These are found in cattle that have fatty humps around their shoulders, and which are called zebu or *indicus* cattle, and placed in the species *Bos indicus*. In reality, *B. taurus* and *B. indicus* are not separate species, but subspecies, and they are more correctly called *Bos primigenius taurus* and *B. primigenius indicus*. Zebu are found in south Asia and parts of Africa, but it is known from historical records that the breeds in Africa descend from cattle that were transported there from Asia, zebu being well adapted to warm environments.

The T and I super-haplogroups have a relatively high degree of genetic difference, as indicated by the lengthy branch that links the T and I parts of the tree shown in Figure 13.5A. With mitochondrial DNA sequences it is possible to use a molecular clock, based on the rate at which polymorphisms accumulate in mitochondrial DNA, to assign a date to the most recent common ancestor of a set of haplotypes. This analysis suggests that the taurine and zebu groups diverged from one another some 200,000 years ago, much earlier than the period when domestication occurred. The mitochondrial DNA data therefore give a clear indication that there were two separate domestications of cattle, one giving rise to the taurine group and one to zebu. The situation is very similar to that discovered by typing STRs in the *indica* and *japonica* types of rice.

13.2.5 Genetic analysis can reveal details of the relationship between domesticated cattle and aurochsen

The wild progenitor of domesticated cattle is the aurochs (plural, aurochsen), *Bos primigenius primigenius*. Aurochsen were at one time common across Eurasia and north Africa and are frequently depicted in prehistoric cave paintings, but their numbers gradually declined, partly through hunting, until the last one died in 1627 in Poland. If we are to use genetic analysis to understand the evolutionary relationships between domesticated cattle and the aurochs, then we will need to sequence ancient DNA from aurochs bones recovered from archaeological sites and elsewhere.

Aurochs bones are not common in the archaeological record; and when found, there is often uncertainty if they are genuinely the remains of wild cattle rather than those of early domesticates. The first sequences were reported in 1996, from two aurochsen from Britain, both of which gave the same mitochondrial DNA haplotype. This sequence, called P, fell on the branch between the T and I super-haplogroups, closer to T rather than I (Figure 13.5B). A more recent survey of aurochs bones from 34 European sites confirmed this initial finding, all but one of these sites yielding P haplotypes, the exception being a single aurochs from Germany whose haplotype was significantly different, though still intermediate between the modern taurine and zebu groups. This unusual haplotype was called E. As we have already established that taurine cattle were domesticated in the Near East and not in Europe, the absence of T haplotypes among European aurochsen is not unexpected. We would, however, predict that at least some aurochsen from the Near East were members of the T haplogroup. Proving this has been difficult due to a paucity of material, but two specimens support the hypothesis. The first is a Bronze Age aurochs from Iran, whose haplogroup was T. Iran is some distance from the supposed center of cattle domestication in the Indus valley, so this result does not necessarily mean that aurochsen with a T haplotype were also present in the Near East, but it is suggestive. The second specimen also gave a T haplotype, and has a better location, coming from the site of Dja'de on the Euphrates River. Unfortunately there is some doubt if this specimen actually is a wild aurochs rather than an early domesticate.

Despite the absence of definite proof that the aurochsen living in the Near East at the time of domestication included members of the T super-haplogroup, there appears to be concordance between the aurochs sequences that have been obtained and the hypothesis that taurine cattle were domesticated in the Indus valley, rather than elsewhere, within Europe for example. But more extensive studies of modern taurine cattle are beginning to cloud the issue. Recently a small number of European cattle have been found to have haplotypes that do not fall within the T super-haplogroup. Some of these make up a new haplogroup, called Q, which is closely related to the T super-group (Figure 13.5C) and probably entered the domesticated population at the same time. In other words, the aurochs that were first taken into captivity in the Near East, although mainly comprising animals belonging to the T super-haplogroup, also included a few with Q mitochondrial DNAs.

The remaining unusual European cattle are more enigmatic. Their haplotypes form another new group, called R, which is located much closer to the P haplogroup of European aurochsen (Figure 13.5C). The evolutionary relationship between P and R is so close that the most likely explanation is that the cattle with the R haplogroup are descended from aurochsen that were taken into captivity in Europe, probably somewhere in Italy, as this is where the R haplogroup is now located. As mitochondrial DNA is inherited only through the female line, these aurochsen must have been female, and presumably were mated with domesticated bulls. There have also been suggestions that the complementary situation also occurred, male aurochsen being mated with domesticated female cattle in Europe. The evidence for this comes for studies of STRs on the Y chromosome, a particular Y chromosome haplotype that is common in northern European cattle also being common in aurochsen bones from the same region. This observation has not, however, been supported by more detailed studies of Y STRs carried out more recently. The possibility that cross-breeding occurred between aurochsen and domesticated cattle has important implications for our understanding of how cattle were managed by early farmers, but at present the extent to which this happened, if at all, remains unclear.

13.3 Biomolecular Studies of the Spread of Agriculture

Once agriculture became established at its points of origin, the technology began to spread to neighboring regions, and in some cases across entire continents. The arrival of agriculture in new parts of the world can be detected in the archaeological record by the first appearance at particular sites of artifacts such as ploughing implements and grindstones, which have a specific association with agricultural activity, and of the bones and preserved remains of domesticated animals and plants. Outside of the centers of origin the wild version of the domesticate is often absent, and so distinguishing the two is not an issue. The key to charting the progress of agriculture across a continent is therefore not so much recognizing its presence as obtaining an accurate date for its arrival at a particular site.

Before development of the accelerator mass spectrometry (AMS) version of radiocarbon dating, relatively large amounts of carbon-containing material were needed in order to obtain a date, possibly requiring that material from different samples be pooled, giving rise to large standard deviations in the resulting data. Often the material would be charcoal from hearths thought to lie in the same stratigraphic layer as the artifacts or biological indicators of agriculture. This indirect method is prone to errors arising from incorrect interpretation of the stratigraphy, so greater reliance is now placed on dates obtained directly from animal bones or charred seeds by the AMS method. These have

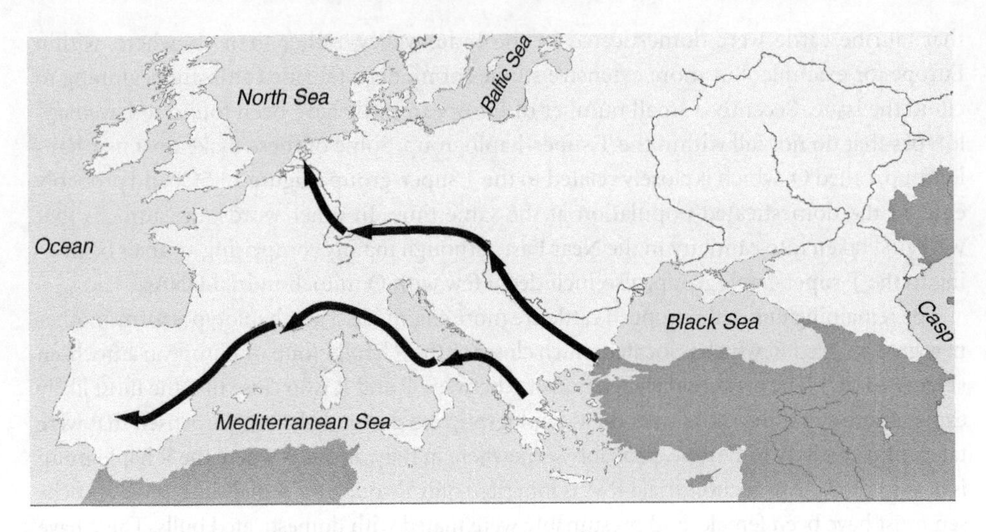

Figure 13.6 Trajectories of the spread of agriculture through Europe.

shown that the first movement of agriculture towards Europe took place around 10,700 BP, very soon after it first became established in southeast Turkey and the Levant. Initially, agriculture spread to southeast Anatolia and Cyprus by sea voyages as well as overland routes. There then appears to have been a relatively rapid movement across the eastern Mediterranean from Cyprus to Crete and mainland Greece by 9,000 BP, and a rather slower expansion into central Anatolia by about the same time. Agriculture then spread through the Danube and Rhine valleys to central Europe and the northern European plain, and along the Mediterranean coast to Italy, Iberia, and thence to northwest Europe (Figure 13.6). These trajectories are associated with two distinct early Neolithic material cultures, the Linearbandkeramik (LBK) and Cardial, respectively.

Biomolecular archaeology can contribute to various aspects of the study of agricultural spread, for example by helping to map trajectories and by giving detailed information about the factors underlying adoption of agriculture at particular places. The latter might include, for example, the genetic adaptation of a crop plant to a colder climate. It can also help us to understand *how* agriculture spread, an issue that is less tractable to other forms of archaeological research. We will therefore begin with the studies that have been made of how agriculture was brought into Europe.

13.3.1 The spread of farming into Europe can be studied by human genetics

How does farming spread through a continent? Does it involve the large-scale migration of farmers from one place to another, taking with them their implements, animals, and crops, and displacing the indigenous pre-agricultural communities that they meet along the way? Or can farming spread without the large-scale movement of people, but instead by the idea of agriculture and the basics of the technology – a few animals, a few plants – being passed from one group to another?

Human genetics has been used to address these questions, especially in Europe. In Section 16.3, we will study in more detail the methods by which the distribution of alleles in modern populations can be used to trace the prehistoric migrations of humans. One of the first attempts at using human genetics in this way was carried out in the 1970s with populations from across Europe. Genetic data were obtained mainly by typing blood groups and other protein polymorphisms, rather than by DNA analysis, which was much less easy to carry out before the invention of PCR. The allele distributions for 95 genes were examined by principal components analysis (PCA), a statistical method that identifies patterns in large and complex datasets by splitting the overall amount of variation into separate, non-overlapping components. The "first principal component" contains the largest amount of the total variation, the "second principal component" is the next largest, and so on. When applied to the European dataset, PCA revealed a first principal component that accounted for 27% of the total variation in the allele distributions. The variation represented by this principal component took the form of a gradation of allele frequencies across Europe, one edge of this gradation in the northwest of the continent, and the other in the southeast. The pattern implies that a migration of people occurred either from southeast to northwest Europe, or in the opposite direction. The former is exactly what we would expect if agriculture had been spread across Europe by a **wave of advance** of farmers moving from their homeland in the Fertile Crescent.

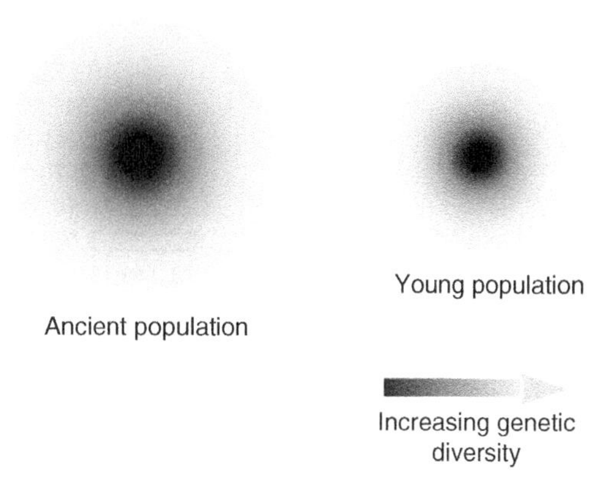

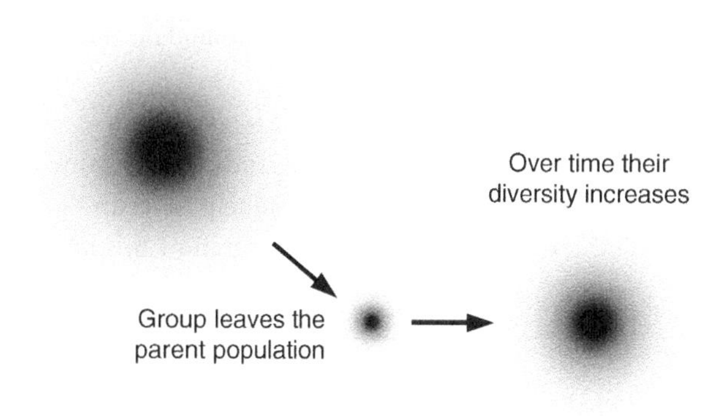

Figure 13.7 The basis to (A) coalescence analysis, and (B) founder analysis.

The analysis looks convincing but has one major limitation. The data provide no indication of when the inferred migration took place, so the link between the first principal component and the spread of agriculture is based only on the allele distribution pattern, not on any complementary evidence relating to the period when this pattern was set up. In contrast, when mitochondrial DNA, rather than nuclear allele frequencies, is studied, it is possible to assign a date to some past events, at least approximate ones. This can be done by using the overall degree of diversity among a set of haplotypes to estimate the **coalescence time** for their haplogroup, this being an estimate of the time that has elapsed since the haplogroup first came into existence (Figure 13.7). The reasoning is that the greater the diversity among the haplotypes, the larger the number of nucleotide substitutions that have occurred, and the more ancient the coalescence time. An extension of this concept is **founder analysis**, which uses haplogroup diversities to estimate the time of a population split, such as occurs when a group of people leave their parent population and migrate to another location. The reasoning is that the haplogroup diversity in the descendents of the new population set up

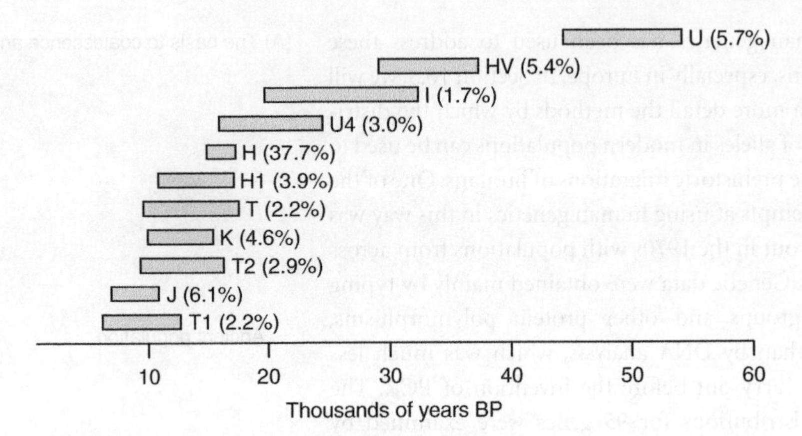

Figure 13.8 Graph showing the results of founder analysis of the 11 main European mitochondrial haplogroups. The bars indicate the deduced periods of origin for each of the haplogroups, with the frequencies of the haplogroups among modern Europeans given in brackets. Only two haplogroups, J and T1, originate around the time when agriculture entered Europe.

by the migrants indicates the time when the migration took place. This is exactly the type of analysis that will help us to understand if farming spread by a human wave of advance.

In the first research of this kind, mitochondrial DNA haplogroups from various populations across Europe were examined. All of these haplogroups are also present in western Asia, so founder analysis was used to work out approximately when the European versions split away from the parent groups. The results failed to confirm the gradation of allele frequencies detected in the nuclear DNA dataset, and instead suggested that European populations have remained relatively static over the last 20,000 years. A subsequent study of almost 4250 individuals led to the discovery that 11 mitochondrial DNA haplogroups predominate in the modern European population. Founder analyses suggest that these haplogroups entered Europe at different times (Figure 13.8). The most ancient haplogroup, called U, first appeared in Europe approximately 50,000 years ago, coinciding with the period when, according to the archaeological record, the first modern humans moved into the continent as the ice sheets withdrew to the north at the end of the last major glaciation. The youngest haplogroups, J and T1, which at 9000 years in age could correspond to the origins of agriculture, are possessed by just 8.3% of the modern European population, suggesting that the spread of farming into Europe was not the huge wave of advance indicated by the principal component study. Instead the data indicate that farming was brought into Europe by a smaller group of "pioneers" who interbred with the existing prefarming communities rather than displacing them.

Equivalent research has been carried out with Y chromosome sequences which, being inherited through the male line, provide a complement to mitochondrial studies, which only reveal the evolution of female lineages. This work has generated a more complicated picture of the spread of farming into Europe. One recent study shows that the commonest Y chromosome haplogroup in Europeans, called R1b1b2, has a Neolithic origin, suggesting that the spread of agriculture involved a large human expansion from the Near East. Studies of other Y haplogroups are more consistent with a small-scale human movement, with agriculture spreading largely by cultural diffusion – the transfer of ideas rather than people. Y chromosome data have to be interpreted carefully because polygamy can lead to a relatively small number of males having a disproportionate effect on Y haplotype frequencies in modern populations, but it is becoming clear that, as many archaeologists have thought for some time, the "domestication of Europe" was not a simple process. So, for example, cultural diffusion might have been responsible for the first expansion of farming into the Balkans, with migration of people responsible for the subsequent, rapid movement of agriculture into the central European plain. Strontium isotope studies additionally suggest that in some parts of Europe hunter-gatherer women moved into farming communities. The one point on which there is agreement is that a more complete picture of the human dimension to agricultural spread will become available when the analyses move away from mitochondrial and Y chromosome haplogroups and take account of polymorphisms across the entire genome. New methods for rapid typing of single nucleotide polymorphisms are gradually being applied to human evolutionary studies, and although the data analysis is complex the work is likely to be fruitful.

13.3.2 Stable isotope analysis suggests that agriculture spread rapidly through Britain

So far all of our discussion of the application of biomolecular archaeology in the study of prehistoric agriculture has been concerned with genetic investigations. DNA is certainly very important in this area of research, but it is not the only type of biomolecular evidence that can be used in agricultural studies. Stable isotope examination of preserved skeletons can also provide important information on shifts in diet that signal a change in subsistence from hunting-gathering to agriculture. In Section 6.2, we learnt how measurement of the $\delta^{13}C$ values in Native American skeletons has been used to date the arrival of maize cultivation in the United States. A similar investigation has suggested that the transition to agriculture occurred with remarkable rapidity in Britain.

All of the domesticated plants introduced into Europe during the Neolithic follow the C_3 photosynthetic pathway, the same as all the indigenous vegetation. A change in subsistence from one dependent on wild plants to one using domesticates cannot therefore be detected by stable isotope analysis of prehistoric European skeletons. The replacement of a diet rich in marine resources to one with a high input from domesticated animals can, however, be distinguished. This is because the high $\delta^{13}C$ values resulting from photosynthesis in the ocean are passed down the food chain to humans who eat fish and other marine produce. When bulk collagen is studied, the typical $\delta^{13}C$ value indicating a diet with high marine input is $-12 \pm 1‰$, whereas that for a typical terrestrial diet, such as provided by farmed livestock, is $-20 \pm 1‰$.

Stable isotope measurements were made for 19 skeletons from British sites dating to the late Mesolithic period, from 9000 to 5200 BP, and these values compared with those from 164 skeletons from the early Neolithic, from 5200 to 4500 BP. The availability of a greater number of Neolithic skeletons reflects the different funerary practices adopted at the onset of that period in Britain, with people interred for the first time in collective burials in visible monuments such as chambered tombs and causewayed enclosures. The distinctiveness of this new burial custom is one of the factors that have led archaeologists to believe that Britain was domesticated by a migration of farmers from continental Europe, which might imply a rapid replacement of the indigenous mode of subsistence.

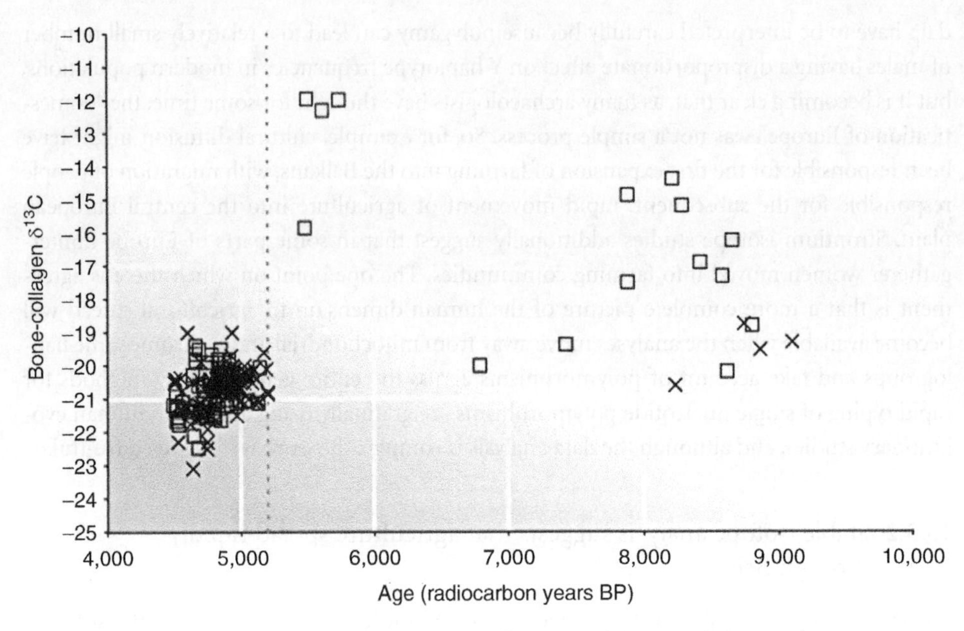

Figure 13.9 Sudden shift in $\delta^{13}C$ values for collagen from British human skeletons after 5200 BP, coinciding with the onset of the Neolithic. Squares are individuals from within 10 km of the coast, and crosses are individuals from further inland. Reprinted from Michael Richards, Rick Schulting, and Robert Hedges, "Sharp shift in diet at onset of Neolithic," *Nature*, 425, 366, copyright 2003, with permission from Macmillan Publishers and the authors.

As expected, most of the Mesolithic skeletons had $\delta^{13}C$ values consistent with a diet with a high marine content, whereas the Neolithic skeletons gave values more typical of a terrestrial diet. The striking feature of the results was the apparent rapidity of the shift from one set of values to the other (Figure 13.9). None of the skeletons from after 5200 BP, even those from very shortly after this date, gives any indication of a marine input. It would appear that marine resources were dropped from the diet very suddenly and reliance placed entirely on terrestrial protein, with no intermediate, such as a part marine, part terrestrial diet, visible in the stable isotope record. These results are controversial and not accepted by all archaeologists, but similar results have been obtained from other areas in the north and west of Europe, as far apart as Portugal and Denmark. In southern Sweden, however, the situation appears to be more complex, with a mixture of marine and terrestrial diets, as well as some intermediates, present either side of the Mesolithic–Neolithic boundary, and subsistence determined more by geography and culture rather than pre- and post-agricultural economies.

13.3.3 Ancient DNA can help resolve the trajectories for maize cultivation in South America

Finally, we will consider an example where ancient DNA has helped to reveal the trajectories taken by agriculture as it spread away from a center of origin. Maize (*Zea mays* ssp. *mays*) is the cultivated form of teosinte (*Zea mays* ssp. *parviglumis*), a grass endemic to southern and western Mexico. Genetic evidence suggests that there was a single domestication, about 9000 years ago, in the Rio Balsas drainage area of western Mexico. By the time that Europeans arrived, maize was being cultivated throughout most of the Americas and had become a central, in some cases indispensable, component of many of the sophisticated cultures occupying the two continents.

The earliest evidence of maize in South America comes from microfossils such as phytoliths and starch granules, which are present in deposits in Ecuador dated to 7150 BP and Peru at 4000 BP. Preserved cobs and other maize macrofossils do not appear until later: according to directly dated specimens, not until after 3500 BP in coastal Ecuador and not until 2000–2250 BP in Argentina and Chile. In most regions, the dates of the macrofossils correspond with the earliest human remains whose stable isotope ratios indicate a diet rich in maize. These discrepancies between the microfossil and macrofossil evidence has led to the interesting suggestion that initially maize was grown as a source of fermentable juice from its stalks (consistent with the early dates for microremains) and might not have developed well-defined cobs and become a nutritional source until later. This is inconsistent, however, with the presence at Wayuna, Peru, of maize starch grains showing damage indicative of processing for food use, dated to 4000 BP.

Unravelling the complexities highlighted by the dating evidence requires a clear understanding of the trajectories by which maize spread into and through South America. This problem was first addressed by the famous geneticist Barbara McClintock in the 1970s as part of a monumental study of the relationships between maize landraces, worked out by comparing the positions of "knobs" on the maize chromosomes. These are segments of tightly wound DNA, which form bulges and projections visible when the chromosomes are stained in dividing cells. Their numbers and precise locations vary in a manner that appears to be inherited, which means that they can be used as a type of genetic marker. One of the many conclusions that McClintock drew from these cytogenetic studies was that maize from central America is more closely related to that in the South American Andes than it is to lowland South American varieties. Her hypothesis was, therefore, that the cultivation of maize spread first to the Andes and from this source eventually to the lowland regions of the continent.

McClintock's model was thought to be correct until a STR study of maize landraces from all over the Americas was carried out. In this project, 99 STRs were typed in 193 maize accessions from all over the Americas and the data used to construct a phylogenetic tree. All of the central and South American landraces were located in one part of the tree, the central American ones closest to the root and the Andean landraces furthermost away, with lowland South American ones in between. This topology implies the reverse of McClintock's model, with maize first being introduced into the lowlands on the east side of South America and reaching the Andes later.

Why do the cytogenetic and STR studies give such different outcomes? One possibility is that the geographical distribution of the landraces used in one or both projects has been altered by the extensive moving around of cultivated maize that has occurred since Europeans arrived in the Americas. This movement might have obscured the original geographic signal and added a degree of insecurity to the conclusions that were reached regarding the earlier prehistoric events. To achieve accuracy we need to strip off the post-Columbian "noise" and get closer to

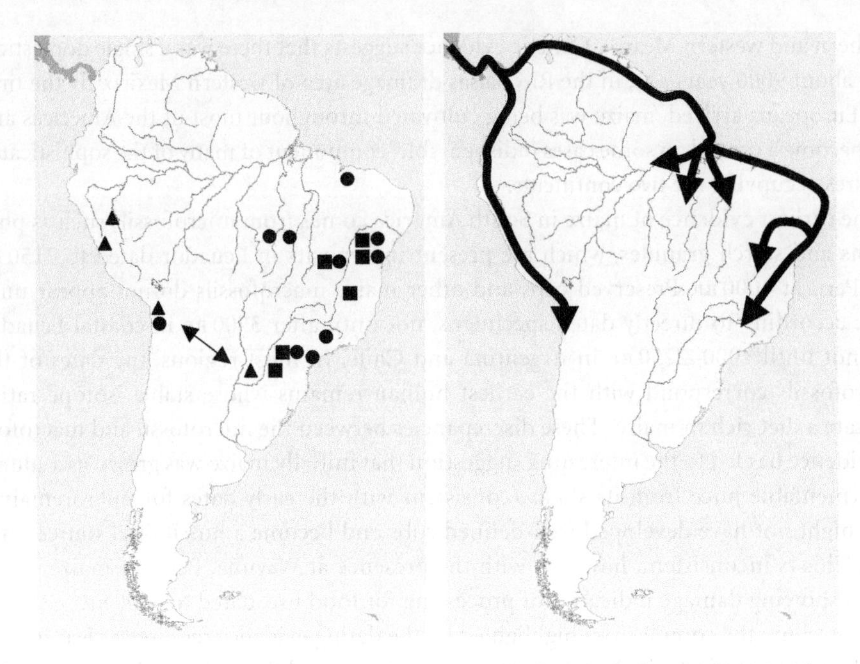

Figure 13.10 STR genotypes of indigenous landraces and archaeological maize from South America. On the left, the distribution of the three forms of the *Adh2* STR is shown. These three versions have the sequences GA_n (shown as triangles), GA_nTA (circles), and $GA_1AA_1GA_n$ (squares). The simple GA_n type predominates in the Andean regions and the more complex types in the eastern parts. On the right, the deduced trajectories by which maize cultivation spread into South America are shown. The double-headed arrow on the map on the left indicates the region between Chile and Paraguay where cultural exchange is known to have occurred across the Andes.

the original geographical signal resulting from the original spread of maize cultivation. There are two ways of doing this. The first is to use primitive landraces obtained from indigenous farming communities who have been largely unaffected by post-Columbian agricultural developments. The second approach is to type ancient DNA in archaeological specimens that predate the arrival of Europeans. Maize cobs, with or without attached kernels, are preserved in the South American archaeological record as both charred and desiccated specimens, and the latter are particularly amenable for ancient DNA analysis (Section 7.3)

Both indigenous landraces of maize and archaeological specimens from Brazil have been used to assess the accuracy of the two models for the spread of maize cultivation into South America. This was achieved by typing part of the *Adh2* gene, which as well as various SNPs contains an STR sequence that exists in three forms. Landraces and archaeological specimens possessing each of the three versions of the STR were found to have distinct distributions within South America. Those with the simplest form of the STR were found mainly in the Andes, whereas the two more complex types were commoner in the lowlands to the east of the continent (Figure 13.10). These distributions support a model in which there were two independent trajectories for the spread of maize cultivation into the continent. One of these was through the Andean regions on the western side of South America, and involved plants with the simpler type of STR. The second was along the lowlands of the northeast coast, probably entering the continent through the river systems, and comprising plants with the more complex STRs. The results therefore suggest that although the two previous models were inaccurate, elements of both were correct.

There are two areas in which the STR distribution expected by this new model breaks down – in northern Chile, where there is an example of a complex STR, and in Paraguay where there is a simple one. In fact, the area of the Andes that separates northern Chile from Paraguay is one of the few points where the mountains can be crossed on foot, and is a point of cultural exchange between east and west. It is therefore entirely possible that there was some mixing of the maize genotypes in these regions.

13.4 Biomolecular Studies of the Development of Agriculture

One of the great advantages of working with archaeological material is the opportunity to examine specimens from different periods, and hence to follow developments over time. The use of stable isotope analysis to follow changes in diet accompanying the arrival of agriculture in northwest Europe is one excellent example of the way in which biomolecular studies of a time series of specimens can be informative. The examination of potsherds from sites of different ages for traces of milk fats in order to follow the spread of dairying (Section 12.3) is a second example. There are many others in which the time series is provided by skeletons or artifacts from sites dating to different periods, or from different layers within a single site. To round off our study of agriculture we will look at two projects of this kind.

13.4.1 Ancient DNA has been used to follow the evolution of domesticated maize

For our first example we stay in the Americas to consider a second investigation making use of ancient DNA in preserved maize cobs. One of the remarkable features of cultivated maize is the huge amount of morphological change that the plant has displayed since its initial domestication from wild teosinte. Maize plants have fewer branches than teosinte, the side branches are shorter, and the flowers at the ends are female rather than male, which means that it is at the ends of these lateral branches that the cobs develop. In teosinte, the cobs are relatively small, with just two rows containing a total of 5–12 kernels. The maize cob, on the other hand, has up to 500 kernels arranged in multiple rows. These changes are in addition to the special phenotypes that we associate with the domesticated versions of other plants, such as increased kernel size and a retention of the kernels on the ripe cob (see Table 13.1). The morphological changes have been accompanied by equally important changes to the plant's biochemistry, including increased synthesis of protein in the kernels, and alterations to the amount of branching in the amylopectin molecules present in the starch grains, giving better baking properties.

The alterations in the visible morphology of maize cobs can be followed through time by examining preserved cobs from sites of different ages, but other parts of the plant are rarely

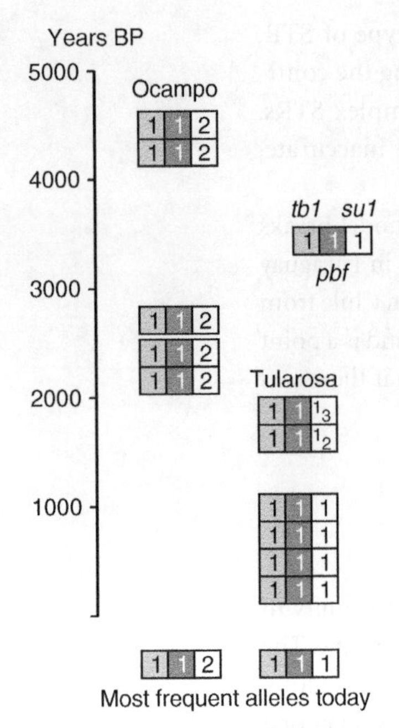

Figure 13.11 Alleles of *tb1*, *pbf*, and *su1* found in maize cobs from Ocampo Cave, Mexico, and Tularosa Cave, New Mexico. The most frequent alleles in modern maize from the local regions are also shown. At Tularosa, the two oldest specimens were heterozygous at *su1*. The allele designated "3," present in the oldest specimen, is unknown in modern maize but present as a rare allele in teosinte.

preserved and morphological study cannot, of course, identify biochemical changes that are occurring with the kernels. We need a more subtle approach if we are to follow in detail the evolution of the traits that distinguish maize from teosinte.

Increased understanding of the genetic basis to some of these traits makes a more detailed study possible. For example, the difference in the branching pattern of the maize plant compared with teosinte is due, at least in part, to the effects of the *teosinte branched 1* (*tb1*) gene, which codes for a transcription factor, a protein that controls the expression of other genes that have the direct effect on the plant's architecture. In modern maize, a single allele of *tb1*, called Tb1-M1, has become **fixed**, meaning that it is possessed by all members of the population. In contrast, Tb1-M1 has a frequency of only 36% in teosinte. We can therefore infer that fixation of this allele was the underlying genetic cause of the change in the branching pattern of cultivated maize. Similarly, a gene called *pbf* codes for a second transcription factor called the prolamin box binding factor, which controls expression of a number of seed protein genes. Only two alleles of *pbf* are found in maize, Pbf-M1 and Pbf-M2, with the former the much commoner with a frequency of 97%. In teosinte, the frequencies of these alleles are 17% and 83%, respectively, the reverse of the situation in maize. Finally, there is the *sugary 1* or *su1* gene, whose product is a starch debranching enzyme (Section 8.5). The Su1-M1 and Su1-M2 alleles have frequencies of 30% and 62% in maize but are both relatively rare in teosinte. The overall picture that emerges is one in which certain alleles have become predominant in cultivated maize, these presumably being the ones that specify the particular traits that we associate with the domesticated plant.

What was the pace of these changes in allele frequencies during and since the domestication of maize? Part of the answer has been provided by studies of ancient DNA in preserved cobs from two archaeological sites, the Ocampo Caves in northeast Mexico, which span the period 4300–2300 BP, and the more recent Tularosa Cave in New Mexico where the remains date to between 1900 and 650 BP. PCRs were directed at those parts of the *tb1*, *pbf*, and *su1* genes containing the diagnostic SNPs that distinguish the different alleles. The results were striking (Figure 13.11). The Tb-M1 and Pbf-M1 alleles, the most common ones for their genes in modern maize, were the only alleles found in any of the archaeological specimens, suggesting that these two had reached frequencies close to those in modern maize by as early as 4405 BP, the age of the oldest specimen. Only with *su1* was the situation at all complex. With this gene, the major allele present in the specimens from the two sites was different, Su1-M1 for the cobs from Tularosa, and Su1-M2 for those from Ocampo. At first this result might not seem to fit in with the allele frequencies in modern maize (30% for Su1-M1 and 62% for Su1-M2), but closer inspection of the geographical distribution of these alleles reveals a correlation. Su1-M1 is much more common in maize from the southwest United States, with a frequency of 72%, whereas in Mexico the Su1-M2 allele is the more common, with a frequency of 59%, increasing to 86% in South America. The ancient DNA results therefore suggest that the geographical partitioning of the *Su1* alleles that we see today had been established by 2000 BP. The research as a whole provides an excellent example of the power of biomolecular archaeology to provide entirely novel information about the development of prehistoric agriculture.

13.4.2 DNA from sediment can chart changes in land usage over time

At a single archaeological site, the traditional way of obtaining a time series of material is to examine different sedimentary layers, these layers built up gradually over time with the oldest at the bottom and the youngest at the top. In conventional archaeology, it is necessary to find artifacts or biological remains in these layers if changes that occurred over time are to be studied. In a remarkable type of biomolecular archaeology, this requirement can be circumvented by obtaining DNA sequences directly from the sediment.

So-called "dirt" DNA was first studied in permafrost cores and in sediments taken from caves. After PCR it has been possible to identify sequences of animals, plants, and microorganisms, enabling information to be obtained on the local ecosystems at different past periods. With Greenland ice cores, DNA from layers dating to 450,000–850,000 years ago have contained DNA from coniferous trees, showing that during that relatively warm period in Earth's history Greenland actually was green, being covered in forest. The exact source of the plant and animal DNA in these deposits is still not understood, as no obvious remains can be seen when samples are examined with the microscope. With plants, it is thought that cell layers from the external surfaces of the roots might become stripped off during plant growth, the DNA from these remaining in the sediments. For animals, faeces and urine are the most likely sources, which raises the question of whether the DNA moves down through the sediment during deposition, which would confuse the stratigraphy, making it appear that animals were present much earlier than actually was the case. In New Zealand, sedimentary layers that pre-date the arrival of Europeans contain sheep DNA, showing that this downward migration does occur, at least in some contexts.

Sedimentary DNA analysis has been used with deposits from a site in southwest Greenland called "The Farm Beneath the Sand." Voyagers from Scandinavia arrived here around AD 1000 and remained for about 400 years, attempting to maintain an existence in an area that today is mostly a cold, sandy desert, and even 600 years ago was probably nothing more than meagre grassland. To see if DNA could be obtained from the sediments at this site, cores were taken down to a depth of 260 mm, sufficient to cover the entire anthropogenic layer – that part laid down while the farm was occupied. The cores were split into seven sections and each one dated by the AMS radiocarbon technique. This showed that a depth of 260 mm corresponded to 903 ± 36 years ago and that the occupation period spanned five of the core sections.

Next the DNA content of each section was examined by carrying out PCRs that would amplify part of the 16S rRNA gene of the mitochondrial genome of most vertebrates. In this type of PCR, the primers anneal to short sequences that are the same in many different species, but amplify a region that contains sufficient nucleotide variation for individual ones, or in some cases closely related groups of species, to be identified. When the PCR products were examined, it was found that they contained sequences from sheep, goat, and cattle, as well as some human ones and a few from reindeer and moose. These were not distributed evenly through the core, and instead suggested the presence of different groups of animals at different times. In the early years of occupation, sheep were the most abundant of the domesticated animals, but as time went by first cattle and then goats became predominant. After AD 1450 there is a sharp shift to sections that contain only human DNA, thought by the researchers to derive from modern contamination. The important point is that the sections corresponding to the time after the site was abandoned do not show any evidence of livestock DNA.

What about the possibility that DNA is moving down through the layers as urine from the livestock soaks into the ground? A downward movement seems unlikely, as this would produce gradients of DNA, gradually decreasing at lower depths, which would give the impression that each animal gradually increased in numbers as time went by. Instead, the sheep DNA is present at approximately the same concentration all through the occupation levels of the core, and the cattle DNA is present in greater quantities at the lower levels, the opposite to the expectation if the cattle DNA is migrating downwards. It therefore seems that the results are authentic and the analysis is giving a true record of the changes in livestock present at The Farm Beneath the Sand during the period when it was settled.

In this project, the ancient DNA results do not add greatly to our overall knowledge of the site, as the fossil remains of animal bones have already enabled the changes in livestock abundance to be recognized. But the work is important as a "proof of concept," showing that sedimentary DNA analysis can contribute to biomolecular archaeology in the same way that it is contributing to paleoecology. Future research will reveal the extent to which this approach can be applied more generally in archaeological settings, possibly in combination with analysis of lipids in the same sediments.

Further Reading

Achilli, A., Bonfiglio, S. & Olivieri, A., *et al.* (2009) The multifaceted origin of taurine cattle reflected by the mitochondrial genome. *PLoS ONE*, 4(8), e5753.

Balaresque, P., Bowden, G.R. & Adams, S.M., *et al.* (2010) A predominantly Neolithic origin for European paternal lineages. *PLoS Biology*, 8(1), e1000285.

Balter, M. (2007) Seeking agriculture's ancient roots. *Science*, 316, 1830–5. [Review of recent archaeological research into the origins of agriculture.]

Battaglia, V., Fornarino, S. & Al-Zahery, N., *et al.* (2008) Y-chromosomal evidence of the cultural diffusion of agriculture in southeast Europe. *European Journal of Human Genetics*, 17, 820–30.

Bentley, R.A., Price, T.D. & Luning, J., *et al.* (2002) Prehistoric migration in Europe: strontium isotope analysis of early Neolithic skeletons. *Current Anthropology*, 43, 799–804. [Evidence that in some parts of Europe hunter-gatherer women moved into farming communities.]

Brown, T.A. (1999) How ancient DNA may help in understanding the origin and spread of agriculture. *Philosophical Transactions of the Royal Society series B*, 354, 89–98.

Brown, T.A., Jones, M.K., Powell, W. & Allaby, R.G. (2009) The complex origins of domesticated crops in the Fertile Crescent. *Trends in Ecology and Evolution*, 24, 103–19.

Colledge, S., Comolly, J. & Shennan, S. (2004) Archaeobotanical evidence for the spread of farming in the Eastern Mediterranean. *Current Anthropology*, 45, S35–S58. [Describes the dating evidence for the establishment and early expansion of farming in the Fertile Crescent.]

Edwards, C.J., Bollongino, R. & Scheu, A., *et al.* (2007) Mitochondrial DNA analysis shows a Near Eastern Neolithic origin for domestic cattle and no indication of domestication of European aurochs. *Proceedings of the Royal Society of London Series B*, 274, 1377–85.

Freitas, F.O., Bandel, G., Allaby, R.G. & Brown, T.A. (2003) DNA from primitive maize landraces and archaeological remains: implications for the domestication of maize and its expansion into South America. *Journal of Archaeological Science*, 30, 901–8.

Garris, A.J., Tai, T.H., Coburn, J., Kresovich, S. & McCouch, S. (2005) Genetic structure and diversity in *Oryza sativa* L. *Genetics*, 169, 1631–8.

Hebsgaard, M.B., Gilbert, M.T.P. & Arneborg, J., *et al.* (2009) "The Farm Beneath the Sand" – an archaeological case study on ancient "dirt" DNA. *Antiquity*, 83, 430–44.

Jaenicke-Després, V., Buckler, E.S. & Smith, B.D., *et al.* (2003) Early allelic selection in maize as revealed by ancient DNA. *Science*, 302, 1206–8. [The evolution of maize followed by ancient DNA analysis.]

Lidén, K., Eriksson, G., Nordqvist, B., Götherström, A. & Bendixen, E. (2004) "The wet and the wild followed by the dry and the tame" – or did they occur at the same time? Diet in Mesolithic-Neolithic southern Sweden. *Antiquity*, 78, 23–33.

Loftus, R.T., MacHugh, D.E., Bradley, D.G., Sharp, P.M. & Cunningham, P. (1994) Evidence for two independent domestications of cattle. *Proceedings of the National Academy of Sciences USA*, 91, 2757–61.

Londo, J.P., Chiang, Y.-C., Hung, K.-H., Chiang, T.-Y. & Schaal, B.A. (2006) Phylogeography of Asian wild rice, *Oryza rufipogon*, reveals multiple independent domestications of cultivated rice, *Oryza sativa*. *Proceedings of the National Academy of Sciences USA*, 103, 9578–83.

Matsuoka, Y., Vigouroux, Y. & Goodman, M.M., *et al.* (2002) A single domestication for maize shown by multilocus microsatellite genotyping. *Proceedings of the National Academy of Sciences USA*, 99, 6080–4.

McClintock, B., Kato, T.A. & Blumenschein, A. (1981) *Chromosome Constitution of Maize*. Colegio de Postgraduados, Chapingo, Mexico. [Data on maize chromosome knobs.]

Menozzi, P., Piazza, A. & Cavalli-Sforza, L. (1978) Synthetic maps of human gene frequencies in Europeans. *Science*, 201, 786–92. [Genetic data suggesting a wave of advance.]

Richards, M.P., Schulting, R.J. & Hedges, R.E.M. (2003) Sharp shift in diet at onset of Neolithic. *Nature*, 425, 366. [The rapid spread of farming in Britain.]

Soares, P., Achilli, A. & Semino, O., *et al.* (2010) The archaeogenetics of Europe. *Current Biology*, 20, R174–83. [Includes a section on genetic studies of the spread of agriculture.]

Stock, F., Edwards, C.J. & Bollongino, R., *et al.* (2009) *Cytochrome b* sequences of ancient cattle and wild ox support phylogenetic complexity in the ancient and modern bovine populations. *Animal Genetics*, 40, 694–700.

Vigouroux, Y., Glaubitz, J.C. & Matsuoka,Y., *et al.* (2008) Population structure and genetic diversity of New World maize races assessed by DNA microsatellites. *American Journal of Botany*, 95, 1240–53.

14

Studying Prehistoric Technology

We tend to look on technology as an attribute of advanced, "developed" societies, but humans first began to manufacture and use objects over 2.5 million years ago, when the first stone tools were produced. Indeed, we identify different periods of prehistory from the nature of the predominant technology – Stone Age, Bronze Age, and so on – and within these periods recognize subdivisions according to the degree of sophistication of the implements that were made. This is especially true of the pre-metal periods where the major subdivisions correspond to the invention of new innovations in stone tool manufacture.

As well as their ability to make implements out of stone and metal, our ancestors also developed technology based around the use of natural biological products. These fall within the sphere of biomolecular archaeology, as often the products are oils or resins, which were used for many different purposes in the past. Identifying these products helps to show exactly which natural resources were being utilized, and with some products the biomolecular analysis can also throw light on the ways in which the resources were processed to yield new biological materials with altered properties and hence a different range of uses. Here we see the beginnings of the chemical industry that influences our modern lives so greatly.

14.1 Illustrations of the Biomolecular Approach to Prehistoric Technology

This chapter is relatively short but this should not be taken as an indication that prehistoric technology is an unimportant area of biomolecular archaeology. The chapter is short because most of the biological products used for technological purposes were lipids, and these are studied in exactly the same way as lipids from food residues in cooking utensils and storage vessels. This means that we are already familiar with the techniques and analytical strategies used to investigate prehistoric technology and can move directly to case studies that illustrate

 © 2011 Terry Brown & Keri Brown

the types of questions being asked and the nature of the information being obtained. We will begin with work that identified the material used as a source of light in prehistoric Crete.

14.1.1 Compound specific residue analysis has identified beeswax in Minoan lamps

Today we take it so much for granted that light is available at the touch of a button that we fail to appreciate the importance of the technological advances that enabled humans to gain control over their sources of illumination. The discovery of biological materials that could be used as illuminants in portable containers broke the dependence on natural light as the determinant of the daily cycle of events and, importantly, allowed activities during the hours of darkness to be expanded away from the confines of the fireside.

Stone or pottery vessels identified as lamps are found at some Neolithic sites, in particular ones associated with caves or mines, but in general are not common in the archaeological record until around 3500 years ago. From then on they begin to appear in increasing numbers, especially at Minoan sites on Crete and elsewhere in the western Mediterranean. According to the pollen record, this increase in the use of lamps occurred at about the same time as an increase in the number of olive trees being grown around the Aegean, leading to the suggestion that the illuminant in the early lamps was olive oil. Written records tell us that olive oil was used in this way some 1500 years later in Classical Greece and Rome, and in the absence of any evidence to the contrary, or indeed any evidence at all, the assumption was made that the discovery of this application of olive oil was the main driver in the development of portable sources of illumination.

Many pottery fragments from lamps and vessels thought to hold illuminants prior to use have been recovered from Bronze Age sites. Some of these potsherds have organic residues either on their surfaces or absorbed into the clay matrix, so attempts could be made to identify the illuminant by applying the standard techniques for residue analysis, in the same way that food residues have been studied in cooking pots and storage vessels (Section 12.2). When this was done with lamps from the Minoan I site at Mochlos on Crete, dating to 1600–1450 BC, the compounds that could be identified bore no relationship to those expected if the residues had been derived from olive oil. Instead the compounds included esters made up of long chain alcohols linked to fatty acids, as well as uncombined versions of the long chain alcohols presumably produced by breakdown of the esters. These compounds are typical of waxes (Section 4.1) rather than plant oils. More specifically, the particular lengths of the carbon chains in the compounds that were recovered from one lamp were characteristic of beeswax.

Although indicative of the presence of beeswax, these initial experiments, carried out using gas chromatography-mass spectrometry, were inconclusive. Waxes are also present on the outer surfaces of the stems and leaves of many plants, and the ones detected in the pottery lamps might conceivably come from this source rather than beeswax. To distinguish between the two, compound specific stable isotope analysis was carried out. Plant epicuticular waxes, being made by the plants themselves, would have $\delta^{13}C$ values typical of C_3 vegetation. Beeswax, on the other hand, is a metabolic product of the bee, and so would be expected to have a higher $\delta^{13}C$ value because of the isotopic shift that occurs as one progresses up a food chain (Section 6.2). When the lamp residues were examined, it was found that the $\delta^{13}C$ values of the waxy esters were 5–10‰ higher than those of typical plant waxes. The presence of beeswax was therefore confirmed.

To maintain a sufficient supply of beeswax to fuel their lamps, the people living in Minoan Crete must have kept bees. In an interesting corollary to the work described above, organic residue analysis has been used to identify ancient beehives, not from Minoan Crete itself but from a later site from mainland Greece dating to 200 BC. A pottery vessel, approximately 300 cm high and 350 cm diameter, with rectangular slits at its base, was suspected of being a beehive because of the presence on its internal surface of grooves thought to provide attachment points for the honeycomb. Absorbed residues were recovered from the inside of the vessel and examined by gas chromatography-mass spectrometry and by compound specific stable isotope analysis. Once again the characteristic fatty acid esters of beeswax were detected, with $\delta^{13}C$ values similar to those measured in the residues from the Minoan lamps.

Organic analysis has been used to identify a number of other ancient commodities recovered as residues from ceramics or occasionally found in glass or metal containers. A particularly interesting example is provided by a metal container from Roman London which contained a white creamy residue, probably a cosmetic used by ladies of the time to lighten their skin. The creamy base was shown to consist of a mixture of starch and animal fat, with the whiteness provided by the mineral cassiterite, a tin oxide that the Romans obtained from mines in Cornwall. Other residues containing fats and waxes have been detected in Roman bottles which might have been used to hold perfumes.

14.1.2 Wood tars and pitches had widespread uses in prehistory

The production of tars and pitches from tree resin by heating wood to high temperatures under anoxic conditions is an ancient technology, and these materials are frequently located in archaeological settings. Tars and pitches appear to have been used for a range of purposes, notably as adhesives to attach stone arrowheads to wooden shafts. This technology therefore pre-dates metalworking, but also survived into Classical times, the same type of tar that was used as a hafting agent in Neolithic Britain still finding applications as an adhesive during the Romano-British period. Tars and pitches were also used to line ceramic vessels in which liquids were stored for long periods, presumably to make the vessels more watertight, and as coatings on ropes.

Most tars and pitches are derived either from coniferous trees such as pine or spruce, or from birch bark, but archaeological tars from other sources are also known. The resins present in coniferous trees are largely composed of diterpenoids, in particular abietic acid and pimaric acid, and those of birch bark are triterpenoids such as betulin and lupeol (Figure 4.9). The source of a tar or pitch can therefore be identified from the presence of one or more of these characteristic compounds or their decay products. The tar used in prehistoric Britain as a hafting agent and centuries later as a glue to repair Roman vessels contains betulin and lupeol and so must have been obtained from birch bark.

Several different technologies for heating wood to produce tar were in existence in historic periods, and some are still used today in traditional settings. Through a particularly

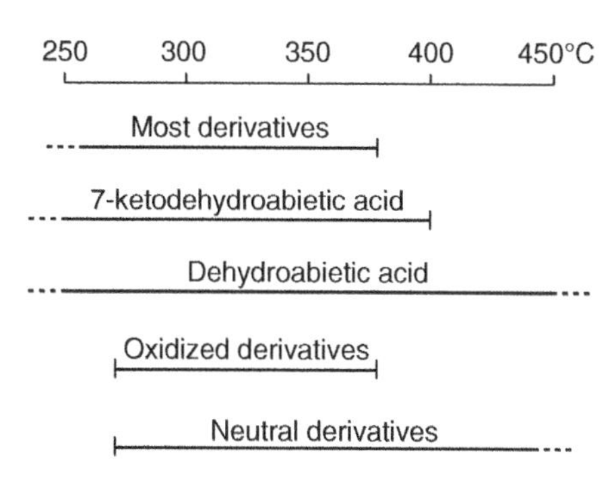

Figure 14.1 Temperature stabilities of various derivatives of pimaric and abietic acid formed during the heating of pine resin.

clever use of organic analysis it is possible to infer some of the details of the prehistoric tar manufacturing process from the identities of the terpenoid derivatives that are formed during the heating process. One important distinction that can be made is between tar that has been produced in the normal way by heating wood, and tar obtained by heating the resin itself. The former production method leads to synthesis of ester derivatives in which the terpenoids have become attached to alcohol compounds derived from wood cellulose, the reaction between the two catalyzed by acids derived from the lignin present in wood. If the tar is obtained directly from resin, then these esters do not form, because the alcohol and acid needed for their production are absent. Identification of terpenoid esters in tar therefore indicates a manufacturing process involving wood distillation.

The second aspect of the technology that can be inferred from the chemical signature of the tar is the temperatures that were used in the tar production process. Experiments with pine resin have shown that the various modified versions of pimaric and abietic acid present in the resin have different temperature stabilities, most being degraded at 350–400°C but two, dehydroabietic acid and the 7-keto form of this compound, remaining at 400°C, with only the former present at 450°C (Figure 14.1). Oxidized derivatives of the acids begin to appear at temperatures between 250 and 300°C and are then degraded if the temperature reaches 400°C. Non-acidic, neutral derivatives such as norabietatriene and tetrahydroretene are also formed as the temperature increases, but at different rates, which means that the relative proportions of these compounds depends on the temperature (Table 14.1). By understanding the thermal stabilities of the pine acids, as well as the synthesis and breakdown temperatures for the oxidized and neutral derivatives, it is therefore possible to predict the suite of compounds that would be present in tars produced at different temperatures. All of these compounds are readily detectable and quantifiable by gas chromatography-mass spectrometry, so analysis of archaeological tars by this method is able to reveal the temperatures reached during the manufacturing process, which in turn can be used to infer the structure and operation of the device within which the tar was made.

14.1.3 Early agritechnology included soil enrichment by manuring

The final example of prehistoric technology that we will examine concerns the use of fecal material as manure to improve the quality of land used for cultivation of crops. Manuring is still widely used today, the addition of animal or human excrement to soils having several benefits. Manure acts as an aggregating agent that improves the structure of the soil, consolidating dry soils and providing aeration to heavy ones with a high clay content. It directly adds organic nutrients to the soil and further improves fertility by promoting the activity of bacteria and fungi, which add to the soil's organic content and also solubilize trace elements so that these become available to the growing plants. All of this is well know to us today, but when was the technology first adopted?

Table 14.1 Presence of neutral derivatives of pimaric and abietic acid at different temperatures during the heating of pine resin.

	% neutral component at			
Compound	*300°C*	*350°C*	*400°C*	*450°C*
18-Norabieta-8,11,13-triene	0.97	3.60	8.65	11.96
19-Norabieta-3,8,11,13-tetraene	1.84	4.24	7.98	4.61
19-Norabieta-8,11,13-triene	0.52	2.40	12.72	22.17
1,2,3,4-Tetrahydroretene	2.07	6.16	11.26	12.59

We learnt in Section 8.4 that a detailed knowledge of the decay pathways for cholesterol has led to identification of biomarkers for fecal material. The bacteria present in animal intestines convert the C_5–C_6 double bond present in the cholesterol molecule into a single bond, at the same time adding a hydrogen group to carbon 5 in the β orientation, giving β-cholestanol or coprostanol (Figure 8.13). This compound can subsequently be converted into 5β-cholestanone and related compounds after excretion. Coprostanol and its derivatives are therefore biomarkers for fecal material, this particular decay pathway for cholesterol only occurring in animal intestines. Confirmation of the presence of fecal material can be obtained by searching for bile acid derivatives, which have the advantage that they enable the species of origin to be identified, with deoxycholic acid characteristic of human and cattle feces, hyodeoxycholic acid being the dominant bile acid in pigs, and cholic acid present in human but not cattle feces (Figure 8.14).

An example of the use of these biomarkers to identify sites where manuring has been used in soil improvement takes us back to the Minoan period, and to the island of Pseira which is just off the western coast of Crete. Pseira was occupied from the Neolithic period until about 1450 BC, when the Late Minoan civilization collapsed. The soils on the island are naturally poor and do not support a great deal of vegetation, but in one attempt to improve productivity by reducing water run-off, a series of terraces were constructed, reaching a peak during the period 2000–1600 BC, when the population on the island was at its highest. Evidence that the soils in these terraces were fertilized with manure was obtained when the site was excavated, the different layers of the terraces containing debris such as pottery fragments, these waste materials possibly finding their way onto the fields along with excrement and decaying food.

The biomolecular investigation of the terraces involved taking soil samples down to a depth of 95 cm to cover the entire occupation period of Pseira. Each sample was then examined for the presence of coprostanol and the other cholesterol decay products characteristic of fecal material. The results indicated that fecal material was indeed present in the layers containing waste material, these dating from the Early to Late Minoan periods, corresponding to 2500 BC up to the time of abandonment in 1450 BC. One interesting observation was the relatively high presence of the coprostanol derivative called epicoprostanol in the upper samples, from the later occupation periods (Figure 14.2). Epicoprostanol is synthesized from coprostanol by the activity of anaerobic microbes. These are present in soil but not in large numbers, because the soil environment contains too much oxygen for their proliferation. They are much more abundant in midden heaps, the interior of these becoming more and more anaerobic over time. The presence of epicoprostanol in the later terrace soil samples is therefore evidence that the manure applied to these was not raw excrement but material that had been stored before being used. The additional decomposition that occurs during storage improves the fertilizing quality of the compost. This fact is well known to farmers and local vegetable growers today, and appears also to have been common knowledge for the ancient Minoans of prehistoric Crete.

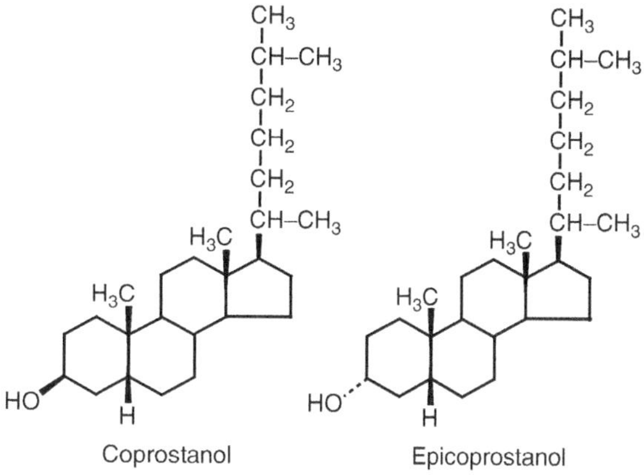

Coprostanol

Epicoprostanol

Figure 14.2 Coprostanol and epicoprostanol.

Further Reading

Aveling, E.M. & Heron, C. (1998) Identification of birch bark tar at the Mesolithic site of Starr Carr. *Ancient Biomolecules*, 2, 69–80.

Beck, C.W., Stout, E.C. & Jänne, P.A. (1997) The pyrotechnology of pine tar and pitch inferred from quantitative analyses by gas chromatography-mass spectrometry and carbon-13 nuclear magnetic resonance spectrometry. In *Proceedings of the First International Symposium on Wood Tar and Pitch* (ed. Brzezinski, W. & Piotrowski, W.), 181–92. State Archaeological Museum, Warsaw.

Bull, I.D., Betancourt, P.P. & Evershed, R.P. (2001) An organic geochemical investigation of the practice of manuring as a Minoan site on Pseira island, Crete. *Geoarchaeology*, 16, 223–42.

Charters, S., Eveshed, R.P., Goad, L.J., Heron, C. & Blinkhorn, P. (1993) Identification of an adhesive used to repair a Roman jar. *Archaeometry*, 35, 91–101.

Evershed, R.P. (2008) Organic residue analysis in archaeology: the archaeological biomarker revolution. *Archaeometry*, 50, 895–924.

Evershed, R.P., Vaughan, S.J., Dudd, S.N. & Soles, J.S. (1997) Fuel for thought? Beeswax in lamps and conical cups from Late Minoan Crete. *Antiquity*, 71, 979–85.

Regert, M., Delacotte, J.M., Menu, M., Pétrequin, P. & Rolando, C. (1998) Identification of Neolithic hafting adhesives from two lake dwellings at Chalain (Jura, France). *Ancient Biomolecules*, 2, 81–96.

15

Studying Disease in the Past

For most people, life in the past was nasty, possibly brutish, and certainly short. We are fortunate to live in an age of antibiotics, anesthesia, sanitation, and medical science, and expect to live to our eighties or longer. Life expectancy for early hominids and Neanderthals was in the twenties, and Neolithic people had an average age of death of just 30 years for men, less for women. The dangers associated with childbirth were one of the leading causes of death for women, and, in the absence of medical understanding, trauma and injury were more likely to have had fatal consequences. Septicemia and tetanus from infected wounds would have led almost inevitably to death. But the primary reason for the shorter life expectancies in the past was almost certainly the greater prevalence of infectious diseases. In the absence of vaccination, epidemics would spread unchecked except for pockets of natural resistance among the population. In the absence of antibiotics and other aspects of modern patient care, a disease once contracted would run its full course, often culminating in death.

Paleopathology is the area of bioarchaeology that investigates diseases in past populations, especially those diseases that leave a signature on the skeleton, either as bone resorption or new bone growth, and hence can be identified by osteoarchaeology. Some diseases also leave a biomolecular signature in the skeleton, examination of which can add to the information provided by osteological study. The study of biomolecular signatures of disease is called **biomolecular paleopathology**, and is the subject of this chapter.

15.1 The Scope of Biomolecular Paleopathology

The osteological approach to paleopathology provides tremendous opportunities for understanding paleodisease, but is not without its limitations. Not all diseases result in a recognizable change to the skeleton, and few of the **lesions** that do occur are entirely

 © 2011 Terry Brown & Keri Brown

diagnostic for a single disease. Biomolecular paleopathology has therefore been looked on as a means of extending the range of diseases that can be typed in human skeletons and making more precise the identification of diseases in skeletons whose lesions are ambiguous. Biomolecular paleopathology also has the potential to address entirely new areas, especially ones concerned with the evolution of disease-causing organisms and the genetic response of humans to exposure to disease. Later in this chapter we will examine the biomolecular research that has been carried out with individual diseases, but first we must consider the overall scope and potential of this area of biomolecular archaeology.

15.1.1 Infectious diseases can be studied by examining the biomolecular remains of the pathogen

To date, biomolecular paleopathology has focussed largely on infectious diseases, ones caused by a pathogenic organism and spread by transfer of that pathogen from person to person. There are various ways of classifying infectious diseases, the official scheme of the World Health Organisation currently recognizing 21 categories and 187 subcategories based on the nature of pathogen, the mode of transmission, and the parts of the body that are affected. For our purposes, we can greatly simplify this scheme and divide infectious diseases into just three groups, depending on the type of organism responsible for the disease (Table 15.1):

- **Diseases caused by bacteria**. The most important of these in modern human health is tuberculosis, which is caused by *Mycobacterium tuberculosis* and is spread through the air by coughing and sneezing. Tuberculosis is preventable and curable but still kills some 1.5 million people per year, mostly in poor parts of the world, and is beginning to re-emerge globally due to the evolution of antibiotic resistant strains of the bacterium and the failure of some western countries to implement national vaccination programmes. As we will see later in

Table 15.1 Main classes of infectious disease.

Type	*Example*	*Causative agent*
Bacterial	Tuberculosis	*Mycobacterium tuberculosis*
	Leprosy	*Mycobacterium leprae*
	Syphilis	*Treponema pallidum*
	Bubonic plague	*Yersinia pestis*
	Stomach ulcers	*Helicobacter pylori*
	Brucellosis	*Brucella* sp.
	Cholera	*Vibrio cholerae*
	Diphtheria	*Corynebacterium diphtheriae*
Viruses	Influenza	Influenza virus
	Common cold	Rhinoviruses, coronaviruses
	AIDS	Human immunodeficiency virus
	Smallpox	Variola virus
Parasitic	Malaria	*Plasmodium falciparum, P. vivax*
	Schistosomiasis	*Schistosoma* sp.
	Amoebic dysentry	*Entamoeba histolytica*

this chapter, tuberculosis is one of the diseases that is most amenable to biomolecular study in archaeological specimens. The same is true for leprosy, a related disease caused by *Mycobacterium leprae*. Other bacterial diseases include syphilis, caused by *Treponema pallidum*, which is sexually transmitted but can cause damage to many parts of the body including the brain; and bubonic plague, caused by *Yersinia pestis*, notorious in history as the cause of the Black Death. Bubonic plague is transmitted by rats and passed to humans initially by fleabite. Infected humans frequently develop a pulmonary version of the disease called pneumonic plague which can spread rapidly by inhalation of bacteria ejected by coughing.

- **Diseases caused by viruses**, of which there are many, some of mild severity, such as the common cold, and others which are almost always fatal, such as AIDS (acquired immune deficiency disease). Some types of cancer are also caused by viruses. Throughout history, the most devastating viral disease has probably been influenza, which regularly re-emerges in modified forms that even today can evade the vaccination and other protective measures used to limit its spread. The Spanish flu epidemic of 1918–20 killed between 20 and 100 million people, the upper estimates putting it on a par with the Black Death of the 14th century, although representing a smaller proportion of the population living at the time.
- **Diseases caused by parasites.** A parasite is an organism that spends at least part of its life cycle within a host to which it contributes nothing and which it might actively harm. Strictly speaking, bacteria and viruses are both types of parasite; but when the term is used in the context of infectious disease, it usually refers to eukaryotic disease-causing organisms. The most important parasitic disease of modern and historic times is malaria, caused by protozoa of the genus *Plasmodium*. The most virulent of these is *P. falciparum*, which is responsible for about 80% of the 250 million new malarial cases that occur each year, and 90% of the 900,000 or so deaths. Schistosomiasis, or bilharzia, caused by microscopic worms of the genus *Schistosoma*, is similarly prevalent, with over 200 million infected people worldwide, but the disease, although debilitating, is not often fatal by itself.

Infectious diseases are studied by searching archaeological remains for biomolecules derived from the pathogenic organism. Most frequently this involves analysis of ancient DNA, with PCRs directed at sequences that are possessed by the pathogen and no other organism, and hence are diagnostic for the presence of the pathogen. Increasingly this work is progressing beyond simply confirming or extending identifications made by osteological analysis to examination of sequences that enable different strains or geographical variants of the pathogen to be distinguished, and there is hope that it might soon be possible to type sequences involved in virulence or other aspects of the disease progression. All of this work is aided by the immense amount of DNA sequence information that is available for many of the important infectious agents. Viral genome sequences, being much shorter than those of free-living organisms, have been available for several years, but these have recently been supplemented by complete sequences for bacterial pathogens, including those causing tuberculosis, leprosy, syphilis, bubonic plague, cholera, and many more. The same is true of the parasitic pathogens of humans, the genomes of both main malaria parasites, *P. falciparum* and *P. vivax*, having been published and that of *Schistosoma mansoni*, the major cause of schistosomiasis, recently completed.

Although most research into biomolecular paleopathology has been with ancient DNA, both lipids and proteins also have potential in the study of past disease, and both have been exploited to a limited extent. Attempts have been made to diagnose tuberculosis and leprosy by identifying lipids called mycolic acids, which are found in the cell walls of the mycobacteria, and an immunological test for the protein II antigen of *P. falciparum* has been used in studies of malaria.

Before going any further, we must appreciate exactly what detection of pathogen biomolecules in an archaeological specimen means. Detection shows that the pathogen was present in the individual at the time of death. It does not mean that the disease caused by the pathogen was the reason for the death of that individual. Infection with a disease-causing pathogen does not always result in death: some diseases, such as bilharzia and malaria, are largely debilitating rather than fatal, and some diseases have lengthy incubation periods when the symptoms might not be fully manifest. And it is, of course, quite possible for a person suffering from terminal illness to die for a completely different reason. These issues must always be kept in mind when interpreting the results of pathogen detection in archaeological specimens.

15.1.2 Not all infectious diseases leave a biomolecular signature in the skeleton

In the early days of biomolecular archaeology, when the excitement generated by the first detections of ancient DNA in bones and other old specimens led to a certain overexuberance about what could be achieved with this new technology, it was suggested that the biomolecular approach would revolutionize paleopathology by enabling many new diseases to be detected and studied in the archaeological record. What was forgotten back in those early times, and is still sometimes forgotten today, is that not all diseases will leave a biomolecular signature in an archaeological specimen. If the specimen is a bone or tooth, then it is necessary for the pathogen to have been present in that bone or tooth at the time of death. Most pathogens are quite specific with regard to the part of the body that they infect, and not all find their way into bones, or into the bloodstream that permeates some types of bone. Even if they do, then the number of copies of the pathogen that are present when death occurs might be too small to leave a detectable ancient DNA or other biomolecular signature. Two examples, tuberculosis and syphilis, will serve to illustrate the relevant issues.

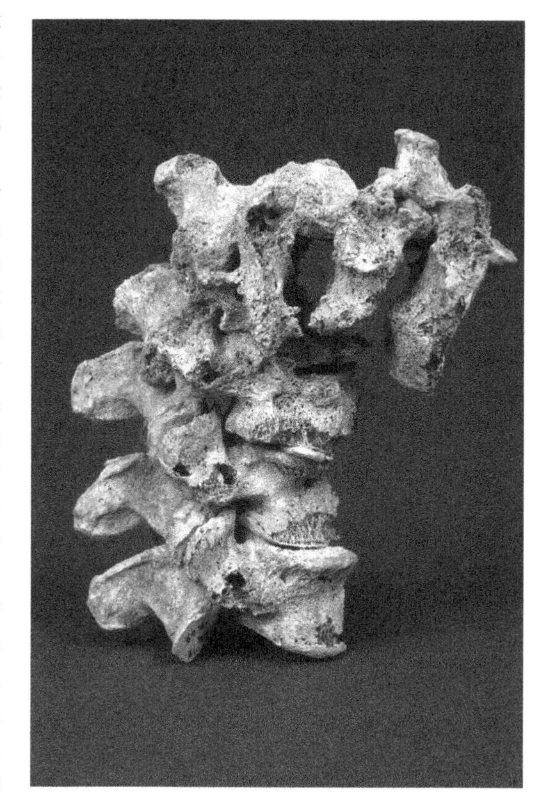

Figure 15.1 Portion of the spine showing Pott's disease, a typical skeletal indicator of tuberculosis, caused by collapse of vertebrae leading to curvature of the spine. Image kindly provided by Charlotte Roberts.

Tuberculosis is the disease that has been most extensively studied by biomolecular examination of skeletons. The relative success of work with ancient tuberculosis is due in no small part to the natural progression of the disease. Although initially infecting the lungs, tuberculosis bacteria can spread via the blood and lymphatic systems to the bones, in particular causing bony lesions on the lower thoracic and lumbar vertebrae (Figure 15.1). Mycobacteria are present both in these bony lesions and also in the bone marrow of patients who have suffered this type of systemic infection. If the patient dies, either from the effects of tuberculosis or through some other cause, then biomolecules from the infecting bacteria might remain in the vertebrae and the marrow-containing bones, and possibly also in other parts of the skeleton that are infiltrated by blood vessels or are close to the primary infection sites, the latter possibly including the inner surfaces of the ribs, which are in close contact with the lungs. Because the bacteria are present in the skeleton at the time of death, their biomolecular remains have the potential to survive in archaeological specimens.

The same is not true of all diseases; in fact, experience is showing that survival of pathogen biomolecules in the skeleton is the exception rather than rule. Our second example is syphilis, the venereal form of which is caused by *T. pallidum*. One of the most interesting paleopathological questions that could be answered through use of ancient DNA is the origin of venereal syphilis in Europe. The first historical records for this disease date to an epidemic which affected all parts of Europe in 1496. The close proximity between this date and the return of the Columbus expedition in 1493 has led to the popular view that venereal syphilis was introduced from the Americas by infected crewmen, but this theory is contradicted by the existence of a few earlier Old World skeletons that appear to show osteological signs of venereal syphilis. If syphilis was present in Europe before 1493, but in a more benign form, then possibly the epidemic was initiated by a mutation that increased the virulence of *T. pallidum*. These alternatives – introduction from the Americas or change in an endogenous Old World disease – are potentially testable by ancient DNA analysis, but the way in which syphilis develops in infected people makes it very difficult to carry out the necessary work. During the early stage of syphilis, when changes to the bones occur, the number of organisms within the host is relatively low and there is little opportunity for pathogen DNA to enter the lesions that are formed. This has been underlined by a study of 46 bones, of various ages, most displaying osteological indicators of syphilis. These bones were tested with nine different *T. pallidum* PCRs, but none of these gave positive results, despite each of the PCRs being sensitive enough to detect the presence of just five or fewer copies of the *T. pallidum* genome, and even though several of the bones contained human DNA, indicating that the burial conditions had been conducive to ancient DNA preservation.

The pathogen content in the bloodstream is in fact higher during the secondary and tertiary stages of venereal syphilis, but during these periods there is very little bone remodelling and the only opportunity for treponemal DNA to enter the bones would be via normal turnover. This might lead to small amounts of ancient *T. pallidum* DNA being present in the bones of people who died while at these stages, but without osteological indicators to point out the bones that should be studied, searching for specimens containing this DNA would be a very hit and miss affair, and it would be very difficult to distinguish genuine detections of *T. pallidum* from spurious results arising from cross-contamination with modern DNA. This means that the interesting paleopathological questions posed by syphilis will probably remain beyond the boundaries of what can be achieved by biomolecular research.

15.1.3 Studies of pathogen evolution can also address archaeological questions

Understanding the evolution of infectious diseases is a central goal in modern medical science, not least because the past evolutionary pathway of a pathogen might provide clues regarding its possible future evolution, enabling clinical microbiologists to stay one step ahead in the search for treatments to combat the emergence of new variants of the disease.

In some cases, these evolutionary studies provide information that is relevant to archaeological questions. Later in this chapter we will see how evolutionary studies of the mycobacteria, making use exclusively of modern strains, has answered a longstanding puzzle regarding the origins of the disease in prehistoric times. It is also possible for studies of modern pathogen DNA to address archaeological issues not directly related to paleodisease. One such example concerns the bacterium *Helicobacter pylori*, the main cause of gastric and duodenal ulcers, leading in some cases to stomach cancer.

Over half the people in the world are infected with *H. pylori*, only 10–15% of these ever developing stomach ulcers. The bacterium is transmitted from parents to their offspring, the latter becoming infected during childhood by a process that is not understood but is possibly encouraged by poor sanitation, infections being more common in the less developed parts of the world. Because of the predominantly vertical transmission, with little or no transfer between members of different families, the movement of *H. pylori* bacteria around the world has mirrored the movement of human populations. This was first demonstrated by examining the sequences of eight genes in 370 strains of the bacterium from all over the Old and New Worlds. The genetic analysis revealed that these modern *H. pylori* strains were derived from five ancestral populations, two located originally in Africa, two in Asia, and one in Europe. The modern distribution of the bacteria derived from these ancestral groups matches known prehistoric and historic migrations, such as the colonization of the New World from Asia some 12,000 years ago, the Bantu expansion into southern Africa about 2000 years ago, and the modern colonial expansion of Europeans. A more detailed analysis of strains from East Asia has revealed that one of the Asian groups predominant in this area gave rise to a distinct subgroup that moved with its human hosts into New Guinea and Australia. Another subgroup located primarily in Austronesia originated about 5000 years ago in Taiwan, again in accord with archaeological data for the source and timing of the first human migrations into the Pacific islands and New Zealand.

15.1.4 Inherited diseases can be studied by typing ancient human DNA

Not all diseases are infectious. Many are caused by defects in the human genome and, like other genetic features, are passed from parents to offspring. These are called genetic or, more correctly, **inherited diseases**. There are over 6000 inherited diseases that result from defects in individual genes, with approximately one in every 200 births being of a child with one or other of these **monogenic disorders** (Table 15.2). The commonest include the lung disease cystic fibrosis, and Huntington's disease, a neurological disorder characterized by uncoordinated body movements. Other diseases are much rarer, with just a few cases reported worldwide, possibly affecting just a few families.

Many of the mutations that cause inherited disease have been identified and could be typed in ancient human DNA extracted from archaeological remains. As an example of this type of work we will look briefly at a project carried out with the skeleton of a child from an Ottoman grave at Akhziv, Israel, dating to the 16th–19th century AD. This child's skull displayed osteological features typical of extreme anemia, including the presence of pores in the upper areas of the eye sockets (cribra orbitalia), small spongy holes elsewhere in the skull (porotic hyperostosis), and a thickening of the cranium. Anemia can result from iron

Table 15.2 Some of the commonest monogenic inherited diseases.

Disease	*Symptoms*	*Frequency (UK births per year)*
Inherited breast cancer	Cancer	1 in 300 females
Cystic fibrosis	Lung disease	1 in 2000
Huntington's chorea	Neurodegeneration	1 in 2000
Duchenne muscular dystrophy	Progressive muscle weakness	1 in 3000 males
Hemophilia A	Blood disorder	1 in 4000 males
Sickle cell anaemia	Blood disorder	1 in 10 000
Phenylketonuria	Mental retardation	1 in 12 000
β-Thalassemia	Blood disorder	1 in 20 000
Retinoblastoma	Cancer of the eye	1 in 20 000
Hemophilia B	Blood disorder	1 in 25 000 males
Tay–Sachs disease	Blindness, loss of motor control	1 in 200 000

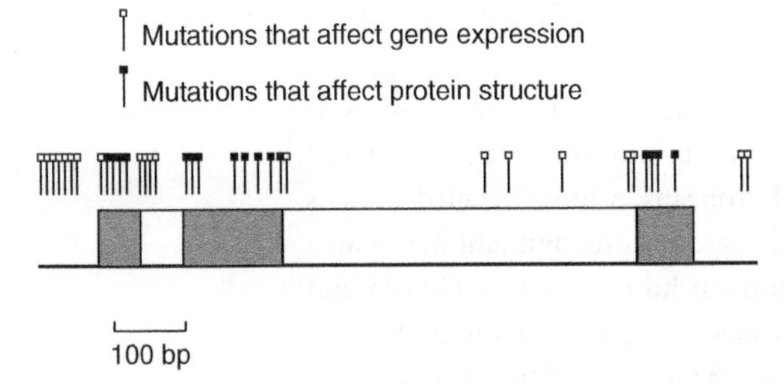

Figure 15.2 Locations of some of the many mutations in the human β-globin gene that result in thalassemia. Shaded boxes are exons, open boxes introns.

deficiency and is therefore a common indicator of malnutrition, but it can also be caused by the inherited disease called thalassemia, in which one of the globin genes is defective. Iron deficiency and thalassemia both reduce the amount of hemoglobin in the blood, giving rise to anemia. To find out whether this child's anemia was dietary or genetic, PCRs were directed at part of the β-globin gene. These revealed a 2 bp deletion, called the FS8 mutation, which is one of the commonest thalassemia defects with present-day frequencies of 2–10% in eastern Mediterranean populations and 2.3% in Arabs. The implication was that the anemia from which this child suffered had a genetic basis and was not due to dietary deficiencies. The result was, however, unexpected because in the homozygous condition, as in this child, this type of thalassemia is fatal with death usually occurring soon after birth. But the child appeared to be eight years or so in age, an exceptional period of survival. Further examination of the β-globin genes identified a second mutation, this one associated with the continued production of the fetal version of hemoglobin. It would therefore appear that the child was able to survive into later childhood, albeit with severe thalassemia, because he or she was still synthesizing fetal hemoglobin.

Ancient DNA was therefore able to specify the cause of the Ottoman child's anemia. In some regards, this was a fortuitous outcome, as identifying an inherited mutation in ancient DNA is not always a straightforward task. Over 250 different mutations are known to give rise to thalassemia, these mutations spread throughout the α- and β-globin gene clusters, several kilobases of DNA having to be searched to locate all of them (Figure 15.2). When ancient DNA is being studied, and each individual PCR can target only 100–150 bp of sequence, checking for all of these mutations is an almost impossible task. The search therefore has to be limited in some way, usually by targeting only those mutations likely to have been present at high frequency in the population from which the skeleton comes. For the relatively recent Ottoman period there is a reasonable degree of likelihood that the most frequent mutations at that time were the same as those in modern eastern Mediterranean populations, but extrapolating from present to past populations becomes less reliable the further back in time we go. The same problem applies to most other inherited diseases. Cystic fibrosis, for example, can be caused by any one of 1400 different mutations, at different positions along the 200 kb length of the cystic fibrosis transmembrane regulator gene. One of these, the ΔF508 mutation, accounts for over two-thirds of all cystic fibrosis cases today, but whether the frequency of this mutation was similarly high in prehistory cannot be guaranteed. Indeed cystic fibrosis illustrates how quickly mutation frequencies can change in a largely inbreeding population, a totally different mutation called W1282X, which has a global frequency of only 1.2%, accounting for 60% of cases of cystic fibrosis in Ashkenazi Jews.

Even when an inherited mutation is identified, there may be no assurance that the person actually suffered from the associated disease. Not all mutations have an impact on a child immediately after birth, some having a delayed onset and only giving rise to the disease later in the individual's life. Others display non-penetrance in some individuals, never being expressed at all. Great care must therefore be taken in both the design and interpretation of any biomolecular paleopathology study aimed at an inherited disease.

15.1.5 Ancient human DNA can indicate exposure to an infectious disease

Thalassemia, as detected by DNA analysis of the Ottoman child, is relevant to paleodisease, not only in its own right as an inherited disorder. A high incidence of thalassemia can also be an indicator that the population is or has been exposed to malaria. The classic illustration of the link between thalassemia and malaria is provided by sickle cell anemia, a type of thalassemia associated with allele *S* of the β-globin gene, which differs from the normal allele *A* by a single point mutation, which converts the 6th codon in the gene from one that specifies glutamic acid in the normal allele to one that codes for valine in the sickle cell version. The resulting change in the electrical properties of the hemoglobin molecule causes the protein to form fibers when the oxygen tension is low. In the homozygous form *SS*, the red blood cells take on a sickle shape and tend to become blocked in capillaries, where they break down. The resulting disease is fatal without treatment. Heterozygotes, with the *AS* genotype, suffer less extensive breakdown of red blood cells and a milder form of thalassemia. This genotype is relatively common in malarial parts of world, because the malaria parasite *P. falciparum* finds it less easy to multiply within the bloodstream of an *AS* person, possibly because the low red blood cell count simply reduces the capacity of the parasite to reproduce. Individuals with the *AS* genotype are therefore more resistant to malaria than *AA* homozygotes, and the *AS* genotype predominates in the population. This is an example of **balancing selection**, where two opposing selective pressures result in a greater number of heterozygotes than expected. The same is true for many other thalassemia mutations, and also for mutations in a different human gene, this one coding for the glucose-6-phosphate dehydrogenase (G6PD) enzyme. A deficiency in G6PD makes red blood cells more susceptible to oxidative stress, damaged cells being filtered out of the bloodstream by the spleen. The increased turnover of red blood cells by the spleen is thought to decrease the ability of the malaria parasite to survive in the bloodstream, reducing the severity of the disease.

Long-term exposure to malaria therefore increases the frequency of thalassemia and G6PD deficiency mutations in human populations, providing a second means of studying malaria in the past. A similar approach is possible with other diseases, including tuberculosis, mutations in the human SP110 gene conferring some resistance to this disease and likely to display greater frequency in populations with a high incidence of tuberculosis. At present, however, the general difficulties in obtaining sequences from human nuclear DNA limit the usefulness of these mutations in biomolecular paleopathology, as we will see in the remainder of this chapter when we examine the work that has been carried out with individual diseases.

15.2 Biomolecular Studies of Ancient Tuberculosis

Tuberculosis is the disease that has been most extensively studied by biomolecular archaeologists. There are several reasons for this. Tuberculosis is looked on as the most important infectious disease today, responsible for some 1.5 million deaths per year and re-emerging as a major disease in developed countries, from which it was largely eradicated during the second half of the 20th century. Some forms of tuberculosis result in bony lesions that are present in archaeological skeletons, showing that the disease has affected humans for many centuries. Until recently it has been uncertain exactly how long humans have lived with tuberculosis, one popular theory, now known to be incorrect, suggesting that the disease initially transferred to humans from infected cattle soon after the latter were domesticated. The presence of tuberculosis lesions enables skeletons likely to be amenable to biomolecular examination to be identified, and the early work that was directed at *M. tuberculosis* DNA appeared to be successful, giving impetus to this particular area of biomolecular archaeology.

The work that has been done on ancient tuberculosis provides an excellent illustration of both the strengths and limitations of biomolecular paleopathology, and also highlights many of the technical challenges that need to be addressed if biomolecular studies are to make a genuine contribution to our understanding of paleodisease. We will therefore spend some time focussing in detail on ancient tuberculosis.

15.2.1 Osteology is not a precise means of identifying tuberculosis in the archaeological record

Prior to the introduction of biomolecular techniques, osteological examination of skeletons was the only way of identifying cases of ancient tuberculosis. A small percentage of people infected with tuberculosis (some 3–5% of the total in modern populations) develop bone changes. These arise when the infecting bacteria spread via the blood and lymphatic systems from the lungs to other parts of the body and invade the bone marrow and other parts of certain bones.

Almost half the people who acquire this skeletal form of tuberculosis develop spinal abnormalities, and the diagnosis of tuberculosis in skeletons is normally based on the presence of destructive lesions in the lower thoracic and/or lumbar vertebrae (Figure 15.1). In severe cases, the vertebrae collapse and fuse together, resulting in curvature of the spine and the condition called Pott's disease. Bony lesions can also occur in the hip and knee joints and elsewhere in the skeleton. It is important to note that none of these lesions is wholly specific to tuberculosis: they are not **pathognomonic** and cannot be taken as definite indicators of the presence of the disease. Similar changes to the vertebrae can be caused brucellosis, fungal infections, septic arthritis, neoplastic disease, and osteoporosis, and changes to the hips and knees can also arise from arthritis. The same is true of various other skeletal changes that are sometimes used as osteological evidence for tuberculosis. Abnormal bone formation on the inner surfaces of the ribs might be indicative of tuberculosis, but these lesions are typically due to inflammatory processes secondary to pulmonary infection, and thus cannot alone be considered pathognomonic of tuberculosis. Other bone changes that have been attributed to tuberculosis include inflammation of the fingers and toes, and hypertrophic pulmonary osteoarthropathy, which results in further abnormalities to the fingers as well as inflammation of joints throughout the body. All these bone changes can also result from other pathological conditions and cannot be used as pathognomonic of tuberculosis.

In view of the lack of precision inherent in the osteological identification of tuberculosis, it is perhaps not surprising that there has been debate about the prevalence of the disease in the archaeological record, in particular in older specimens. Although claims have been made for the identification of tuberculosis in *Homo erectus*, these are controversial and the earliest accepted example of skeletal tuberculosis derives from Italy, dated to 7800 ± 90 BP, with the first evidence from the New World much later than this, not until AD 700. In Europe, tuberculosis does not seem to become particularly frequent until after AD 1600, when historic medical records first begin to ascribe a substantial proportion of deaths to "consumption" and various other ailments that we now recognize as tuberculosis. In London, the Bills of Mortality suggest that up to 25% of all deaths in the 1780s and 1790s were due to this cause. But only limited trust can be placed in the older medical records, as there is confusion over the names given to diseases that might or might not have been tuberculosis, and the records are generally inaccurate because of the poor training giving to the people (often local townsfolk) hired by the authorities to compile the death records, and sometimes paid per disease reported. Biomolecular studies therefore have a clear potential as a means of confirming osteological identifications and making more accurate our understanding of the prevalence of tuberculosis in the past.

15.2.2 Early biomolecular studies were aimed simply at identifying tuberculosis in archaeological specimens

The first biomolecular studies of ancient tuberculosis were carried out in the mid 1990s and were aimed primarily at confirming and extending identifications of the disease made by osteological methods. In these early projects, PCRs were often directed at two **insertion sequences** present in the *M. tuberculosis* genome. These are mobile sequences that can move about within a genome and also be transferred between bacteria, possibly taking genes, such as those for antibiotic resistance, with them. The *M. tuberculosis* insertion sequences are called IS6110 and

IS1081. Both have copy numbers of up to 20 in a single bacterium, aiding detection by giving greater sensitivity to the PCRs, but IS6110 has the drawback that it is absent from some strains of *M. tuberculosis.* In most of the world, strains lacking IS6110 are uncommon, but their frequency is greater than 10% in parts of Asia, and the prevalence of such strains in the ancient world is unknown. PCRs directed at this sequence are therefore not discriminatory for the presence or absence of *M. tuberculosis*, as a negative result is compatible with both alternatives.

When detections of *M. tuberculosis* are attempted it is essential to be certain that the PCRs are specific for this bacterium, or to understand what other species might give a positive result if there is any lack of specificity. *M. tuberculosis* is one member of a complex of closely related mycobacteria, all with very similar genomes. The other members include *Mycobacterium bovis*, which is responsible for tuberculosis in various animals including cattle, and can sometimes be caught by humans; *M. africanum* and *M. canettii*, which cause versions of human tuberculosis in parts of Africa; and *M. microti* and *M. pinnipedii*, which cause the disease in voles and seals, respectively. All of these species are thought to harbor IS6110 and IS1081 elements, so PCRs directed at these sequences merely show that one member of the *M. tuberculosis* complex is present. More specific markers that enable members of the complex to be distinguished are therefore needed to supplement these insertion sequence PCRs. From the archaeological standpoint, the most critical requirement is to separate *M. tuberculosis* and *M. bovis*, so that skeletons displaying human tuberculosis can be identified without the possible confusion that the individuals concerned had actually contracted the bovine form of the disease. A distinction between *M. tuberculosis* and *M. bovis* first became possible with identification of a single nucleotide polymorphism (SNP) in the *oxy*R pseudogene, at which an A indicates *M. bovis* and a G indicates one of the other members of the complex. Shortly afterwards two SNPs were discovered in the *kat*G and *gyr*A genes, which, in combination, achieve the same result, and in more recent years the list has gradually been added to.

One of the first biomolecular archaeology projects to exploit these new, more diagnostic, markers was carried out in 1999 with bones from the cemetery of the Abbey of St Mary Graces in London, which was in use between 1350 and 1538. Two lumbar vertebrae showing signs of Pott's disease were sampled from one burial, along with a wrist bone from a second skeleton showing possible indicators of tuberculosis, and a lumbar vertebra showing no signs of tuberculosis, the last used as a negative control. The extracts were tested with a set of PCRs with varying degrees of discriminatory power within the *M. tuberculosis* complex. These included the multicopy but non-specific IS6110 sequence, the *oxy*R pseudogene mentioned above, the mtp40 region of the phospholipase C gene, which allows *M. tuberculosis* to be distinguished from most *M. bovis* strains, and a randomly identified fragment of the *M. bovis* genome thought to be absent in the other complex members. The *rpo*B gene was also tested because, although it does not allow discrimination between the complex members, it contains an SNP that indicates resistance to the antibiotic rifampicin, and so provides a control for modern contamination. All three of the bones from the two skeletons with tuberculosis lesions gave positive results with the IS6110 and *rpo*B PCRs, none of the latter displaying the rifampicin resistance mutation, suggesting that *M. tuberculosis* complex DNA was present and that this did not derive from a modern *M. tuberculosis* contaminant. The mtp40 fragment was amplified from all three bones, consistent with the presence of *M. tuberculosis*, and the *M. bovis* genome fragment could not be detected in either of the samples.

Only one of the three samples yielded an *oxy*R PCR product, but this product contained the version of the SNP associated with *M. tuberculosis*. The bone from the skeleton lacking any indications of tuberculosis did not yield products in any of the PCR tests. Overall the results showed that the causative pathogen in these samples was *M. tuberculosis*, from the pre-antibiotic era. The project as whole was an excellent demonstration of the use of biomolecular techniques to confirm an osteological identification of paleodisease.

15.2.3 Biomolecular studies have detected tuberculosis in skeletons with no bony lesions

There have been several studies similar to that described above at St Mary Graces, where ancient DNA has been used to check the accuracy of a tuberculosis identification made by osteological examination. Can biomolecular work also take identification of tuberculosis further than is possible by osteological means? We noted above that only a small minority of people who contract pulmonary tuberculosis actually go on to develop bony lesions. It would therefore be interesting to know the extent to which osteological identification of tuberculosis underestimates the actual prevalence of the disease in past societies. This is something that ancient DNA could address, at least in theory. If a person was infected with *M. tuberculosis* when he or she died, then perhaps the bacterial DNA could be detected in the skeleton even if the latter shows no osteological signs of the disease.

In practice, attempts to identify *M. tuberculosis* DNA in skeletons without lesions raises many of the problems and questions that are inherently tied up with biomolecular paleopathology. If the bacterium has not begun to spread systemically, then will it be present in the bones at the time of death? Experimental studies with rabbits have shown that during the pulmonary stage the vast majority of bacteria are, not surprisingly, located in the lung tissue, with very few elsewhere in the body and virtually none in the bone marrow. We might therefore predict that skeletons that do not display the bony indications of tuberculosis will not yield *M. tuberculosis* DNA even if the person had tuberculosis at the time of death. Armed with this prediction we would be tempted to dismiss any reports of the presence of *M. tuberculosis* DNA in such skeletons as arising from contamination with modern DNA.

Unlike the equivalent situation with syphilis, where comprehensive analysis of bones with the appropriate lesions failed to detect DNA of the causative bacteria, those most careful studies that have been reported have tended to suggest that *M. tuberculosis* DNA can sometimes be recovered from skeletons that lack bony indicators of tuberculosis. One project was carried out with skeletons from two Lithuanian cemeteries, at Alytus and Krazai, that were in use during the 15th–17th centuries AD. Tuberculosis was clearly a major cause of death in Lithuania at this time, as many of the skeletons in the cemetery had the expected pathology. Six individuals from Alytus and one from Krazai were analyzed, three of those from Alytus showing no osteological signs of tuberculosis. All samples taken from the four skeletons with bone lesions gave positive results with the IS6110 PCR. Samples taken from the three skeletons without lesions also gave positive results with these PCRs. A typical example of cross-contamination? The intriguing feature of this report is that the soil from the areas around the burial was also tested for the presence of *M. tuberculosis* DNA, to see if migration of DNA was occurring between the burials, possibly explaining why the bones without lesions were giving positive results. These soil PCRs were negative, except for one, which, on microscopic examination, was found to contain minute fragments of soft tissue from one of the burials. As well as showing that DNA was not migrating through the soil, the soil results also show that rampant cross-contamination between samples was not occurring in the laboratory, suggesting that the positive PCRs from the bones without lesions were not themselves the result of contamination.

There have been other similar reports of *M. tuberculosis* DNA in skeletons that show no osteological signs of tuberculosis. The difficulty in understanding the meaning of these detections—whether they are genuine or, despite all indications to the contrary, result from contamination—illustrates the problems in interpreting any ancient DNA result that goes against one's expectations.

15.2.4 Phylogenetic studies using modern DNA have shown that *M. tuberculosis* is not derived from *M. bovis*

In recent years, much work has been carried out to understand the sequence variations not just between the species of the *M. tuberculosis* complex, but also between different strains of these species. This work, which was prompted by publication of the complete genome sequence of *M. tuberculosis* in 1998, has enabled the evolutionary relationships between the members of the *M. tuberculosis* complex to be unravelled in increasing detail. The results have forced a reappraisal of the conventional archaeological view that human tuberculosis is descended from the bovine disease.

The first comprehensive evolutionary scheme to be published was based on SNP variations at the *kat*G, *gyr*A, *oxy*R, *pnc*A, and *mmp*16 loci, on the presence or absence of a deletion sequence called TbD1, and on the sequences within 10 "regions of difference" (RDs) previously identified in the *M tuberculosis* genome sequence. All of these loci were typed in 100 isolates of *M. tuberculosis* bacteria from all over the world. The resulting evolutionary tree (Figure 15.3) has a number of interesting features, the most notable of these being the clear demonstration that *M. bovis* is not ancestral to *M. tuberculosis*. The hypothesis that human tuberculosis is derived from the bovine form had arisen largely because *M. tuberculosis* infects only humans, whereas *M. bovis* has a broad host range, suggesting that *M. tuberculosis* is a relatively young species that has not yet adapted to infect other hosts. The supposed transfer from cattle to humans was thought to have taken place in southwest Asia soon after cattle were domesticated, humans at that time living in close association with their livestock, providing the opportunity for pathogens to transfer from one species to the other. This idea has now been rejected with the realization that *M. bovis* is probably descended from *M. africanum*, which in turn is derived from an ancestor of *M. tuberculosis*.

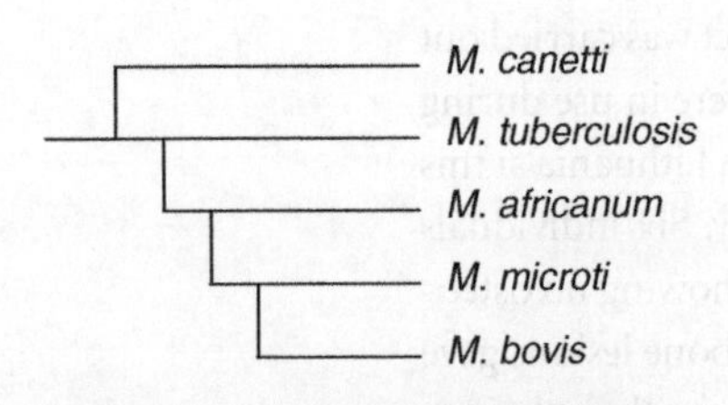

Figure 15.3 Evolutionary relationships between members of the *M. tuberculosis* complex as revealed by typing variable loci in 100 isolates, and showing that *M. tuberculosis* is not derived from *M. bovis*.

The ancestral nature of *M. tuberculosis* has been confirmed by further studies of modern strain variations, using more and more loci and greater numbers of isolates. The species probably originated 2–3 million years ago in East Africa and has been a human pathogen since them, initially affecting early hominids. Analysis of 875 *M. tuberculosis* strains from 80 different countries has divided the species into six main lineages and 15 sublineages, each of the main lineages associated with a specific geographical area. All six of these are also found in Africa, supporting the hypothesis that the disease spread out of Africa with the first migrations of *Homo erectus* almost 2 million years ago, and probably also again when *Homo sapiens* left Africa some 70,000 years ago.

Although these phylogenetic studies show that human tuberculosis is not derived from the bovine disease, this does not mean that all cases of tuberculosis in the archaeological record must be due to *M. tuberculosis*. It is quite possible that there have been localized transfers of the bovine form of the disease in the past, which might not have been transmitted very far within the human population, but which occurred nonetheless. This has been suggested by a study of four skeletons from the Iron Age site of Aymyrlyg in southern Siberia. The *oxy*R PCR product obtained from extracts of these skeletons consistently showed an A nucleotide, indicative of *M. bovis*, and the TbD1 deletion was absent, also consistent with *M. bovis*. The Iron Age society at Aymyrlyg was pastoralist, providing the opportunity for these individuals to acquire bovine tuberculosis through close contact with infected animals or through ingesting infected meat and milk.

15.2.5 The relationship between Old and New World tuberculosis could be solved by biomolecular analysis

A combination between the evolutionary insights from large-scale sampling of modern *M. tuberculosis* DNA and the temporal depth provided by ancient DNA raises hopes that longstanding questions about the origins of tuberculosis in the New World can be settled. Tuberculosis was already present in the New World, both in North and South America, before the arrival of Columbus. Pre-Columbian skeletons with Pott's disease have been described from several parts of North America, the earliest dated around AD 1000, and South American skeletons and mummies with indicators of tuberculosis have been found from even earlier periods, certainly present at AD 700.

How did tuberculosis get to the New World? There are three possibilities:

- It could have been brought into North America by the first colonizers who crossed the Bering straits 12,000 years ago, or earlier (Section 16.3). If this is the case, then we should expect to see Asian *M. tuberculosis* lineages in pre-Columbian human remains.
- It could be have been passed to humans from infected wild animals such as bison. Some skeletons of these animals from paleoindian killing sites dating to 15,000–20,000 years ago have bone lesions that could indicate the presence of tuberculosis, or which equally could be due to some other disease such as brucellosis.
- It is possible that pre-Columbian contact by Europeans was responsible for the introduction of tuberculosis into the New World. Bearing in mind that tuberculosis was present in the New World by AD 1000, the only possible candidates would be the Vikings, who were present in Newfoundland at the site of L'Anse aux Meadows at about AD 1000. If European contact was the source of New World tuberculosis, then we would expect to find European *M. tuberculosis* strains in the pre-Columbian native populations.

Ancient DNA studies of pre-Columbian remains might enable these various possibilities to be tested. *M. tuberculosis* complex DNA has been detected in Andean mummies from before AD 1000 and in human skeletal remains from AD 1000–1200 in North America, but these projects made use primarily of insertion sequence PCRs and hence provide no information on the geographical affinities of the strains with which the individual were infected. Future work should be more revealing.

Although tuberculosis was present in the New World, the disease caused immense destruction in most Native American populations following contact with Europeans. At one time, before archaeological examples of Pott's disease had been discovered, the inability of Native Americans to resist the effects of European tuberculosis (which now dominates the Americas) was looked on as strong evidence that the disease originated in the Old World, not reaching the New World until the arrival of Europeans. Now we know that the disease was already present in the Americas, we can speculate that there must have been differences between the types of tuberculosis present in the Old and New Worlds at the time of Contact. These differences must have related to the virulence or immunological features of the particular strains of *M. tuberculosis* present either side of the Atlantic at that time. Clinical microbiologists are becoming aware that an interrelationship exists between modern hosts and mycobacteria whereby particular lineages of *M. tuberculosis* are adapted to specific human populations in particular geographical locations, and the Contact situation represents an example of what happens when these lineages and populations mix. At present, identifying the important differences between the Old and New World strains of *M. tuberculosis* is just beyond the limit of biomolecular paleopathology, but advances in understanding the genetic basis to the bacterium's virulence, and the development of better means of acquiring sequence data from ancient DNA, will make these issues more tractable in the future.

15.2.6 Mycolic acids have also been used in attempts to identify ancient *M. tuberculosis*

Ancient DNA is not the only biomarker than can be used for studying tuberculosis in human skeletons. Some success has also been achieved in the detection of characteristic lipid components of the *M. tuberculosis* cell wall. These lipids, called **mycolic acids**, are synthesized by all mycobacteria and also by members of related genera such as *Corynebacterium* and *Nocardia*. Like most types of lipid, mycolic acids have a number of structural variations, and the particular combination of molecules produced by a bacterium is characteristic of that species. If these combinations are retained during diagenesis, then mycolic acid typing might provide an alternative means of identifying *M. tuberculosis* in a specimen.

Mycolic acids are a type of fatty acid, and hence are long-chain hydrocarbons with a terminal carboxylic acid group (Figure 4.1). They differ from the standard fatty acid structure in that they are branched molecules (Figure 15.4), the branch point being the first carbon adjacent to the carboxylic group. The hydrogen normally attached to this carbon is replaced by a second long hydrocarbon chain, the first carbon of which carries a hydroxyl (–OH) group. The total length of the

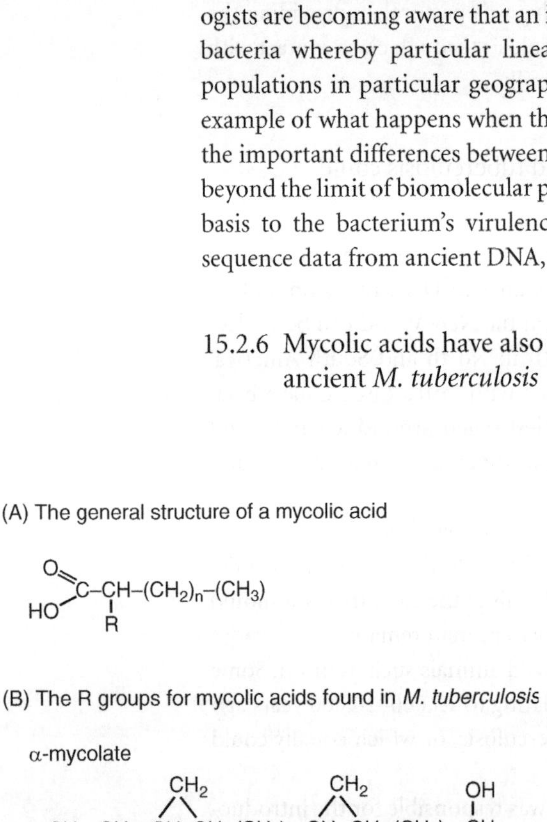

(A) The general structure of a mycolic acid

$HO(O{=})C–CH(R)–(CH_2)_n–(CH_3)$

(B) The R groups for mycolic acids found in *M. tuberculosis*

α-mycolate

$CH_3–CH_2–CH–CH–(CH_2)_x–CH–CH–(CH_2)_y–CH(OH)–$ (each CH–CH pair bridged by CH_2)

keto-mycolate

$CH_3–CH_2–CH(CH_3)–C(=O)–(CH_2)_x–CH–CH–(CH_2)_y–CH(OH)–$ (CH–CH bridged by CH_2)

methoxy-mycolate

$CH_3–CH_2–CH(CH_3)–CH(OCH_3)–(CH_2)_x–CH–CH–(CH_2)_y–CH(OH)–$ (CH–CH bridged by CH_2)

Figure 15.4 Mycolic acids.

two hydrocarbon chains varies between 60 to 90 carbons in *Mycobacterium* species, and is generally shorter in other genera. Mycolic acids are divided into categories depending on the nature of the chemical groups attached to the second hydrocarbon chain. There are at least 10 different categories in the mycobacteria as a whole, a single species usually making versions of two or three of these. In *M. tuberculosis*, over 70% of the molecules are α-mycolic acids. Each of these contains one or more cyclopropane units, in which two adjacent carbons are linked via a methyl ($-CH_2$) group. The remainder are made up of roughly equal amounts of methoxy-mycolates, in which one of the carbons has a methoxy ($-OCH_3$) group, and keto-mycolates, which have a carbonyl (=O) group. Methoxy- and keto-mycolates usually also have cyclopropane units and at least one carbon carrying a methyl ($-CH_3$) group.

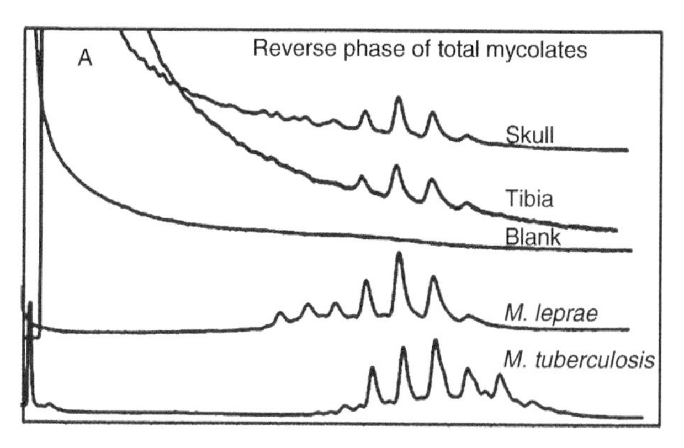

Figure 15.5 Reverse phase HPLC separation of mycolic acids. The two lower traces show the profiles from *M. tuberculosis* and *Mycobacterium leprae*, the causative agent of leprosy. Species-specific differences in the mycolic acid peaks can be seen. The upper two traces are from a skeleton dating to the 1st–4th centuries AD from Uzbekistan. The comparison with the controls reveals that this skeleton contains mycolic acids from *M. leprae*. Reprinted from Michael Taylor, Soren Blau, Simon Myas, Marc Monot, Oona Lee, David Minnikin, Gurdyal Besra, Stewart Cole, and Paul Rutland, "*Mycobacterium leprae* genotype amplified from an archaeological case of lepromatous leprosy in Central Asia," *Journal of Archaeological Science*, 36, 2408–14, copyright 2009, with permission from Elsevier and the authors.

Mycolic acids extracted from bacterial cultures can be examined by thin layer chromatography (TLC) or high performance liquid chromatography (HPLC), the latter giving a characteristic trace that identifies the species of origin (Figure 15.5). Whether these traces are truly species specific has not yet been established, simply because there are over 100 different species of *Mycobacterium* and not all have been tested, but from what is known so far the assumption seems reasonable. A greater problem in the use of mycolic acids as archaeological biomarkers is the absence of information on the way these molecules break down over time, which means that the mycolate structures expected in a bone that contained *M. tuberculosis* at the time of death are not known. Those studies that have been carried out so far with archaeological specimens have yielded HPLC traces identical to those obtained with modern *M. tuberculosis* extracts, the only difference being lower peak heights indicating that less material is present. These results suggest that the mycolic acid structures do not change during diagenesis, the molecules simply decreasing in frequency with different types of mycolic acid degrading at similar rates. If it can be established that these results are in accord with the way that the molecules actually break down during diagenesis, then mycolic acids are likely to become popular biomarkers for ancient tuberculosis.

15.2.7 Difficulties in the study of ancient tuberculosis

The establishment of mycolic acids as alternative biomarkers for ancient tuberculosis is an important step forward, as it provides a means of obtaining independent confirmation of identifications made with ancient DNA. Independent confirmation is needed because, in some respects, the use of ancient DNA is even more problematic in paleopathology than it is in other areas of biomolecular archaeology.

Principal among these problems are the additional challenges that must be met in ensuring that supposed ancient DNA detections are not due to contamination. In Section 9.1, we examined the various precautions that are taken to ensure that extracts of archaeological material do not become contaminated with modern DNA, either from one of the people who handle the material or from PCR products that have been generated in previous experiments. It is very unlikely that a bone would become contaminated with *M. tuberculosis* DNA by handling, but cross-contamination with old PCR products is always a possibility. When human DNA is being studied, the sequence of the PCR product that is obtained sometimes indicates that this type of contamination has occurred. If this sequence is from the mitochondrial hypervariable region, or some variable region of the nuclear genome, and is identical to a sequence recently obtained from a previous specimen, then it is immediately apparent that cross-contamination might have occurred, and the necessary steps can be taken to determine if this has been the case or not. The problem when a disease is being studied is that most of the DNA biomarkers used to detect the presence of the pathogen are invariant. With *M. tuberculosis*, for example, the IS6110 and IS1081 PCRs frequently used in an initial screen for ancient DNA always give the same sequence, because the sequence of the amplified region is identical in all copies of the insertion element. The same is true of most of the other PCRs routinely used to detect *M. tuberculosis*, including the valuable TbD1 amplicon that distinguishes this species from other members of the complex. As these PCRs always give the same sequence, it is impossible, simply from looking at the sequence, to know if it has been obtained from genuine ancient DNA present in the specimen under study, or if it arises from cross-contamination with a previous PCR product. Biomolecular paleopathology therefore places extreme demands on the measures taken to prevent cross-contamination with old PCR products, these involving the use of isolated laboratories with filtered air supplies, because without these facilities it is impossible to have any assurance that the DNA that is detected is genuinely ancient.

Studies of ancient tuberculosis present a second contamination problem, one we have not encountered previously. The genus *Mycobacterium* includes over 100 species, most of which are found in soil and water and only a few of which are pathogenic. The expectation is therefore that any archaeological material that has come into contact with soil and/or water is likely to contain some of these environmental mycobacteria. This has been borne out by a study of 12 Andean mummies, 7 of which were found to contain DNA from environmental mycobacteria, as well as work with skeletons from London, dating to AD 1350–1550, which have shown that these contain DNA from a variety of environmental mycobacteria. Could a PCR directed at *M. tuberculosis* also amplify DNA from one of these environmental species? This is almost certainly not the case for the IS6110 and IS1081 PCRs, as the distribution of these insertion sequences has been studied and it is known that they are present only in the *M. tuberculosis* complex. Similarly the specificity of the TbD1 PCR for *M. tuberculosis* can be assured because of the phylogenetic studies that have shown that this deletion occurred relatively recently in the *M. tuberculosis* lineage. But a problem might arise when attention moves to the more interesting SNPs in the *M. tuberculosis* genome, the ones that must be typed in order to study the particular strains of the bacterium present in skeletons from different places and periods. Most of these SNPs lie within essential genes whose overall sequences are likely to be very similar in many different mycobacteria, raising the possibility that a PCR designed to amplify the region containing an SNP in *M. tuberculosis* might also amplify the equivalent region from an environmental mycobacterium whose DNA is present in the bone. The sequence of this amplicon might be identical to that of *M. tuberculosis*, so the error that has been made will not be recognized, but the resulting SNP data will be erroneous. The only way to circumvent this problem is to be aware of the sequences of the relevant regions of the genomes of all the environmental mycobacteria likely to have contaminated the bone, so that PCRs that are truly specific for *M. tuberculosis* can be designed. Solving this problem is perhaps the major challenge for future biomolecular research into ancient tuberculosis.

15.3 Biomolecular Studies of Other Diseases

The research that has been carried out on ancient tuberculosis illustrates many of the strengths and limitations of biomolecular paleopathology, and we have therefore spent some time focussing on this disease. Now we will survey the work that has been done on other ancient pathogens.

15.3.1 Leprosy is a second mycobacterial disease

Leprosy might be thought of as a medieval disease of no relevance in the modern world, yet there are about half a million new cases each year. It is a disease of poverty and war, and of places where there is little medical organization and support. Leprosy can be cured with antibiotics, but the people who need them cannot obtain them. Leprosy is spread by close contact, and hence is found in families, with genetic susceptibility thought to play a role in whether a person acquires leprosy after exposure. Leprosy was first documented in India around 600 BC, and seems to have been replaced by tuberculosis in medieval times, leading to the suggestion that tuberculosis might confer some immunity against leprosy, although exactly how is not clear.

There are two forms of the disease, the milder tuberculoid leprosy, and lepromatous leprosy, which is the form associated with the most severe manifestation of symptoms, including loss of extremities and facial disfigurement. The loss of extremities and resorption of bone seen in living victims and skeletal remains is associated with numbness, as the disease attacks the nervous system. Characteristically the fingers and toes are damaged and lost, and the nose and mouth areas of the skull are subject to bone resorption. The skull damage is similar to that seen with syphilis, so it is the pattern of damage to the extremities that provides the identification of this disease in skeletons.

Leprosy is caused by *Mycobacterium leprae*, and many of the issues relevant to the study of ancient tuberculosis are relevant to leprosy also. There have been fewer biomolecular studies of ancient leprosy, but those that have been published suggest that the disease is amenable to ancient DNA investigation. One such project concerned a human skeleton from Orkney dating to the 13th or 14th centuries AD. The skull of this individual showed classic signs of leprosy, with resorption of the nasal bones and premaxilla, and also a large lesion through the upper palate, but no bones from the lower limbs were preserved, so the diagnosis could not be definite. PCRs were designed to amplify part of a repetitive sequence called RLEP, which has a copy number of at least 28 in the *M. leprae* genome and is thought to be specific for this bacterium. Amplicons were obtained with bone samples taken from the skull, but samples from the clavicle, left scapula, and other bones gave negative results. In a second project, PCRs were also directed at short tandem repeats and longer repeat sequences in the *M. leprae* genome, the results suggesting that different strains of the bacterium were present at two medieval sites in Britain, one at Wharram Percy in North Yorkshire and the other at Ipswich, Suffolk. These and other projects have emphasized that a knowledge of the locations of the active disease centers at the time of death is required in order to select the appropriate parts of the skeleton to sample, the best parts being those where bone resorption or growth has taken place, but not the center of these lesions, as these appear to lack *M. leprae* DNA. The demonstration that *M. leprae* DNA can be obtained from the soil in areas inhabited by leprosy cases indicates that special care must be taken to ensure that a putative ancient DNA detection is genuinely endogenous to the skeleton being studied.

As with tuberculosis, modern strains of *M. leprae* are providing information on the evolution and past spread of the disease. The species has a very low level of genetic diversity compared to other bacteria, and so far only 215 polymorphic sites, mostly SNPs, have been identified in genomes from strains taken from different parts of the world. These have enabled 16 genotypes of *M. leprae* to be identified and the geographical distributions of these have been studied in both modern and archaeological samples. The global distribution of these genotypes is consistent with a single origin for leprosy in East Africa, followed by its spread into Asia and Europe when modern humans left Africa some 70,000 years ago, with a much later transmission of more recent strains to the Americas by European colonists and African slaves, and to East Asia along the Silk Road. The proposed prehistoric spread of leprosy has, however, been questioned on the grounds that modern leprosy is transmitted by close contact and is associated with urban conditions, whereas the dispersal of humans out of Africa probably involved small numbers of highly mobile hunter-gatherer groups that had little contact with each other. Future work on leprosy in archaeological remains may throw more light on these questions.

15.3.2 Malaria has been detected in some archaeological bones, but not in others

Malaria has global importance and there are fears that with climate warming its distribution could spread, possibly including re-entry into Europe. The disease was widespread in the Mediterranean region, until drainage works and insecticide spraying in the mid-20th century eradicated the mosquito responsible for its transmission. It was widespread in classical times, both the ancient Greeks and Romans being aware of its existence, though at that time the role of the mosquito vector was not understood, it being thought until the late 1800s that the disease was caused by the "bad air" or *mal aria* associated with marshes and stagnant water.

Whether or not malaria leaves its mark on the human skeleton is debatable. The osteologist Lawrence Angel proposed that the skeletal indicators of chronic anemia – porotic hyperostosis and cribra orbitalia – seen at high frequencies in Neolithic Greek remains could be attributed to the presence of malaria. This link is logical because, as described above (Section 15.1), some types of genetic anemia or thalassemia provide a degree of protection against malaria and so are more frequent in populations in which the disease is endemic. It is not a secure link, however, because porotic hyperostosis can also result from chronic anemia caused by other factors, including iron-deficient diets. This presents a problem for the biomolecular archaeologist, as any attempt to use a biomarker to detect a disease pathogen in a skeleton becomes a hit and miss affair if there are no firm skeletal indicators to point out the best material to examine. In view of this, it is not surprising that there are few positive reports of the detection of ancient malaria. In the first to be published, a segment of the *Plasmodium falciparum* 18S ribosomal RNA gene was successfully amplified from a 60 year old rib bone from an individual believed to be infected with malaria at the time of death, although other bones from the same collection failed to give PCR products. Negative results were also obtained with samples from an Egyptian mummy dating to 700 BC, even though the PCR system could amplify as little as 20 femtograms of DNA. Similarly incomplete results were obtained with the infant and fetal bones from the 5th century AD cemetery at Lugnano in Teverina (Section 10.4), thought to be the victims of a malaria epidemic because the deaths seemed to have occurred in a short time in one year, probably during the summer. Forty-seven skeletons were excavated, 22 of which were premature births. Only one of these yielded *P. falciparum* PCR products, this being the oldest child in the cemetery, a female of 2–3 years. Indirect evidence for the presence of malaria was also sought by examination of the G6PD gene, a deficiency in this enzyme arising as a response to endemic malaria in a manner similar to thalassemia (Section 15.1). One male infant possessed the most common G6PD mutation found in the Mediterranean, known as the G6PD Med variant. Detection of this variant indicates that malaria was present in Roman Italy at this time, but tells us nothing about the cause of death of the infant who carried the mutation.

The difficulties in using biomolecular methods to study ancient malaria are further emphasized by attempts to detect *Plasmodium* DNA in skeletons from medieval England. Malaria was endemic in many parts of England until the 19th century, particularly in the marshy coastal districts of Kent and Essex and in the East Anglian Fens. This type of malaria was caused by *P. vivax*, the most widely distributed of the four malaria parasites, with a lower temperature requirement for reproduction in the mosquito than the more dangerous *P. falciparum*. A study of 159 skeletons from four sites in England located close to or within marshy areas, all thought to be likely locations of vivax malaria, failed to detect the parasite DNA in any sample, despite the use of optimized PCRs directed at multicopy targets and the detection of human ancient DNA in several of the skeletons. The failure of this particular project underlines the difficulty in searching for pathogen DNA in the absence of skeletal indicators of the disease, and also raises questions about the feasibility of the biomolecular approach to this disease. The number of malaria parasites in the bloodstream of an individual who is suffering from malaria but not actually dying of the disease might be too low to leave a biomolecular signature that can subsequently be detected in archaeological remains. This was the explanation given by the researchers who carried out the Lugnano study – that the single infant who yielded *P. falciparum* PCR products might have had a high parasite load in her blood because she was suffering from a massive infection at the time of death. The less virulent vivax malaria might never give such a high parasite load and hence might simply be undetectable in the archaeological record.

15.3.3 *Yesinia pestis* has been detected in archaeological teeth

The bacterium *Yersinia pestis* causes bubonic plague, the Black Death of 14th century Europe (1347–52), which is believed to have killed up to one-third of the population, and persisted with episodic outbreaks for the next few hundred years. Earlier epidemics of plague, such as the Justinian (AD 451–700) and the plague of Athens (430 BC), have been retro-diagnosed as bubonic plague by medical historians, although discrepancies between historical accounts and the known progression of *Y. pestis* infection have led some to propose that these ancient plagues were caused by a different pathogen. A third pandemic emerged in mid-19th century China. The ancient, medieval and 19th century diseases have been linked to three variants of *Y. pestis*:

- Antiqua, the main type, which originated in Central Asia and entered Europe from Africa, is thought to have caused the Justinian plague.
- Medievalis, a variety of Antiqua, also originated in central Asia, and then spread to the Crimea and caused the Black Death in Europe.
- Orientalis is thought to be derived from Medievalis, and arose in the 19th century to cause the third pandemic.

These variants have genetic differences, and so could be distinguished by ancient DNA typing, if it can be established that *Y. pestis* DNA does indeed survive in human skeletal remains. Bubonic plague does not cause any diagnostic bone changes in the skeletons of its victims but, at least for the medieval period, there are recorded "plague pits" where the victims of the disease were buried. Any skeleton from such a site can be assumed, with a reasonable degree of confidence, to belong to a person who died of bubonic plague.

The presence of *Y. pestis* DNA in human remains was first reported for individuals buried in two mass graves in Provence in the 16th–17th centuries AD. Dental pulp was used as the source material, experimental studies having shown that infected animals harbor plague bacteria in their teeth. PCRs were directed at two *Y. pestis* genes: *pla*, coding for the plasminogen activator protein associated with virulence, and the RNA polymerase gene *rpoB*. The first of these PCRs is expected to give a 300 bp product, quite long for ancient DNA detection, but 6 of the 12 individuals that were tested gave positive results. The *rpoB* PCR, whose product is a more reasonable 133 bp, was positive for four extracts, two of those testing positive for *pla*, and two others. No *Y. pestis* amplification products were obtained with teeth from a medieval cemetery in Toulon, which had no signs of plague, but human nuclear DNA could be detected in all the teeth that were tested, from both the Provence and Toulon sites. In more recent projects, the same researchers have used information from the complete genome sequences of the Medievalis and Orientalis strains of *Y. pestis* to design PCRs that enable these variants to be distinguished. Application of these PCRs to teeth from six burial sites in France, from the 5th to 18th centuries, gave sporadic results, but in all cases the strain that was detected was

Orientalis, despite this variety being associated more specifically with the later, 19th century, pandemic in the Far East.

Typing of *Y. pestis* DNA in human teeth therefore appears to be providing interesting new information on the nature of the bubonic plague that was present in Europe at various times in the past. Standing against these results, however, is the outcome of a large project that used material from seven sites from across Europe, dating to the 13th–17th centuries AD, five of these sites looked on as possible plague cemeteries. Extracts of 108 teeth from 61 skeletons were tested with four human and eight different *Y. pestis* PCRs. Although human mitochondrial DNA could be detected in some of these specimens, only a few positive results were obtained with the *Y. pestis* PCRs, and none of these could be reproduced when attempts were made to replicate the results in a second laboratory. The discrepancy between the outcome of this project and the positive results obtained with specimens from France is a depressingly familiar illustration of the difficulties that arise when attempts are made to use ancient DNA in biomolecular paleopathology.

15.3.4 Though not archaeology, studies of the 1918 influenza virus indicate a future goal for biomolecular paleopathology

We conclude this chapter by looking at an ancient DNA project that does not, strictly speaking, fall within the realms of archaeology, but that indicates ways in which archaeological material might be used in the future. The 1918 influenza virus killed between 20 and 100 million people in the "Spanish flu" epidemic of 1918–20. Was this virus particularly virulent, or does the high death rate reflect a decreased level of resistance among the stressed and, in some cases, starving populations of Europe immediately after World War I? There have also been questions about the origins of the 1918 influenza virus, alternative theories being that it resulted from reassortment between human and swine flu, or was a form of avian flu that "jumped" to humans. We are overdue another large flu pandemic, and a better knowledge of the origins and virulence of the 1918 virus might help us prepare for its inevitable arrival.

The influenza genome is made of RNA rather than DNA, so before ancient extracts can be examined by PCR any RNA that they contain must be copied into DNA. This is carried out with the enzyme called **reverse transcriptase**, which is an RNA-dependent DNA polymerase, and the process as a whole is called **reverse transcriptase PCR** or **RT-PCR** (Figure 15.6). Samples of the 1918 virus have been obtained from tissue sections that were taken from flu victims and then preserved by fixing in formalin, and also from the lungs of a female victim in Alaska who became buried in permafrost. Initially PCRs were directed at eight viral genes, coding for the hemagglutinin protein, neuraminidase, three RNA polymerase subunits, two proteins from the core of the virus capsid, and an RNA-binding protein associated with the viral genome. Comparisons of the sequences of the PCR products with the equivalent genes of modern flu viruses have provided some interesting insights

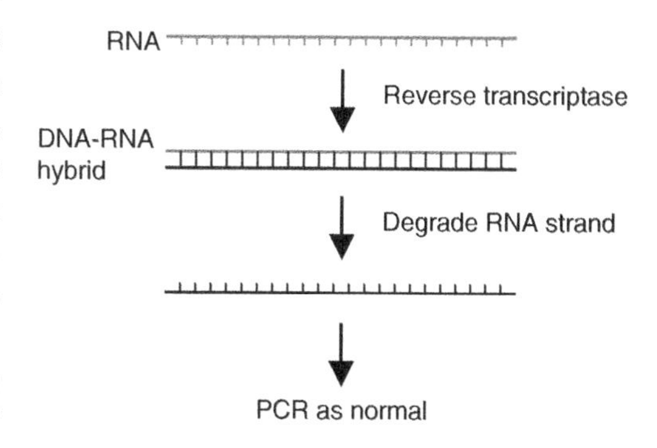

Figure 15.6 Reverse transcriptase PCR.

into the origins and extreme virulence of the 1918 disease. The RNA polymerase subunits of the 1918 virus are very similar to avian flu proteins, with only a few amino acid changes, suggesting that 1918 flu was caused by an avian virus that adapted to humans, rather than a reassorted human-swine virus. The hemagglutinin PCRs showed that in the 1918 virus this protein had largely retained its avian structure, but with small alterations that enabled the virus to bind to surface receptors on human cells. These are unusual features not seen in modern influenza strains, and could well explain the high virulence of the 1918 virus.

The sequence information obtained from preserved specimens has also enabled partial and complete copies of the 1918 virus genome to be reconstructed. Initially 1918 sequences were inserted into a modern virus genome and mice inoculated with the resulting hybrids. The mice suffered severe lung damage when the 1918 hemagglutinin gene was present in the virus construct, much more so than when the virus carried a modern hemagglutinin. But the hemagglutinin was not on its own responsible for the extreme virulence of the 1918 virus. Experiments with macaque monkeys, whose immune system is very similar to that of humans, showed that the 1918 virus elicits a stronger immune response than contemporary strains of flu. Paradoxically, this enhanced response might benefit the virus, as the resulting tissue inflammation facilitates entry of the virus particles into the host's lung cells. This could explain one of the unusual features of the 1918 pandemic – the high mortality among younger people of 15–45 years in age whose relatively robust immune systems should have made them better able to withstand the effects of the virus. The strength of their immune response was the cause of death among these people.

Sequence analysis of ancient RNA from the 1918 influenza virus has given clinical virologists a number of leads that are being pursued in efforts to prepare us for the next severe influenza pandemic. In the long term, work of this kind is likely to be one of the major outcomes of biomolecular paleopathology. Rather than using biomolecular data solely to study the archaeological implications of disease, we will see the resulting information used more frequently as a means of understanding the pathogenicity of the causative organism and hence of combating the disease in its modern form. The gradual move from PCR to high throughput sequencing procedures (Section 2.5), which can yield partial or even complete genome sequences of ancient pathogens, will move biomolecular paleopathology even more rapidly in this new direction.

Further Reading

Bouwman, A.S. & Brown, T.A. (2005) The limits of biomolecular archaeology: ancient DNA cannot be used to study venereal syphilis. *Journal of Archaeological Science*, 32, 691–702.

Brosch, R., Gordon, S.V. & Marmiesse, M., *et al.* (2002) A new evolutionary scenario for the *Mycobacterium tuberculosis* complex. *Proceedings of the National Academy of Sciences USA*, 99, 3684–9.

Drancourt, M., Aboudharan, G., Signoli, M., Dutour, O. & Raoult, D. (1998) Detection of 400-year-old *Yersinia pestis* DNA in human dental pulp: an approach to the diagnosis of ancient septicemia. *Proceedings of the National Academy of Sciences USA*, 95, 12637–40.

Faerman, M., Janauskas, R. & Gorski, A., *et al.* (1997) Prevalence of human tuberculosis in a medieval population of Lithuania studied by ancient DNA analysis. *Ancient Biomolecules*, 1, 205–14.

Filon, D., Faerman, M., Smith, P. & Oppenheimer, A. (1995) Sequence analysis reveals a β-thalassaemia mutation in the DNA of skeletal remains from the archaeological site of Akhziv, Israel. *Nature Genetics*, 9, 365–8.

Gilbert, M.T., Cuccui, J. & White, W., *et al.* (2004) Absence of Yersinia pestis-specific DNA in human teeth from five European excavations of putative plague victims. *Microbiology*, 150, 341–54.

Konomi, N., Lebwohl, E., Mowbray, K., Tattersall, I. & Zhang, D. (2002) Detection of mycobacterial DNA in Andean mummies. *Journal of Clinical Microbiology*, 40, 4738–40. [Showing contamination with environmental mycobacteria.]

Minnikin, D.E., Minnikin, S.M., Parlett, J.H., Goodfellow, M. & Magnusson, M. (1984) Mycolic acid patterns of some species of *Mycobacterium*. *Archives of Microbiology*, 139, 225–31.

Monot, M, Honoré, N. & Garnier, T., *et al.* (2009) Comparative genomic and phylogeographic analysis of *Mycobacterium leprae*. *Nature Genetics*, 41, 1282–9.

Moodley, Y., Linz, B. & Yamaoka, Y., *et al.* (2009) The peopling of the Pacific from a bacterial perspective. *Science*, 323, 527–30. [*Helicobacter pylori*.]

Roberts, C.A. & Buikstra, J. (2008) *The Bioarchaeology of Tuberculosis: A Global View on a Reemerging Disease*. University Press of Florida, Gainesville.

Sallares, R. & Gomzi, S. (2001) Biomolecular archaeology of malaria. *Ancient Biomolecules*, 3, 195–203.

Salo, W.L., Aufderheide, A.C., Buikstra, J. & Holcomb, T.A. (1994) Identification of *Mycobacterium tuberculosis* DNA in a pre-Columbian Peruvian mummy. *Proceedings of the National Academy of Sciences USA*, 91, 2091–4.

Sreevatsan, S., Pan, X. & Stockbauer, K.E., *et al.* (1997) Restricted structural gene polymorphism in the *Mycobacterium tuberculosis* complex indicates evolutionarily recent global dissemination. *Proceedings of the National Academy of Sciences USA*, 94, 9869–74.

Taylor, G.M., Goyal, M., Legge, A.J., Shaw, R.J. & Young, D. (1999) Genotypic analysis of *Mycobacterium tuberculosis* from medieval human remains. *Microbiology*, 145, 899–904. [The Abbey of St Mary Graces in London.]

Taylor, G.M., Murphy, E., Hopkins, R., Rutland, P. & Christov, Y. (2007) First report of *Mycobacterium bovis* DNA in human remains from the Iron Age. *Microbiology*, 153, 1243–9.

Taylor, G.M., Rutland, P. & Molleson, T. (1997) A sensitive polymerase chain reaction method for the detection of *Plasmodium* species DNA in ancient human remains. *Ancient Biomolecules*, 1, 193–203.

Taylor, G.M., Watson, C.L., Bouwman, A.S., Lockwood, A.N.J. & Mays, S.A. (2006) Variable nucleotide tandem repeat (VNTR) typing of two paleopathological cases of lepromatous leprosy from Mediaeval England. *Journal of Archaeological Science*, 33, 1569–79.

Taylor, G.M., Widdison, S., Brown, I.N. & Young, D. (2000) A mediaeval case of lepromatous leprosy from 13th–14th century Orkney, Scotland. *Journal of Archaeological Science*, 27, 1133–8.

Thierry, D., Cave, M.D. & Eisenbach, K.D., *et al.* (1990) IS*6110*, an IS-like-element of *Mycobacterium tuberculosis* complex. *Nucleic Acids Research*, 18, 188.

16
Studying the Origins and Migrations of Early Modern Humans

The evolutionary history of our species holds a fascination not just for academics but also for people in all walks of life. At one time the preserve of paleoanthropology – the study of human fossils – over the last 30 years human evolution has increasingly been studied by geneticists and biomolecular archaeologists. Three related questions have dominated this area of research. The first concerns the geographical origins of *Homo sapiens* and has been answered by evolutionary studies of DNA samples from living people, which show that modern humans originated in Africa some 200,000 years ago. The second question asks how modern humans emerged from Africa and spread around the planet – when did these migrations take place and what routes did our ancestors follow as they travelled through Asia and into Europe and the New World? Again, studies of modern DNA are addressing these issues. The final question concerns the relationships between the first modern humans to enter Europe and the Neanderthals who occupied the continent prior to our arrival. How different were Neanderthals from modern humans and did the two types interbreed? The Neanderthal genome sequence, being built up from ancient DNA preserved in Neanderthal fossils, is helping us to understand these relationships.

Before looking at the research that is addressing these questions we must set the context by stepping back beyond our own species and following the fossil record through to the first appearance of *H. sapiens*.

16.1 The Predecessors of *Homo sapiens*

Humans are **hominids**, part of the taxonomic family called the Hominidae. The living members of this family are divided into four genera:

- *Homo*, the humans, containing the single species *H. sapiens*.
- *Pan*, the chimpanzees, of which there are two species, the common chimpanzee (*P. troglodytes*) and the bonobo (*P. paniscus*).

 © 2011 Terry Brown & Keri Brown

- *Gorilla*, again with two species, the western and eastern gorillas (*G. gorilla* and *G. beringei*, respectively).
- *Pongo*, the orangutans, once more with two species, the Borneo orangutan (*P. pygmaeus*) and the Sumatra orangutan (*P. abelii*).

The Hominidae are thought to have split from the related Hylobatidae family, which gave rise to the modern gibbons, some 14 million years ago. Since this split, the Hominidae evolutionary tree has included a number of extinct genera and species known only from the fossil record. Often these remains are fragmentary and poorly preserved, making it difficult to be certain of the true relationship between an extinct species and the living ones. The further back in time we go, the more uncertain the relationships become. We will therefore restrict ourselves to the branch of the Hominidae tree that leads specifically to humans.

16.1.1 Bipedalism defines the evolutionary branch leading to humans

One of the key distinctions between humans and the other living members of the Hominidae is our mode of locomotion. Uniquely, we are bipedal, walking upright on our two hind legs, in contrast with the knuckle walking locomotion of chimpanzees, gorillas, and orangutans. Bipedalism is a criterion of the types of hominid now called **hominins**, which include not just our own species but also those extinct taxa that preceded us in the evolutionary branch leading to *H. sapiens*, which split away from the one leading to chimpanzees some 6 million years ago.

The earliest known fossil showing the ability to walk upright is called *Ardipithecus ramidus*, represented by a single remarkably intact skeleton approximately 1.2 m in height, as well as the fragmentary remains of at least 35 other specimens, all found near the village of Aramis in northeast Ethiopia. The skeleton dates to 4.4 million years ago, placing *A. ramidus* very soon after the split between the human and chimpanzee lineages (Figure 16.1). Not surprisingly, the species has some features more typical of chimpanzees than of modern humans, especially in the dentition, but also including opposable big toes, which chimpanzees use to cling to branches. The implication is that *A. ramidus* spent at least part of its time in trees, which complicates the earlier theories that bipedalism evolved when our ancestors first moved out of the forests onto the grasslands of the African savannahs.

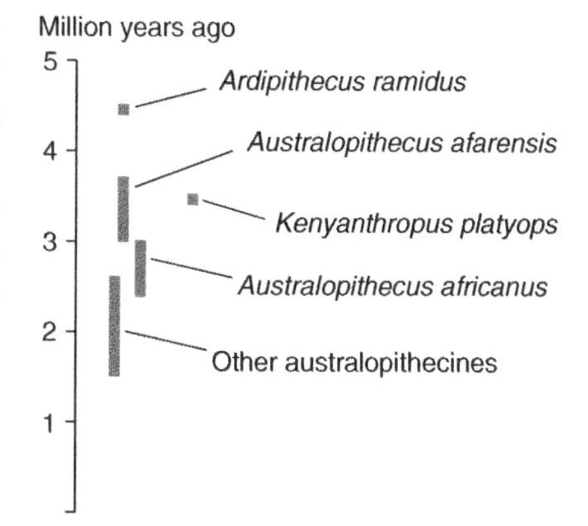

Figure 16.1 Timelines for some of the important pre-*Homo* hominins.

The placement of the earliest bipedal homininin the genus *Ardipethecus* indicates its considerable difference not only from *Homo* but also from the various species of *Australopithecus* that come between *A. ramidus* and us. Exactly how many australopithecine species there were depends on whether the paleoanthropologist studying the remains is a "lumper" or a "splitter," the former preferring as few hominin species as possible and so placing similar fossils in the same species, and splitters emphasizing the differences between individual fossils and so recognizing a greater proliferation of species. Both types of researcher recognize the importance of *Australopithecus afarensis*, known from the famous fossil skeleton called Lucy, dated to 3.6 million years ago and again found in Ethiopia, in the Afar region about 250 km from Aramis. Lucy is about the same height as *A. ramidus*, and her skull with a capacity of 375–550 cm^3, indicates that her brain was no bigger than that of *A. ramidus*, but her skeletal features are much more similar to those of modern humans. In particular, her pelvis and leg bones have modern features, suggesting that she walked much like we walk today.

At about 3 million years ago, *Australopithecus africanus* appears in the fossil record, slightly taller than *A. afarensis* and with a bigger brain, and with teeth that are more similar to humans than are those of the earlier species. But whether *A. africanus* is on the direct line leading to *Homo*, or represents a side branch that became extinct, is still the subject of intense debate, especially after the recent discovery in Kenya of a skull, about 3.5 million years old, with a large flat face and small teeth that is looked on as more human in character. This skull is so different from the other fossils that have been found that it is given not only its own species, but also its own genus – *Kenyanthropus platyops*. It is too early to understand the relationships between *K. platyops*, the australopithecines, and *Homo* – as is always the case with paleoanthropology, more fossils are needed.

16.1.2 There were at least four extinct species of *Homo*

Our own genus first appears in the fossil record about 2.4 million years ago (Figure 16.2). The first species is *Homo habilis*, which is also the earliest hominin at the famous prehistoric site Olduvai Gorge, part of the African Rift Valley in northern Tanzania. The name *H. habilis* means "handy man" and was given to the species because it is the first to be associated with stone tools. These "Olduwan" tools were manufactured by splitting flakes from the surfaces of pebbles of flint or basalt to leave a crude chopping or scraping implement. The Olduwan industry was prominent for about 1 million years and then was gradually replaced by the more advanced Acheuleun technology characterized by more carefully designed tools such as oval handaxes.

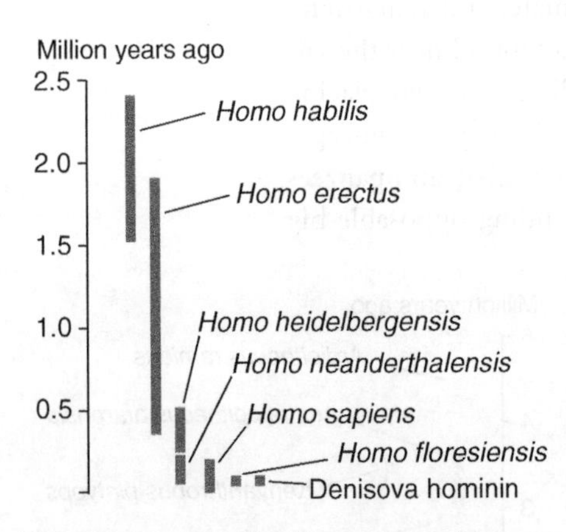

Figure 16.2 Timeline for members of the genus *Homo*. As well as the species mentioned in the text, the figure also shows *H. floresiensis*, known from a few fossils in Indonesia and dating until as recently as 13,000 years ago, and the "Denisova hominin," from 48,000 to 30,000 years ago, known only from the DNA sequences obtained from a finger bone (see Section 16.4).

Homo habilis survived for 900,000 years, until 1.5 million years ago, but during this period australopithecines still existed in Africa and the ranges of the various species probably overlapped. These australopithecines were larger and more robust than *A. afarensis* and *A. africanus*, and some paleoanthropologists believe that there are no significant differences between these and *H. habilis*, and that the latter should be reclassified as a type of *Australopithecus*. Others point out that the *H. habilis* brain, with an average size of 650 cm^3, is 100 cm^3 larger than that of any australopithecine, and that detailed studies of the inside of the cranium suggest that the feature of the brain called the bulge of Broca, which is essential for speech, was more developed in *H. habilis*. As speech is looked on as a uniquely human feature, the placement of this species in *Homo* seems warranted. It is clear, however, that unlike the later humans,

H. habilis was not a great hunter, using its Olduwan tools for scavenging meat from carrion, and itself sometimes becoming the prey of the more successful hunters of the African plains, such as the big cats.

The final disappearance of the australopithecines occurs shortly after the emergence of the first fossils that are unarguably ascribed to *Homo*. These belong to the species *Homo erectus*, which existed from 1.9 million to 300,000 years ago. Reaching 1.8 m in height, more robust than modern humans, with brains of up to 1225 cm^3 (the average for *H. sapiens* is 1350 cm^3), and armed with the advanced Acheulean tools, these were formidable creatures who dominated the African ecosystem. They were also the first hominins to move to other parts of the Old World, fossils of *H. erectus* being found in Georgia and China (Peking Man) from periods very soon after the first appearance of the species in Africa. The earliest hominin fossil in Europe is a juvenile specimen found in a cave at Atapuerca, Spain and dated to 800,000 years ago. This single fossil is placed in its own species, *Homo antecessor*, by some paleoanthropologists, and by others as a member of *Homo heidelbergensis*, which was first identified from a 500,000-year-old jaw from Heidelberg, Germany, a tibia of the same age being discovered more recently at Boxgrove, England. Other fossils of *H. heidelbergensis* have been found in Greece and Ethiopia, suggesting that this was a widespread but variable species. Their skulls have a mixture of features seen in *H. erectus* and *H. sapiens*, with a brain size intermediate between the two. About 200,000 years ago they disappear from the fossil record, replaced over much of their range by *Homo neanderthalensis*, sometimes called *H. sapiens neanderthalensis* to stress the similarity with our own species (according to this terminology we are *H. sapiens sapiens*). Male Neanderthals had an average height of 1.7 m and slightly larger brains than *H. sapiens*, probably a reflection of their greater robustness. They displayed many features typical of cold adaptation, such as large sinuses for warming the air entering through their nostrils. They lived throughout Europe until 30,000 years ago, by which time they shared the continent with members of our own species.

Homo sapiens sapiens first makes its appearance in Africa some 195,000 years ago with fossils found at Omo in Ethiopia. This specimen has a mixture of archaic and modern human features, the latter including a rounded skull and protruding chin. By this time, *H. erectus* had disappeared from Africa and the Omo specimen is looked on as a fairly advanced transition stage on the line eventually leading to anatomically modern humans. Other transition fossils have been found at Herto, also in Ethiopia, from 160,000 years ago, and more modern ones at Laetoli in Tanzania from 120,000 years ago, at Border Cave, South Africa, about 110,000 years ago, and at Klasies River Mouth in South Africa at 90,000 years ago. Around this time, anatomically modern humans were also present at Skhul and Qafzeh caves in the Levant, probably as a result of the African population expanding during a relatively warm period shortly after the beginning of the last glacial cycle.

16.2 The Origins of Modern Humans

Prior to the emergence of *H. sapiens sapiens* we have to rely solely on the fossil record as the source of information on the evolution of hominins. But with our own species we have the opportunity of using DNA to approach our evolutionary history from a different angle.

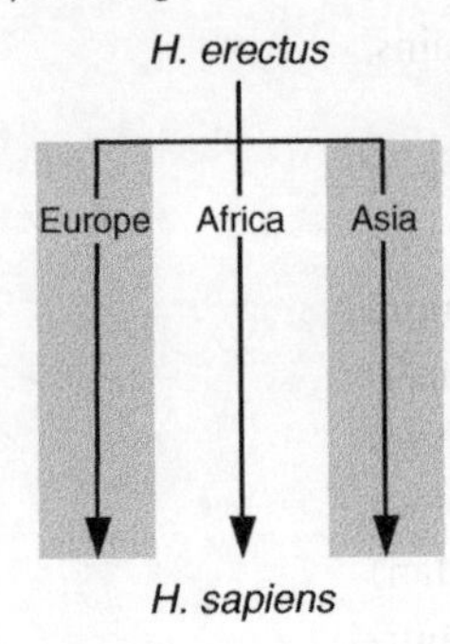

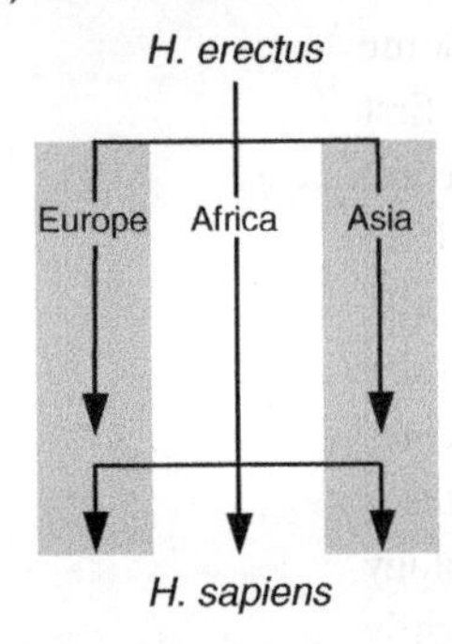

Figure 16.3 The multiregional and Out of Africa hypotheses for the origins of modern humans.

The first attempts to do this were in the late 1980s, and the results immediately challenged some of the views widely held in paleoanthropology.

16.2.1 There have been two opposing views for the origins of modern humans

Although the paleontological record appears to place the origins of anatomically modern humans in Africa, not all paleoanthropologists agreed with the assumption that our species originated solely in Africa. They saw similarities between the features of archaic and modern humans in different parts of the world, such as the relatively prominent brow ridges shared by modern Europeans and Neanderthals, which they believed pointed to a **multiregional** process of evolution. According to this model, modern humans emerged in parallel in different parts of the Old World, evolving directly from the archaic *Homo* species that occupied those regions before them (Figure 16.3A). Such a process is theoretically possible if there is a certain amount of interbreeding between populations from different parts of the world, so these are not reproductively isolated and the entire world population evolves as a single entity. One outcome of the multiregional hypothesis is that the geographical variations that we see today among people from different parts of the world, and which geneticists believe largely reflect adaptation to different climates and other environmental conditions, are relatively ancient, having evolved since the first arrival of *H. erectus* almost 2 million years ago.

The second hypothesis, first proposed in the 1980s, has a very different view of the origins of modern humans. Rather than evolving in parallel throughout the world, as suggested by the multiregional hypothesis, the **Out of Africa** or replacement hypothesis states that *H. sapiens* originated solely in Africa, members of our species then beginning to move into the rest of the Old World less than 100,000 years ago, displacing the descendants of *H. erectus* that they encountered (Figure 16.3B). Strictly speaking we should refer to this scenario as "Out of Africa II" and the previous migration by *H. erectus* as "Out of Africa I." According to this model, the differences between modern human populations are relatively recent variations that evolved mostly during the last 50,000 years.

The multiregional and Out of Africa hypotheses are amenable to testing by genetic analysis because the models make two different predictions regarding our genetic ancestry. The first of these predictions centers on the time when our most recent common ancestor (MRCA) was alive. Our most recent common ancestor is the person from whom all people alive today are descended. The MRCA would not have been the only male or female alive at the time; indeed the population of which they formed a part could have been quite large. But he or she was the only person from that period whose descendants are alive today: all of the lineages derived from everyone else alive at that time have now died out. According to the multiregional hypothesis, our MRCA must have lived almost two million years ago, as after that time our ancestors were distributed across the Old World and no single person could give rise to all of the present day human population. In contrast, if the Out of Africa hypothesis is correct, then our MRCA could have lived much more recently, possibly just 100,000–200,000 years ago, before modern humans left Africa.

The second prediction that distinguishes the multiregional and Out of Africa models concerns the relationship between modern Europeans and Neanderthals. According to the multiregional hypothesis, the Neanderthals that lived in Europe until 30,000 years ago are the direct ancestors of the *H. sapiens* populations living in Europe today. According to the Out of Africa hypothesis, Neanderthals were replaced by modern humans that moved into Europe as they migrated out of Africa. Multiregional evolution therefore predicts that there is genetic continuity between Neanderthals and modern Europeans, whereas the Out of Africa hypothesis suggests that the two *Homo* species are only distantly related.

We will now look at how these differing predictions of the multiregional and Out of Africa hypotheses have been tested by genetic studies.

16.2.2 Molecular clocks enable the time of divergence of ancestral sequences to be estimated

The first question we will address is the time of the MRCA of the modern human population. To estimate this time we must obtain DNA sequences from a representative sample of people and apply a molecular clock to those sequences in order to estimate the time needed for the diversity displayed by the population as a whole to have evolved.

The molecular clock hypothesis, first proposed in the early 1960s, states that nucleotide substitutions (or amino acid substitutions if protein sequences are being compared) occur at a constant rate. This means that the degree of difference between two sequences can be used to assign a date to the time at which their ancestral sequence diverged. However, to be able to do this the molecular clock must be calibrated so that we know how many nucleotide substitutions to expect per million years. Calibration is usually achieved by reference to the fossil record. For example, fossils suggest that the most recent common ancestor of humans and orangutans lived 13 million years ago. To calibrate the human molecular clock we therefore compare human and orangutan DNA sequences to determine the amount of nucleotide substitution that has occurred, and then divide this figure by 2, followed by 13 to obtain a rate of substitution per million years (Figure 16.4).

At one time it was thought that there might be a universal molecular clock that applied to all genes in all organisms. Now we realize that molecular clocks are different in different organisms and are variable even within a single organism. The differences between organisms might be the result of generation times, because a species with a short generation time is likely to accumulate DNA replication errors at a faster rate than a species with a longer generation time. This probably explains the observation that rodents have a faster molecular clock than primates. Within a single genome the variations arise because different parts of the genome evolve at different rates. In general, the sequences of the intergenic regions change more rapidly than those of the genes, because most of the substitutions that can occur in the intergenic regions have no effect on expression of the genome and so have no impact on the organism as a whole. A substitution

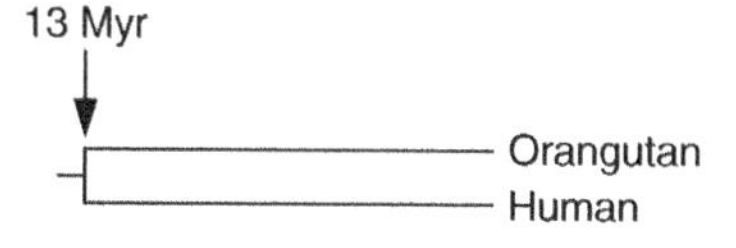

Number of substitutions when orangutans and humans are compared = n

Number of substitutions per lineage = $n/2$

Number per lineage per million years = $n/(2\times13)$

Figure 16.4 Using the date of the human–orangutan split to calibrate the molecular clock.

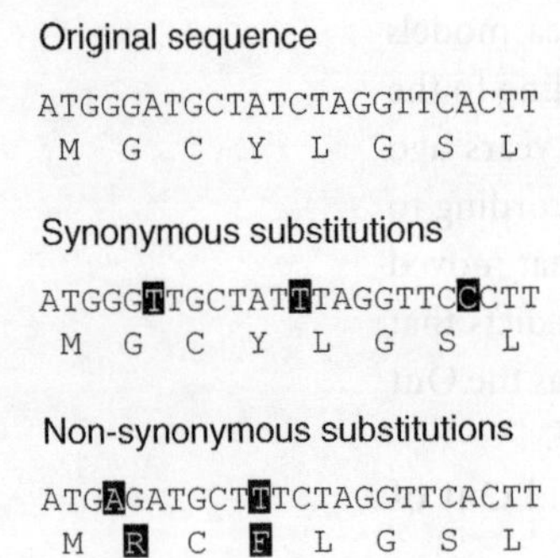

Figure 16.5 Synonymous and non-synonymous substitutions. The amino acids coded by each nucleotide sequence are indicated using the one-letter code (see Table 3.1).

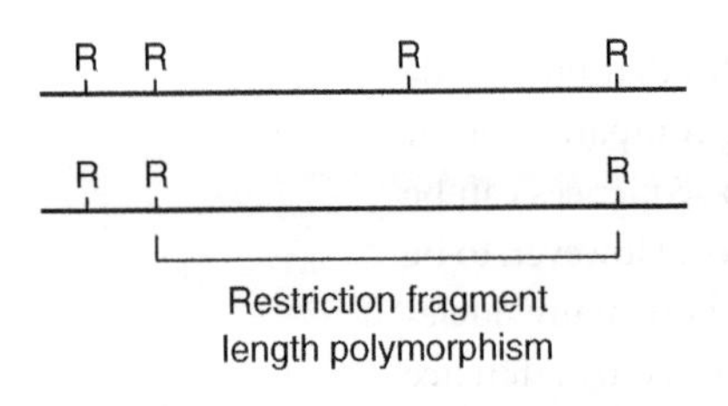

Figure 16.6 A restriction fragment length polymorphism. A substitution in the lower DNA molecule has changed the sequence of the third restriction site so that it is no longer recognized by the enzyme.

within a gene, on the other hand, might result in a deleterious change in the amino acid sequence of a protein. The accumulation of these substitutions in the population will be reduced by the processes of natural selection. Even within a single gene the molecular clock is variable, because of the difference between **non-synonymous substitutions**, ones that change the amino acid sequence of the protein coded by the gene, and **synonymous** ones, which do not affect the amino acid sequence (Figure 16.5). This means that when gene sequences from two species are compared, there are usually fewer non-synonymous than synonymous substitutions.

In mammals such as humans, the molecular clock for mitochondrial DNA is faster than that for DNA in the nucleus. This is probably because mitochondria lack many of the DNA repair systems that operate in the nucleus, enabling a greater proportion of the mutations that occur to be propagated in the population. Mitochondrial DNA therefore evolves relatively rapidly, with the hypervariable regions being the fastest of all. This is one of the reasons why mitochondrial DNA is often used when studies are being made of genetic diversity within a single species – not just of humans but domesticated animals and also many wild species. But caution is needed because a slight error in the value used for the mitochondrial molecular clock will have a significant effect on the date assigned to the most recent common ancestor of modern humans. The latest estimate of the clock for hypervariable region I in human mitochondrial DNA is 1.62×10^{-7} substitutions per nucleotide per year, corresponding to one change every 17,343 years in this region, and for hypervariable region II, 2.23×10^{-7} substitutions per nucleotide per year, or one change every 22,388 years.

16.2.3 The molecular clock supports the Out of Africa hypothesis

A molecular clock was first applied to human mitochondrial DNA in 1987. This project took place before PCR had become established as the means of obtaining sequences from specified regions of DNA, and instead made use of the less direct method of obtaining sequence data called **restriction fragment length polymorphism (RFLP) analysis.** RFLPs result from substitutions that change the sequence of a recognition site for a restriction endonuclease, so that the DNA is no longer cut at that position when treated with the enzyme (Figure 16.6). The substitution results in two fragments remaining joined together and is revealed by examining the banding pattern that is obtained when the restriction digest is analyzed by agarose gel electrophoresis. RFLPs were scored in the mitochondrial genomes of 147 humans, from all parts of the world and the data used to construct a phylogenetic tree (Figure 16.7). The root of this tree represents a woman (remember, mitochondrial DNA is inherited only through the female line) whose mitochondrial genome is ancestral to all the 147 modern mitochondrial DNAs that were tested. This woman has been called **mitochondrial Eve**, but of course she was not equivalent to the biblical character and was by no means the only woman alive at the time: she is simply our female MRCA, the person who carried the ancestral mitochondrial DNA that gave rise to all the mitochondrial DNAs in existence today. It was deduced that mitochondrial Eve lived in Africa because the tree divides into two segments either side of her sequence, one of these segments being made up solely of African mitochondrial DNAs. Because of this split, it was inferred that the ancestral sequence was also located in Africa.

The all important question is the date that the mitochondrial molecular clock assigns to mitochondrial Eve. This figure was estimated at between140,000 and 290,000 years ago. The result is therefore compatible with the prediction made by the Out of Africa hypothesis, but incompatible with the multiregional model, which requires that the MRCA dates to around two million years ago.

Inevitably, such an important result was subject to close scrutiny, and this led to one important problem being uncovered. When the RFLP data were examined by other molecular phylogeneticists, it became clear that the original computer analysis had been flawed, and that several quite different trees could be reconstructed from the data, with greatly varying dates for the MRCA, some old enough to agree with the multiregional hypothesis. The problem arose from the immense amount of computer power needed to construct a phylogenetic tree, which causes difficulties even today, and was much more acute with the limited hardware available back in the 1980s. To reduce the computational requirement, the early programs employed various shortcuts, not all of which are now looked on as sound.

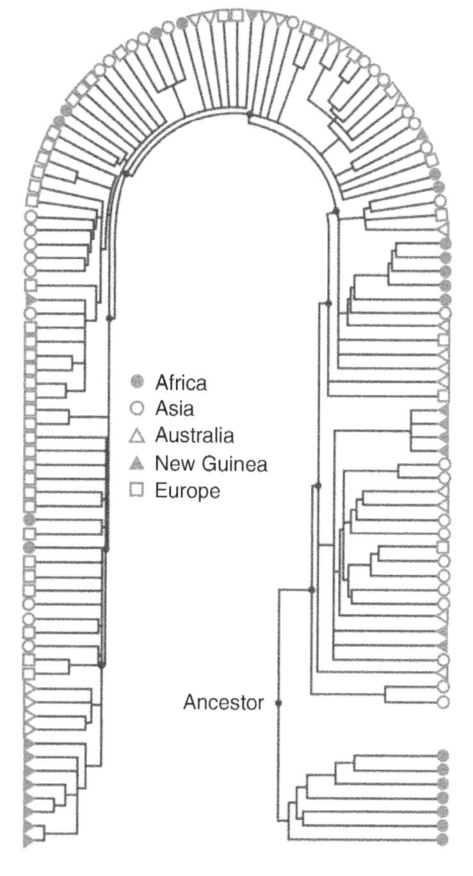

Figure 16.7 Tree depicting the evolutionary relationships between the mitochondrial DNAs of 147 humans from various parts of the world. The position of the ancestor—mitochondrial Eve—is marked. Adapted from Rebecca L. Cann, Mark Stoneking, and Allan C. Wilson, "Mitochondrial DNA and human evolution," *Nature*, 325, 31–6, copyright 1987, by permission from Macmillan Publishers.

So was the discovery of mitochondrial Eve correct or not? Since 1987 there have been numerous additional projects examining mitochondrial DNA in modern humans, many using PCR to amplify parts of the genome so that DNA sequences can be obtained, rather than depending on RFLPs as used in the original project. All of these later studies confirm the relatively recent date for mitochondrial Eve. To take one example, when the complete mitochondrial genome sequences of 53 people, again from all over the world, were compared, a date of 220,000 to 120,000 years for the MRCA was obtained. An interesting complement has been provided by studies of the Y chromosome, which, of course, descends exclusively through the male line. This work has revealed that the paternal MRCA – Y chromosome Adam – also lived in Africa between 40,000 and 140,000 years ago. The wide confidence limit reflects the greater difficultly in applying the molecular clock to the range of markers that are typed on the Y chromosome.

DNA therefore provides strong support for one of the predictions of the Out of Africa hypothesis by showing that our maternal and paternal MRCAs were still living in Africa over a million years or so after the spread of *Homo erectus* into the rest of the Old World.

16.2.4 Neanderthals are not the ancestors of modern Europeans

What answer does DNA provide to the second prediction of the Out of Africa hypothesis, that modern Europeans are not the direct descendants of the Neanderthals who inhabited the continent up to 30,000 years ago? This question has been addressed by studies of ancient DNA from Neanderthal bones. The youngest Neanderthal fossils are just within the time frame for ancient DNA preservation, although as these fossils are found in cool environments in Eurasia rather than permafrost, the study of Neanderthal ancient DNA has consistently pushed the technology to its very limits.

The first breakthrough came in 1997 when ancient DNA was extracted from a Neanderthal type specimen, believed to be between 30,000 and 100,000 years old. A series of nine overlapping PCRs enabled a 377 bp sequence to be built up from hypervariable region I of the Neanderthal mitochondrial genome. A phylogenetic tree was then constructed to compare this sequence with the equivalent sequences from 994 modern humans. The Neanderthal sequence was positioned on a branch of its own, connected to the root of the tree but not linked directly to any of the modern human sequences. Next, a detailed analysis was made of the differences between the Neanderthal sequence and the modern human ones. The results were striking. The Neanderthal sequence differed from the modern sequences at an average of 27.2 ± 2.2 nucleotide positions, whereas the modern sequences, which came from all over the world, not just Europe, differed from each other at only 8.0 ± 3.1 positions.

Sequences of hypervariable region I have now been obtained for 15 different Neanderthal fossils, and hypervariable region II has also been sequenced in two of these. In all cases, the degree of difference between Neanderthal and modern European DNA is incompatible with the notion that modern Europeans are descended from Neanderthals. This conclusion was brought home most forcefully when a complete mitochondrial DNA sequence was obtained, as an offshoot of the larger project aimed at sequencing the entire nuclear genome of Neanderthals. This project is using high throughput methods (Section 2.5), in which millions of short sequences, most of them shorter than 75 bp, are generated at random from an ancient DNA extract, in this case from three small pieces of Neanderthal bone, dating to 38,000 years ago, found in a cave at Vindija in Croatia. We will look at this project in more detail when we examine the Neanderthal nuclear genome later in this chapter (Section 16.4). What interests us at the moment is the mitochondrial sequence and how this compares with the mitochondrial genomes of modern humans. When compared with the **Cambridge Reference Sequence (CRS)**, the modern human mitochondrial DNA sequence, corresponding to haplogroup H, which is used as the standard against which all other human sequences are compared, the Neanderthal mitochondrial genome was found to have a total of 206 differences. This is much greater than the degree of variation displayed by all the known modern mitochondrial DNA haplogroups, the two most divergent of these having only 118 differences. The Neanderthal mitochondrial genome therefore falls outside of the modern human range, which would be unexpected if it is ancestral to modern mitochondrial DNAs as predicted by the multiregional theory. Instead the sequence comparisons suggest that Neanderthals and modern humans form two separate branches of the *Homo* evolutionary tree, a scenario entirely in keeping with the Out of Africa hypothesis.

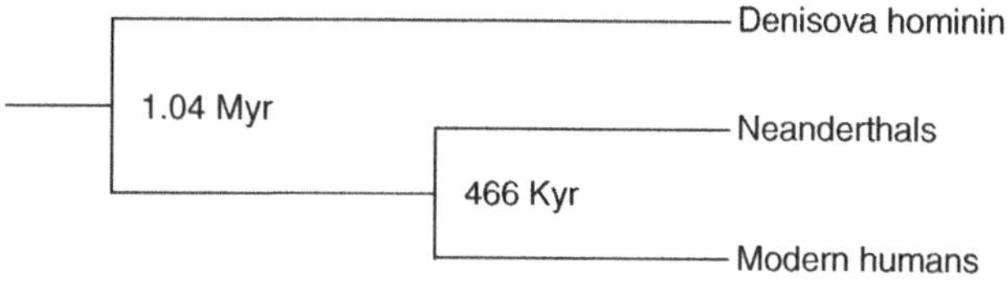

Figure 16.8 Evolutionary tree showing relationships between modern humans, Neanderthals, and the Denisova hominin, according to mitochondrial DNA comparisons. Abbreviations: Kyr, thousand years ago; Myr, million years ago.

16.2.5 It now seems likely that there were several migrations out of Africa

The mitochondrial DNA data suggest that the most recent common ancestor of modern humans and Neanderthals lived between 520,000 and 800,000 years ago. More detailed analyses using sequences from the Neanderthal nuclear genome date the split a little later, at about 466,000 years ago (Figure 16.8). Whichever date is correct, the implication is that the ancestors of Neanderthals also migrated out of Africa, at a period between the departures of *H. erectus* and our own species. The reasoning is quite simple. If *H. sapiens* evolved in Africa, as appears certain from the fossil record, then all the direct ancestors of *H. sapiens* must also have lived in Africa. These direct ancestors include the one we share with Neanderthals, shown by the genetic evidence to have lived about 500,000 years ago. If 500,000 years ago an ancestor of Neanderthals lived in Africa, then Neanderthals cannot be the descendents of the *H. erectus* populations who were living in Eurasia at that time. Instead, their ancestors must have arrived in Eurasia via their own separate migration. So now we have three "Out of Africas" – the first 1.8 million years ago involving the first *H. erectus* populations to colonize Eurasia, then the ancestors of the Neanderthals 500,000 years ago, and finally the first modern humans less than 100,000 years ago.

In fact, there must have been even more than three migrations out of Africa. This is one of the conclusions of the remarkable discovery of an unknown type of hominin who lived in southern Siberia between 48,000 and 30,000 years ago. This hominin has been identified from a single broken finger bone found in Denisova Cave in the Altai mountains. The discovery of this hominin has been remarkable not just because of what it has told us about the human past. For the first time a new hominin was identified not by morphological examination of its fossil bones – which, of course, would be impossible with just a broken finger – but from its DNA sequence.

Denisova Cave was first occupied by hominins some 125,000 years ago, and has been a rich source of stone tools and bone implements, but has yielded very few human bones, most of these being isolated finds like the finger bone used in this study. Ancient DNA was extracted from this bone in the expectation that it would turn out to be a modern human or Neanderthal fossil, but the sequences showed that it was neither. The mitochondrial DNA had 385 differences compared with the consensus modern human sequence, which, as well as showing that clearly it is not a modern human, also places it well outside of the range of Neanderthal mitochondrial genomes that have been sequenced. The molecular clock places the common ancestor of modern humans and the Denisova hominin at 1.04 million years (Figure 16.8), almost twice as ancient as the Neanderthal-to-modern-human split.

The same logic that deduced a migration of Neanderthal ancestors out of Africa also leads to the conclusion that the Denisova forebears also left Africa independently of *H. erectus*. If the Denisova-modern human ancestor was alive 1.04 million years ago, then at that time it was in Africa, and so could not be a direct descendent of the *H. erectus* populations that were already in Eurasia. So now we have four "Out of Africas," and there is no reason to believe that there were not others. Although only connected by a small land link, the environments of North Africa and southwest Asia are not greatly different, and a hominin adapted to the former would have had little difficulty surviving if its population expanded across what is now the Sinai desert. There may well have been successive waves of hominin advance out of Africa.

16.3 The Spread of Modern Humans Out of Africa

As well as establishing that our common ancestor lived in Africa relatively recently, archaeogenetics is also helping to trace and date the migrations by which modern humans colonized the rest of the planet. We have already touched on this issue in the context of the spread of agriculture (Section 13.3). We learnt that studies of blood groups and protein polymorphisms had suggested that agriculture spread into Europe as a large-scale wave of advance of farmers from their original homelands in the Fertile Crescent, but that analysis of mitochondrial and Y chromosome DNA has questioned this view, providing a more complex interpretation of the human dynamics underlying the domestication of the continent. Now we will look more closely at the use of DNA to trace past human migrations by examining two additional questions. These concern the route taken by modern humans when they first left Africa, and the timing and pattern of colonization of the New World.

16.3.1 Many studies of migrations have begun with mitochondrial DNA

The archaeogenetic approach to studying past human migrations has been based on studies of both mitochondrial and Y chromosomal DNA, and is now beginning to utilize the much greater wealth of information present in the genome as a whole. In general, however, the initial studies of migrations into a particular part of the world have relied on mitochondrial DNA, which has therefore been responsible for setting the models that later, broader genetic studies have tested. We will therefore focus on mitochondrial DNA in our short review of the subject,

The use of mitochondrial DNA haplogroups in modern populations to study prehistoric migrations has three basic premises. The first is that the evolutionary relationships between haplogroups can be inferred by comparisons between the nucleotide polymorphisms that are present in the hypervariable regions and elsewhere in the mitochondrial genome. These various differences enable a network displaying the evolutionary relationships between different mitochondrial haplogroups to be constructed (Figure 16.9).

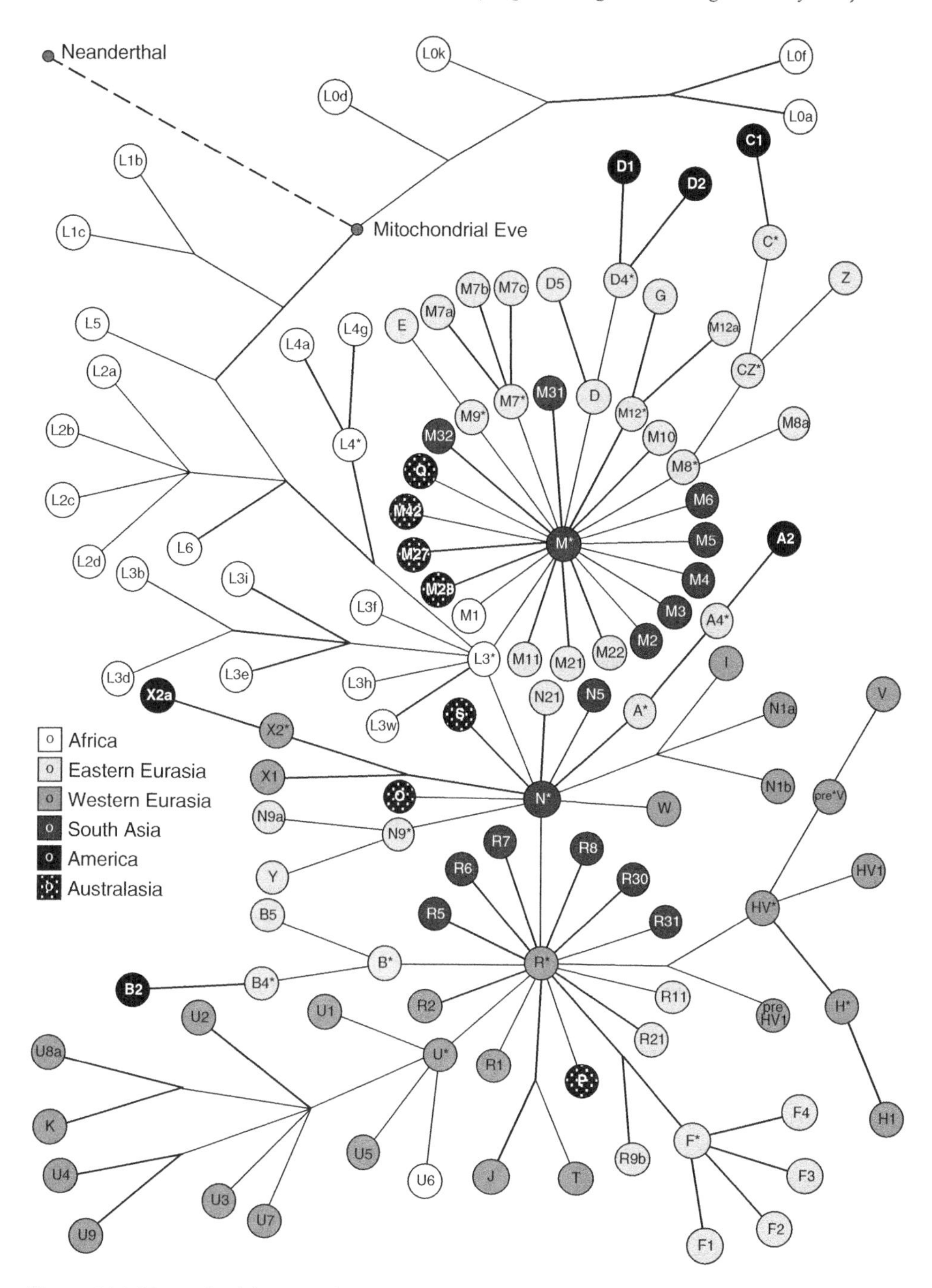

Figure 16.9 Network of the most frequent human mitochondrial haplogroups, showing their geographical associations. Network kindly provided by Martin Richards.

The second premise is that in the modern population, individual haplogroups are associated with different geographical regions. In Section 13.3, we learnt that 11 haplogroups predominate in the modern European population. These are present also in other populations, but are much commoner in Europe and are looked on as characteristic of modern Europeans. Similarly there are haplogroups that are typically found in Africans, others that are Asian, and others that are found mainly in the New World. These geographical associations are marked on Figure 16.9. As we will see when we examine the migrations out of Africa and into the New World, the information in Figure 16.9 enables us to identify haplogroups that have close evolutionary relationships but that are associated with two different parts of the world. The inference is that the populations living in these two regions are also closely related, which is most easily explained if we assume that there has been a movement of people from one of those regions to the other.

The final premise is that the origin of a haplogroup can be dated by coalescence analysis. This is achieved by measuring the overall degree of diversity among the individual haplotypes that make up the haplogroup, and using the molecular clock to estimate the coalescence time for the group as a whole. If a haplogroup is associated with a particular geographical region, then its coalescence time might be equivalent to the date at which the first people possessing the haplogroup arrived in that area (Figure 13.7).

Mitochondrial DNA has proved to be a powerful tool for studying the relationships between human populations and for inferring the migrations that set up the current geographical distributions of these populations. It has, however, been criticized, on two grounds. The first is that, in genetic terms, the mitochondrial genome is a single marker. This is because there is no recombination between maternal and paternal mitochondrial DNAs, so the genome is inherited as a single unbroken unit. The absence of recombination and the uniparental pattern of inheritance both simplify analysis of the evolutionary relationships between haplogroups, but they do mean that, in general terms, the amount of evolutionary information contained in the mitochondrial DNA is quite small. In essence, it is equivalent to just one of the 169 short tandem repeats that were used to establish the evolutionary relationships between different types of rice (Figure 13.2). The second criticism that has been made is that detailed examination of mitochondrial DNA sequences, not just from the hypervariable regions but also from the genes present on the molecule, have revealed that the mitochondrial genome does not evolve in an entirely neutral fashion, as was first assumed, but instead is subject to some natural selective pressures. It has been suggested that these are involved in adaptation to cold environments, but it is possible that the pressure is simply in the form of **purifying selection**, which gradually removes deleterious mutations from the genome. A failure to recognize that mitochondrial DNA is subject to selection could mean that the molecular clock has been calibrated incorrectly.

The problem posed by selection has been solved by a recent project in which the mitochondrial molecular clock has been recalibrated with careful account taken of the selective pressures acting on the genome. Application of this new clock gives coalescence times slightly younger than had been calculated in the past, but the recalibration does not significantly affect any of the major conclusions about the origins and migrations of modern humans that had previously been drawn. Mitochondrial Eve, for example, is placed at 192,400 years ago. The limited evolutionary information provided by the mitochondrial genome is more of a problem as it means that conclusions based only on mitochondrial DNA studies might be incorrect, simply because of the chance inaccuracies that can arise when just a single genetic marker is examined. Y chromosomal haplogroups provide complementary information on the male lineage, but the Y chromosome is itself just another single marker. This is why it so important to harness the full evolutionary content of the nuclear genome to understand with greater accuracy the recent evolution of humans, this work requiring comparisons of single nucleotide polymorphisms from all the chromosomes. Such data are gradually being acquired but new analytical tools will be needed to mine the evolutionary information that the data contain.

16.3.2 One model holds that modern humans initially moved rapidly along the south coast of Asia

The direction taken by modern humans when they first left Africa is still a subject of intense debate. The earliest modern human fossils found outside of Africa, dated to 90,000–100,000 years ago, were discovered in the Qafzeh and Skhul caves near Nazereth in the Levant. These fossils were initially thought to indicate that modern humans migrated out of Africa via modern Suez, and thence northwards into Asia and Europe. The view has now changed and it is generally believed that these early fossils are probably not the remains of people who were involved in the first migration into Eurasia. The caves were reoccupied by Neanderthals at a later date and modern humans are not seen there again for some time. It appears that the Qafzeh and Skhul caves were occupied either as an expansion of the African population into Eurasia during a warm period, or as part of a "failed" attempt at colonization. This northern route did not seem to be sustainable, and modern humans did not colonize Europe until much later, about 45,000 years ago according to the earliest archaeological evidence.

It now seems probable that the first modern human migration out of Africa left from Ethiopia, further south than the area of modern Suez that forms the physical link between Africa and Asia, and then followed a "beachcombing" route along the southern coast of Asia. The earliest evidence for the use of marine resources by modern humans dates to 125,000 years ago at Abdur on the Eritrean coast of the Red Sea, so the first modern humans leaving Africa would have been well aware of the nutritional richness of the coastal regions. Although the fossil remains for early modern humans are scarce or non-existent for much of Asia, comparisons of stone tools found in the Indian subcontinent with those from East Africa are thought to support such a migration. These Indian tools have forms suggesting that they were produced by people who lacked some of the raw materials needed to manufacture the full repertoire of types found in Africa, and who possibly had lost the knowledge of certain aspects of stone tool technology.

The genetic evidence is also compatible with a southern Asian route out of Africa. Most modern African mitochondrial DNAs fall into three haplogroups, called L1, L2, and L3. When the relationships between these and non-African haplogroups are examined, we see that the non-Africans all derive from L3 (Figure 16.9), implying that the initial migration out of Africa involved individuals predominantly of this haplogroup. The immediate non-African descendents of haplogroup L3 are M and N, which are found mainly in Asia,

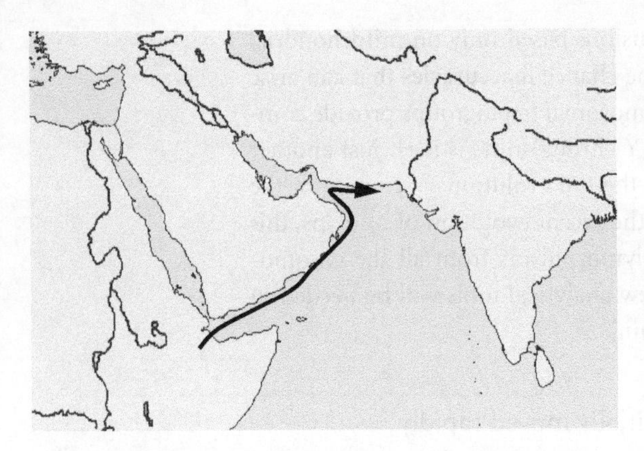

Figure 16.10 One possibility for the route taken by the first migration of modern humans out of Africa.

suggesting that the initial migration was into Asia. The modern Africans whose L3 haplotypes have the greatest similarity to M and N live in the Ethiopian region of East Africa, so the migration probably originated in this area, possibly by boat across the entrance of the Red Sea to the southern coast of Arabia (Figure 16.10). The sea levels might have been lower at that time, making the crossing less arduous than it would be today.

When did this migration begin? The coalescence time for L3 is approximately 70,000 years, and for M and N 50,000–70,000 years. The implication is that humans left Africa sometime during this period. It has been suggested that the massive eruption of a super-volcano on Lake Toba in Sumatra, which the geological record suggests occurred about 74,000 years ago, caused climatic changes that triggered the movement of humans out of Africa, possibly in search of more productive hunting and plant gathering areas. According to the genetic evidence, the Toba eruption was slightly too early to have influenced the subsequent migration into Asia, which at its earliest would have been 5000–10,000 years later.

What did the first migrants do once they had reached southern Asia? It appears that they moved rapidly, in relative terms, along the coast towards Australia, which archaeologists believe might have been occupied by 50,000–60,000 years ago. Again there is DNA evidence to support this model. The Andaman Islanders, who live in the Bay of Bengal to the east of India, have M haplotypes that coalesce at approximately 60,000 years ago, as do the Orang Asli of the Malayan peninsula. These particular populations therefore appear to have originated around that time, not long after the initial migration from Africa and early enough for humans to continue their coastal journey to reach Australia during the period suggested by archaeologists. Interestingly the M haplotypes in India coalesce to 49,000 years ago, which might indicate that the internal parts of the subcontinent were not colonized until several millennia after the coastal regions. Combining various pieces of evidence it has been estimated that the migration along the south coast of Asia moved at a rate of 0.7–4.0 km per year and involved an initial population including some 500–2000 females – remember, mitochondrial DNA tells us only about our female past.

16.3.3 Into the New World

There is no evidence for the presence of *H. erectus* anywhere in the Americas. Possibly this reflects the difficulty in reaching the continent without a sea voyage, which *H. erectus* could not or would not attempt. The obvious point of entry is across the Bering Strait, which separates Siberia from Alaska. The Bering Strait is quite shallow and if the sea level dropped by 50 m it would be possible to walk across from one continent to the other. The sea was 50 m or more below its current level for most of the last Ice Age, between about 60,000 and 11,000 years ago, but for a large part of this time the route would have been impassable because of

Table 16.1 Pre-Clovis sites in the Americas.

Site	*Location*	*Reported date (years BP)*
Cactus Hill	Virginia, USA	15,000–17,000
Channel Islands	California, USA	13,000
Guitarrero Cave	Peru	12,560
Manis Mastodon Site	Washington State, USA	14,000
Meadowcroft Rockshelter	Pennsylvania, USA	16,000–19,000
Monte Verde	Southern Chile	30,000
Paisley Cave	Oregon, USA	14,300
Pedra Furada	Brazil	32,000
Tipacoya	Mexico	25,000
Topper	South Carolina, USA	16,000, possibly 50,000

the build-up of ice, not on the land bridge itself but in the areas that are now Alaska and northwest Canada. Also, the glacier-free parts of northern America would have been arctic during much of this period, providing few game animals for the migrants to hunt and very little wood with which they could make fires. But for a brief period around 14,000 to 12,000 years ago the Beringian land bridge was open at a time when the climate was warming and the glaciers receding such that there was an ice-free corridor leading from Beringia to central North America. These considerations have led to the suggestion that modern humans first reached the Americas some 13,000 years ago, in agreement with the conventional archaeological view that the first paleoindians were the Clovis people, who made large spearheads called Clovis points, named after the site in New Mexico from which they were first extensively recovered. The first Clovis points date to around 13,000 years ago, so the Clovis people may well have entered the Americas across the Bering Strait at about this time.

The supposition that humans entered the New World via Beringia is supported by mitochondrial DNA analysis, as the vast majority of Native Americans have one of four haplogroups, A, B, C, and D. These haplogroups are widely scattered in the mitochondrial DNA network (Figure 16.9), but all four are common in East Asia, suggesting that East Asia is the source of the populations that colonized the Americas. This is exactly what we would expect if these people entered the New World across the Bering Strait. But the genetic data also suggest that this initial migration occurred earlier than 13,000 years ago, as each of the four Native American haplogroups has a coalescence date that falls before this time. Some analyses place the coalescence dates at 20,000 years or earlier, and others at around 14,000 years. Whichever date is favored, the origins of the Native Americans appear to predate the earliest Clovis sites. This debate is still ongoing, and centers largely on the archaeological interpretation of sites in the Americas that might have been occupied by humans earlier than 13,000 years ago. A growing body of archaeologists are questioning the "Clovis-first" hypothesis, arguing that there is evidence of human occupation at sites that date much earlier (Table 16.1). The problem is that skeletons are rarely found, so the evidence that humans were present is indirect, including such things as the possible discovery of the remains of a hearth where a fire was tended and food cooked.

The increasing indications of a pre-Clovis occupation of the New World include the results of an ancient DNA study. Coprolites – fossilized excrement – dating to over 14,000 years ago were discovered in a cave in Oregon. Judging by the size, shape, and color of the coprolites, they could be human in origin, or they might be from wild dogs. Excrement contains DNA from the animal or person responsible for the material, and coprolites have previously been shown to be a good source of ancient DNA, so could DNA be extracted from these specimens to determine whether they are human or canine in origin? The answer is yes, DNA was present, and it was clearly human mitochondrial DNA. This discovery pushes the date for the first migration into the Americas back to 15,000 years or so. As the ice-free corridor was almost certainly closed at that time, it is now thought that the first humans to reach the Americas followed a Pacific coastal route, along the south of Beringia and then down the east of North and South America. Taking into account the locations of the major glaciers, this journey is thought to have been possible at any time after 15,000 years ago.

16.4 Studying the Complete Genome Sequences of Prehistoric People

The last decade has seen a revolution in the technology used to sequence genomes, with the introduction of high throughput methods that enable millions of short sequences, totalling several Mb, to be obtained in a single experiment (Section 2.5). These "next generation" methods are making it relatively easy to sequence the entire genomes of different living people, the current goal being to accumulate 1000 genome sequences from populations from all over the world. The information on human genetic diversity that this project will provide will be valuable not only for evolutionary studies, but also as a means of identifying genome variations that are linked to susceptibility to different genetic diseases.

16.4.1 Next generation sequencing methods are ideal for ancient DNA

The new sequencing technologies are also stimulating a revolution in biomolecular archaeology. Obtaining nucleotide sequences from ancient DNA has, in the past, been a time consuming and difficult process, requiring a specially designed PCR for each different segment of the genome that is being studied, with none of these segments exceeding 100–150 bp in length. Longer sequences have to be built up from a series of overlapping PCRs. Not surprisingly, the longest contiguous ancient DNA sequence that has ever been obtained by the conventional approach is just a few kb in length. But with the new methods, entire genomes have been sequenced from the ancient DNA in preserved human remains.

Although there are several types of next generation sequencing method, as described in Section 2.5, each works in a similar way. The DNA extract is broken into fragments no more than 500 bp in length and adaptors attached to either end of every fragment (Figure 2.18). These adaptors enable the fragments to be attached to metallic beads, with the ratio of DNA fragments to beads set so that, on average, a single fragment becomes attached to each bead. The beads are then placed in water droplets in an oil emulsion, again with one bead per droplet, and the attached DNA fragment amplified by PCR with primers that anneal to the adaptor sequences. This means that every fragment can be amplified by the same pair of primers, and many thousands of PCRs can be carried out in parallel. The amplicons are then sequenced by the pyrosequencing method, which again can be carried out in parallel with many thousands of sequences generated at the same time.

Two features of the next generation sequencing method make it ideal for studying ancient DNA. The first is that the sequence reads are rarely longer than 100 bp, even with modern DNA, so the short lengths of ancient DNA molecules become immaterial. In fact, they become an advantage because they mean that ancient DNA preparations do not require any pre-treatment before being sequenced, whereas modern DNA has to be broken down into lengths suitable for the sequencing methodology. The second advantageous feature is that the next generation methods sequence DNA in an entirely random fashion: in essence, all of a genome is sequenced at once, rather than individual genes and other DNA sequences being targeted one after the other. This is perfect for sequencing an ancient genome, because it means the sequence can be obtained directly from the ancient DNA extract, without the need to prepare overlapping PCR amplicons.

Next generation sequencing of ancient genomes has been pioneered with the mammoth, using DNA from specimens preserved in the permafrost of Siberia. Mammoths fall outside the realm of biomolecular archaeology, but within our discipline there are two genome sequencing projects that we must consider. The first is aimed at obtaining the complete Neanderthal genome.

16.4.2 The Neanderthal genome is being sequenced from 38,000-year-old females

The Neanderthal sequence is being generated from ancient DNA extracted from small pieces of bone from the Vindija Cave in Croatia. These bones are dated to 38,000 years ago and come from three different females, meaning that the genome being assembled is a composite of three people. Although Neanderthal fossils are known from all over Eurasia, these particular specimens were chosen because initial tests suggested that they are free from contamination with modern DNA. Various criteria can be used to judge the extent of contamination, the simplest and most effective being to carry out PCRs that amplify both Neanderthal and modern human DNA, but generate amplicons whose source can be distinguished from their internal sequence features. The ratio of modern to Neanderthal sequences in the PCR products gives a measure of the amount of contamination. Of the 89 different Neanderthal bones that initially were examined, from 18 different sites, the three that are being studied had the lowest amounts of modern contamination. Using a specimen that contains only Neanderthal DNA is important because many regions of the Neanderthal genome will have sequences that are identical to that of modern humans. It will not therefore be possible to screen out contaminating modern DNA simply by inspecting the sequences.

In early 2010 it was reported that 4.1 Gb of Neanderthal DNA sequence had been obtained. The genome is 3.2 Gb in length, the same as for modern humans, so some parts have been sequenced more than once. This is important in order to identify sequence errors resulting from miscoding lesions, but a coverage of five times or more for all parts of the genome is probably needed before all such errors can be excluded. The genome is therefore a draft rather than the finished version.

Despite its draft status, the Neanderthal genome is already providing a mass of new information about the relationships between *H. neanderthalensis* and *H. sapiens*. In particular, the intriguing possibility that Neanderthals and modern humans interbred is being addressed. Modern humans arrived in Europe around 45,000 years ago, and Neanderthals did not become extinct until about 30,000 years ago, so the two types of hominin must have lived in the same regions for about 15 millennia. Although the mitochondrial DNA comparisons suggest that there has been no female Neanderthal contribution to the modern human genome, this does not exclude the possibility of interbreeding involving male Neanderthals. Tentative evidence for interbreeding has been obtained both from archaeological and genetic studies. In 1998, a skeleton of a 4-year-old child was found at Abrigo do Lagar Velho in Portugal. This child is claimed by some archaeologists to represent a Neanderthal and modern human hybrid, largely because of the robustness of the skeleton. The burial has been dated to 24,500 years ago, about 5000 years after the last definite evidence for Neanderthals in Europe, so its status as a hybrid seems doubtful.

The genetic evidence comes from examination of genes in the modern human genome whose inferred evolutionary history might be consistent with transfer of allelic variants from Neanderthals. One intriguing example is provided by the microcephalin gene, which is thought to influence brain size and may even be involved in the development of human cognitive abilities. In the present day population, there are several haplogroups of the microcephalin gene, but one of these, haplogroup D, is predominant, being present in 70% of modern humans. According to an analysis of the gene in 89 people from around the world, haplogroup D has a coalescence time of 37,000 years ago, which means that it must have undergone a rapid expansion in order to reach the frequency that it displays today. This implies that the haplogroup has been under strong positive selection. But haplogroup D appears to have split from the other microcephalin haplogroups around 1.7 million years ago, which leads to the unlikely proposition that it existed in the human gene pool for 1.7 million years, undergoing very little evolutionary change, before suddenly exploding 37,000 year ago. A more parsimonious explanation is that haplogroup D was a Neanderthal variant that was absent from the human lineage until its transfer by interbreeding 37,000 years ago. Once in the modern human gene pool it experienced a positive selective pressure that did not act on Neanderthals, resulting in its present day frequency.

So it is possible to weave archaeological and genetic stories that support the notion that Neanderthals and modern humans interbred. What does the Neanderthal genome reveal? The initial analyses suggest that a small amount of interbreeding did occur. The evidence comes from comparisons between the Neanderthal genome and that of modern humans. These show that the degree of divergence between Neanderthals and modern

Europeans is slightly less than that between Neanderthals and modern Africans. If there had been no interbreeding, then modern Europeans and Africans should be indistinguishable when compared with Neanderthals.

16.4.3 The complete genome sequence of a 4000-year-old paleo-Eskimo has been obtained

The second genome sequencing project relevant to biomolecular archaeology has been carried out with ancient DNA extracted from the hair of a 4000-year-old paleo-Eskimo. The hair came from a tuft that had been preserved in permafrost in Greenland, from a site associated with the Saqqaq culture, one of the first groups of people to live in Greenland, between 4700 and 2500 years ago. In some respects, hair is a better source of ancient DNA than bone. The outer surface of a hair can be rigorously decontaminated to remove modern human contamination, and hair is less subject to invasion by bacteria and fungi than is bone. This means that a greater proportion of the resulting sequence reads are human rather than microbial, reducing the amount of work that is needed to obtain a complete genome sequence.

The ancient DNA in the paleo-Eskimo hair was so well preserved that 79% of the individual's genome could be sequenced, with an average of 20 times coverage. In other words, each of the sequenced regions was represented by an average of 20 reads, meaning that most sequencing errors and miscoding lesions could be identified and excluded. It was therefore possible to compare the sequence with that of other modern human populations in the expectation that any single nucleotide polymorphisms that were identified would be genuine rather than artifacts.

SNPs were identified in a number of protein-coding genes whose functions are known, enabling several aspects of the paleo-Eskimo's physical and biochemical features to be deduced. He was blood group A+, had brown eyes and dark skin, his thick dark hair was prone to baldness, and he had shovel graded front teeth typical of Asians and Native Americans. Of greater scientific value were the comparisons made between the SNPs present in the paleo-Eskimo genome and those known to be characteristic of other populations to whom the Saqqaq people might have been related. The origins of the Saqqaq have been rather mysterious, many archaeologists assuming that they were related to other North American groups such as the Inuit peoples. The SNP analysis suggested that this view is incorrect, the greatest similarities being with populations from arctic Siberia, called the Nganasans, Koryaks, and Chukchis. If this result is correct, then the ancestors of the Saqqaq must have entered the Americas across the Bering Strait, not with the original Clovis or pre-Clovis migrants, but much more recently, probably about 5500 years ago.

The paleo-Eskimo project shows that obtaining a near-complete genome sequence from ancient DNA is feasible, if the DNA is sufficiently well preserved, and is free from modern contamination, or can be decontaminated as is the case with hair. At present, these requirements seem to limit this type of work to permafrost specimens, or at least ones from cold latitudes, such as the Denisova hominin from the Altai mountains of Siberia. There will not be many opportunities to exploit the benefits of hair as a source of ancient DNA, because hair is not common in the archaeological record. But as with all areas of biomolecular archaeology, the technical developments are progressing rapidly. Most of the research we have examined in this book has been carried out since 2000, much of it since 2005, and almost every month brings a new headline report of the outcomes of a new important project. This is an exciting time to be involved in biomolecular archaeology, and we hope that reading this book has inspired you to delve further into the subject.

Further Reading

Bromham, L. & Penny, D. (2003) The molecular clock. *Nature Reviews Genetics*, 4, 216–24.

Cann, R.L., Stoneking, M. & Wilson, A.C. (1987) Mitochondrial DNA and human evolution. *Nature*, 325, 31–6. [The first discovery of Mitochondrial Eve.]

Dorit, R.L., Akashi, H. & Gilbert, W. (1995) Absence of polymorphism at the ZFY locus on the human Y chromosome. *Science*, 268, 1183–5. [One of the reports of Y chromosome Adam.]

Evans, P.D., Mekel-Bobrov, N., Vallender, E.J., Hudson, R.R. & Lahn, B.T. (2006) Evidence that the adaptive allele of the brain size gene *microcephalin* introgressed into *Homo sapiens* from an archaic *Homo* lineage. *Proceedings of the National Academy of Sciences USA*, 103, 18178–83.

Fagundes, N., Kanitz, R. & Eckert, R. (2008) Mitochondrial population genomics supports a single pre-Clovis origin with a coastal route for the peopling of the Americas. *American Journal of Human Genetics*, 82, 583–92.

Gilbert, M.T.P., Jenkins, D.L. & Götherström, A., *et al.* (2008) DNA from pre-Clovis human coprolites in Oregan, North America. *Science*, 320, 786–9.

Green, R.E., Malaspinas, A.-S. & Krause, J., *et al.* (2008) A complete Neandertal mitochondrial genome sequence determined by high-throughput sequencing. *Cell*, 134, 416–26.

Hublin, J. (2009) The origin of Neandertals. *Proceedings of the National Academy of Sciences USA*, 106, 16022–7.

Krause, J., Fu, Q. & Good, J.M., *et al.* (2010) The complete mitochondrial DNA genome of an unknown hominin from southern Siberia. *Nature*, 464, 894–7.

Mellars, P. (2006) Going East: new genetic and archaeological perspectives on the modern human colonization of Eurasia. *Science*, 313, 796–800.

Rasmussen, M., Li, Y. & Lindgreen, S., *et al.* (2010) Ancient human genome sequence of an extinct Palaeo-Eskimo. *Nature*, 463, 757–62.

Relethford, J.H. (2008) Genetic evidence and the modern human origins debate. *Heredity*, 100, 555–63.

Soares, P., Ermini, L. & Thomson, N., *et al.* (2009) Correcting for purifying selection: an improved human mitochondrial molecular clock. *American Journal of Human Genetics*, 84, 740–59.

Stringer, C. & Andrews, P. (2005) *The Complete World of Human Evolution*. Thames and Hudson, London.

Tamm, E., Kivisild, T. & Reidla, M., *et al.* (2007) Beringian standstill and spread of Native American founders. *PLoS ONE*, 2, e829.

Templeton, A.R. (1993) The "Eve" hypothesis: a genetic critique and reanalysis. *American Anthropologist*, 95, 51–72. [Criticism of the work of Cann *et al.* 1987]

White, D., Asfaw, B. & Beyene, Y., *et al.* (2009) *Ardipithecus ramidus* and the palaeobiology of early hominids. *Science*, 326, 75–86.

Glossary

α-helix One of the commonest secondary structural conformations taken up by segments of polypeptides.

β-*N*-glycosidic bond The linkage between the base and sugar of a nucleotide.

β-sheet One of the commonest secondary structural conformations taken up by segments of polypeptides.

2′-deoxyribose The type of sugar in a DNA polynucleotide.

3′-OH terminus The end of a polynucleotide that terminates with a hydroxyl group attached to the 3′-carbon of the sugar.

30 chromatin fiber A relatively unpacked form of chromatin consisting of a possibly helical array of nucleosomes in a fiber approximately 30 nm in diameter.

5′-P terminus The end of a polynucleotide that terminates with a monophosphate, diphosphate, or triphosphate attached to the 5′-carbon of the sugar.

Abasic site A site in a polynucleotide from which the nucleotide base as been removed.

Accession A variety of a crop.

Acylation The attachment of a lipid sidechain to a polypeptide.

Adaptor A synthetic, double-stranded oligonucleotide used to attach sticky ends to a blunt-ended molecule.

Adenine A purine base found in DNA and RNA.

Admixture Refers to members of a population whose genotypes have affinities with two or more groups.

Agarose gel electrophoresis Electrophoresis carried out in an agarose gel and used to separate DNA molecules between 100 bp and 50 kb in length.

Aldehyde A compound containing a carbonyl group.

Aldohexose An aldose sugar with six carbon atoms.

Aldopentose An aldose sugar with five carbon atoms.

 © 2011 Terry Brown & Keri Brown

Aldose A sugar containing a carbonyl group.

Aldotetrose An aldose sugar with four carbon atoms.

Aldotriose An aldose sugar with three carbon atoms.

Allele One of two or more alternative forms of a gene.

Allelic dropout When one of two or more PCR products fails to be synthesized in sufficient amounts to give a visible band in a gel.

Amelogenin A protein found in tooth enamel whose gene is sexually dimorphic and is used in DNA-based sex identification.

Amino acid One of the monomeric units of a protein molecule.

Amino acid racemization Conversion between the two different enantiomers of an amino acid.

Amino terminus The end of a polypeptide that has a free amino group.

Amphipathic Refers to a molecule with one highly hydrophilic end and one highly hydrophobic end.

Amplicon A DNA molecule produced by PCR.

Amyloplast Structures related to chloroplasts in roots, developing seeds and storage organs such as fruits and tubers.

Ancient biomolecule A biomolecule preserved in ancient biological material.

Ancient DNA DNA preserved in ancient biological material.

Anomer A pair of cyclic monosaccharides that differ in the positioning of the hydroxyl group attached to carbon 1.

Antibody A protein synthesized by animals that binds to and helps degrade foreign molecules.

Antigen A substance that elicits an immune response.

Antiparallel Refers to the arrangement of polynucleotides in the double helix, these running in opposite directions.

Archaeobotany The study of preserved plant remains in order to address archaeological questions.

Archaeogenetics The area of biomolecular archaeology that uses analysis of DNA sequences from living organisms to address archaeological questions.

Archaeology The study of the human past.

Atomic number The number of protons in the nucleus of an atom.

Autosome A chromosome that is not a sex chromosome.

Balancing selection The process where two opposing selective pressures result in a greater number of heterozygotes than expected.

Base pair The hydrogen-bonded structure formed by two complementary nucleotides. When abbreviated to "bp," the shortest unit of length for a double-stranded DNA molecule.

Base pairing The attachment of one polynucleotide to another, or one part of a polynucleotide to another part of the same polynucleotide, by base pairs.

Biallelic Refers to a gene that exists as a pair of alelles.

Bile acid A type of sterol synthesized in the liver.

Bioapatite The inorganic component of bone.

Biochemical proxy A straightforward analysis which, when applied to an archaeological specimen, would provide an indication of the degree of preservation of a biomolecule such as DNA or protein.

Biomolecular archaeology The study of biomolecules in order to address archaeological questions.

Biomolecular paleopathology The use of biomolecules to study disease in the past.

Biotin A small protein with a strong affinity for streptavidin, used as a label on DNA and other molecules, for example as an aid to purification.

BLAST An algorithm frequently used in DNA database searching.

Blocking lesion A type of chemical modification in ancient DNA that blocks progress of the *Taq* DNA polymerase during PCR.

Bog body A type of mummy that has been preserved in a bog,

Cambridge reference sequence (CRS) The first version of the human mitochondrial genome to be sequenced.

Cancellous bone A less dense type of bone permeated with channels.

Carbohydrates The group of biomolecules comprising sugars and their derivatives.

Carboxyl terminus The end of a polypeptide that has a free carboxyl group.

Carboxylic acid A compound made up of a central carbon atom attached to an oxygen radical, a hydroxyl group and an R group.

Cementum The surface covering of the part of a tooth that is attached to the jaw.

Chain termination method A DNA sequencing method that involves enzymatic synthesis of polynucleotide chains that terminate at specific nucleotide positions.

Chiral A central carbon atom attached to four different chemical groups.

Chromosome One of the DNA–protein structures that contains part of the nuclear genome of a eukaryote. Less accurately, the DNA molecule(s) that contain(s) a prokaryotic genome.

Cloning Insertion of a fragment of DNA into a cloning vector, and subsequent propagation of the recombinant DNA molecule in a host organism.

Coalescence time An estimate of the time that has elapsed since a haplogroup first came into existence, based on the degree of divergence of the haplotypes in that haplogroup.

Codon A triplet of nucleotides coding for a single amino acid.

Collagen fibril Fibrils made of tropocollagen units.

Collagenase An enzyme that specifically breaks down collagen.

Column chromatography Any type of chromatography that is carried out in a glass, metal, or other type of column.

Compact bone Bone made up of parallel bundles of osteons.

Competent Refers to a culture of bacteria that have been treated, for example, by soaking in calcium chloride, so that their ability to take up DNA molecules is enhanced.

Complementary Refers to two nucleotides or nucleotide sequences that are able to base-pair with one another.

Complex triglyceride A triglyceride in which all three fatty acids are not identical.

Compound specific isotope study A procedure in which a detailed analysis is made of the isotope ratios in individual biomolecules.

Consensus sequence A nucleotide sequence that represents an "average" of a number of related but non-identical sequences.

Contig A contiguous set of overlapping DNA sequences.

Coprolite Preserved excrement.

Coprostanol A sterol biomarker for fecal material.

Cortical bone Bone made up of parallel bundles of osteons.

Crossover immunoelectrophoresis (CIP, CIEP) An immunoassay in which the precipitin reaction is viewed at the crossover point between antigen electrophoresis and antibody electroendosmosis in a gel.

Crystallinity Refers to the crystalline content of an archaeological bone.

C-terminus The end of a polypeptide that has a free carboxyl group.

Cytosine One of the pyrimidine bases found in DNA and RNA.

Degeneracy Refers to the fact that the genetic code has more than one codon for most amino acids.

Denaturation Breakdown by chemical or physical means of the non-covalent interactions, such as hydrogen bonding, which maintain the secondary and higher levels of structure of proteins and nucleic acids.

Dental microwear Small forms of damage on the surfaces of teeth often indicative of diet.

Dentine The main structural component of a tooth.

Deoxyribonucleic acid One of the two forms of nucleic acid in living cells; the genetic material for all cellular life forms and many viruses.

Derivatization A chemical process that modifies hydrophilic and reactive groups in compounds prior to gas chromatography so that these compounds are more readily volatile.

Diagenesis The general breakdown and decay of dead biological material over time.

Diastereomer Isomers that differ from one another only in the relative orientations of their chemical bonds, but which are not enantiomers.

Dideoxynucleotide A modified nucleotide that lacks the 3′ hydroxyl group, and so terminates strand synthesis when incorporated into a polynucleotide.

Diploid A nucleus that has two copies of each chromosome.

Disaccharide A sugar comprising two linked monosaccharide units.

Discontinuous gene A gene that is split into exons and introns.

Distance matrix A table showing the evolutionary distances between all pairs of nucleotide sequences in a dataset.

Disulfide bridge A covalent bond linking cysteine amino acids on different polypeptides or at different positions on the same polypeptide.

DNA Deoxyribonucleic acid, one of the two forms of nucleic acid in living cells; the genetic material for all cellular life forms and many viruses.

DNA ligase An enzyme that synthesizes phosphodiester bonds as part of DNA replication, repair and recombination processes.

DNA packaging The system, involving attachment to proteins, that fits a DNA molecule into its chromosome.

Dominance Refers to the allele that is expressed in a heterozygote.

Double helix The base-paired double-stranded structure that is the natural form of DNA in the cell.

Eicosanoid A type of lipid derived from the fatty acid arachidonic acid.

Electroendosmosis Electrically stimulated diffusion of a compound, distinct from electrophoresis.

Electrospray ionization A mild ionization procedure used in liquid chromatography-mass spectrometry.

ELISA (enzyme-linked immunosorbent assay) An immunoassay in which antibody–antigen binding is detected by measuring the activity of a reporter enzyme that is conjugated to the antibody.

Enamel The surface covering of the part of a tooth that is exposed in the mouth.

Enantiomer Two compounds that are mirror images but non-identical due to the arrangement of their chemical groups around a chiral carbon atom.

Endogamy Marriage between two individuals who belong to the same social group.

Enzyme A protein with catalytic properties.

Epitope A surface feature recognized by an antibody.

Ethanol precipitation Precipitation of nucleic acid molecules by ethanol plus salt, used primarily as a means of concentrating DNA.

Ethnographic analogy Interpretation of past societies via observation and analysis of diverse modern societies.

Eukaryote An organism whose cells contain membrane-bound nuclei.

Exogamy Marriage between two individuals who belong to different social groups.

Exon A coding region within a discontinuous gene.

Faraday collector The component of an isotope ratio mass spectrometer that catches the separated ions, generating a current that is measured in order to quantify the amount of each isotope in the starting sample.

Fatty acid A type of lipid that is a carboxylic acid whose R group is a long-chain hydrocarbon.

Fixation index A statistic indicating the degree of difference between the allele frequencies present in two populations.

Fixed Refers to an allele that is possessed by all members of a population.

Founder analysis An analysis that uses haplogroup diversities to estimate the time of a population split.

Free radical A chemical group containing unpaired electrons and hence highly reactive.

F_{ST} value A statistic indicating the degree of difference between the allele frequencies present in two populations.

Furanose The cyclic form of a ketose.

Gamete A reproductive cell, usually haploid, that fuses with a second gamete to produce a new cell during sexual reproduction.

Ganglioside A type of sphingolipid containing a complex sugar.

Gas chromatography A method for separating compounds according to their differential partitioning between a carrier gas and a liquid stationary phase.

Gas chromatography combustion isotope ratio mass spectrometry (GC-C-IRMS) A type of analytical procedure used to measure stable isotope ratios in the material under study.

Gel filtration chromatography A chromatography system that separates molecules according to their rate of diffusion through a gel matrix.

Gender A system of social categorization that uses the differences between the biological sexes as a way of structuring thought and practice.

Gender archaeology The discipline that investigates the cultural construction of gender in past societies.

Gene A DNA segment containing biological information and hence coding for an RNA and/or polypeptide molecule.

Genetic code The rules that determine which triplet of nucleotides codes for which amino acid during protein synthesis.

Genetic profiling A method based on typing of short tandem repeats that produces a genotype that can be used to identify kinship relationships.

Genome expression The series of events by which the biological information carried by a genome is released and made available to the cell.

Genome The entire genetic complement of a living organism.

Genotype A description of the genetic composition of an organism.

Globular protein A protein with a non-fibrous structure.

Glycerophospholipid A lipid that resembles a triglyceride, but with one of the fatty acids replaced by a hydrophilic group attached to the glycerol moiety by a phosphodiester bond.

Glycosylation The attachment of sugar units to a polypeptide.

Grave goods The materials buried with a skeleton, possibly indicative of the latter's gender, status, and/or wealth.

Guanine One of the purine nucleotides found in DNA and RNA.

Half-life The time taken for half of the atoms of a radioactive element to decay.

Haplogroup One of the major sequence classes of mitochondrial DNA present in the human population.

Haploid A nucleus that has a single copy of each chromosome.

Haplotype An individual mitochondrial DNA sequence.

Haversian canal The central component of an osten containing blood vessels and nerves.

Helix–turn–helix motif A common structural motif for attachment of a protein to a DNA molecule.

Heteroplasmic Possessing two haplotypes of the mitochondrial genome.

Heteropolysaccharide A polysaccharide in which all the monosaccharide units are not identical.

Heterozygous A diploid nucleus that contains two different alleles for a particular gene.

High-performance liquid chromatography (HPLC) A column chromatography method with many applications in biochemistry.

Hilum The formation center for a starch grain.

Histone One of the basic proteins found in nucleosomes.

Hominid A member of the taxonomic family called the Hominidae, comprising humans, chimpanzees, gorillas, and orangutans.

Hominin A bipedal hominid, including modern humans and those extinct taxa that preceded us in the evolutionary branch leading to *Homo sapiens*.

Homopolysaccharide A polysaccharide in which all the monosaccharide units are identical.

Homozygous A diploid nucleus that contains two identical alleles for a particular gene.

Hydrocarbon A compound made up of carbon and hydrogen atoms.

Hydrogen bond A weak electrostatic attraction between an electronegative atom such as oxygen or nitrogen and a hydrogen atom attached to a second electronegative atom.

Hypervariable region One of the two very variable regions of the non-coding part of the mitochondrial genome.

Immunoassay A quantitative test that makes use of an immunological reaction.

immunoelectrophoresis An immunoassay involving electrophoresis of antibody and antigen.

Immunoglobulin The type of protein that acts as antibodies.

Immunological methods Detection and analysis methods that make use of the immunological reaction between antibody and antigen.

Inductively coupled plasma mass spectrometry (ICP-MS) A type of mass spectrometry in which the material to be studied is ionized by plasma exposure.

Infrafemale A sexually underdeveloped female.

Inherited diseases A disease caused by a defect in the genome.

Initiation codon The codon, usually but not exclusively 5′–AUG–3′, found at the start of the coding region of a gene.

Insertion sequence A short transposable element found in bacteria.

Intergenic DNA The regions of a genome that do not contain genes.

Interspersed repeat A sequence that recurs at many dispersed positions within a genome.

Intron A non-coding region within a discontinuous gene.

Ion exchange chromatography A method for separating molecules according to how tightly they bind to electrically charged particles present in a chromatographic matrix.

Isoelectric focusing Separation of proteins in a gel that contains chemicals which establish a pH gradient when the electrical charge is applied.

Isoelectric point The position in a pH gradient where the net charge of a protein is zero.

Isoprene A small hydrocarbon molecule that forms the basic chemical component of a terpene.

Isotope One of two or more atoms that have the same atomic number but different atomic weights.

Isotope fractionation A fractionation that results in a change in the relative proportions of the isotopes in the products of a reaction compared to the proportions in the initial substrates.

Isotope ratio mass spectrometry (IRMS) A type of analytical procedure used to measure stable isotope ratios in the material under study.

Jumping PCR A process that can lead to chimeric products when PCR is carried out with damaged DNA molecules.

Keratin A fibrous protein found, for example, in hair.

Ketone A compound containing a C=O group.

Ketose A sugar containing a C=O group.

Kilobase pair (kb) 1000 base pairs.

Kinetic fractionation Isotope fractionation occurring during a unidirectional physical or chemical reaction.

Kinship A family relationship.

Kinship analysis Any of several methods used to identify kinship relationships between individuals.

Lactase persistence Continued activity of the lactase gene into adulthood, conferring the ability to metabolize lactose and hence digest dairy products.

Lamellar bone Bone in which the collagen fibrils are aligned.

Laser ablation A procedure that makes use of a laser beam to remove very small amounts of material from the surface of a material.

Lesion A change to the skeleton caused by disease or trauma.

Linker histone A histone, such as H1, that is located outside of the nucleosome core octamer.

Lipids A broad group of hydrophobic compounds that include the fats, oils, waxes, steroids, and various resins.

Liquid chromatography A method for separating compounds according to their differential partitioning between a liquid mobile phase and a solid stationary phase.

Macromolecule The large organic compounds (nucleic acids, proteins, lipids, and carbohydrates) found in living organisms and sometimes present, usually in a partly degraded state, in the remains of those organisms after their death.

Magnetic sector mass spectrometer A type of mass spectrometry in which the mass analyzer is a single or series of magnets through which the ionized molecules are passed.

Maillard reaction A reaction occurring between sugars and amino acids, and hence able to link together polynucleotides and peptides.

Marker A degradation product whose detection indicates the former presence of its parent molecule in an archaeological specimen, or any biomolecule whose detection is linked with a particular process or activity, such as dairying.

Mass number The number of protons and neutrons in a nucleus.

Mass spectrometry An analytical technique in which ions are separated according to their charge-to-mass ratios.

Mass spectrum A plot showing the m/z values for the molecular and fragment ions of the compound being analyzed.

Mass-to-charge ratio The relationship between the mass of an atom and its electrical charge, which forms the basis of ion separation during mass spectrometry.

Matrilocal A marriage system in which the man moves and the new couple live within or close to the woman's family.

Matrix-assisted laser desorption ionization time-of-flight (MALDI-TOF) A type of mass spectrometry used in proteomics.

Maximum parsimony A method for constructing a phylogenetic tree in which it is assumed that the most likely tree is the one that requires the least evolutionary change.

Megabase pair (Mb) 1000 kb; 1 000 000 bp.

Melting temperature (T_m) The temperature at which the two strands of a double-stranded nucleic acid molecule or base-paired hybrid detach as a result of complete breakage of hydrogen bonding.

Mercury intrusion porosimetry A method that enables the total volume of pores of different sizes to be measured in a material.

Messenger RNA (mRNA) The transcript of a protein-coding gene.

Micelle A spherical structure formed by a soap, with the carboxyl groups on the surface and the hydrocarbon chains embedded within the structure.

Microsatellite A type of simple sequence length polymorphism comprising tandem copies of, usually, dinucleotide, trinucleotide, or tetranucleotide repeat units. Also called a simple tandem repeat (STR).

Microvariant An STR allele in which the repeat motif is not entirely regular.

Miscoding lesion A type of DNA damage that leads to an error when the molecule is sequenced.

Mitochondrial Eve The woman who lived in Africa between 140 000 and 290 000 years ago, and who carried the ancestral mitochondrial DNA that gave rise to all the mitochondrial DNAs in existence today.

Mitochondrial genome The genome present in the mitochondria of a eukaryotic cell.

Mobile phase The mobile component of a column chromatography system.

Molecular clock A statistical test used to assign a date to the time of the most recent common ancestor of two or more DNA sequences based on the degree of divergence between them.

Molecular phylogenetics A set of techniques that enable the evolutionary relationships between DNA sequences to be inferred by making comparisons between those sequences.

Monoclonal antibody A preparation containing a single immunoglobulin and hence specific for a single epitope.

Monogenic Refers to a characteristic that is specified by a single gene.

Monogenic disorder An inherited disease caused by a defect in an individual gene.

Monolith A chromatography column comprising a single solid block of polymer.

Monosaccharide The monomeric unit of a polysaccharide, such as glucose.

Multiallelic Refers to a gene with three or more alleles.

Multigene family A group of genes, clustered or dispersed, with related nucleotide sequences.

Multiple alignment An alignment of three or more nucleotide sequences.

Multiplex PCR PCR with two or more primer pairs in a single reaction.

Multiregional hypothesis A hypothesis that holds that modern humans in the Old World are descended from *Homo erectus* populations that left Africa 1.8 million years ago.

Mummy An archaeological specimen in which some of the soft tissues have been preserved.

Mutation An alteration in the nucleotide sequence of a DNA molecule.

Mycolic acid A type of modified fatty acid synthesized by all mycobacteria and also by members of related genera such as *Corynebacterium* and *Nocardia*.

Natron A mixture of salts used in Egyptian mummification.

Neighbor-joining method A method for construction of phylogenetic trees.

Next generation sequencing Recently developed methods for DNA sequencing that enable hundreds of thousands of short sequences to be obtained in a single experiment.

Nitrogenous base One of the purines or pyrimidines that form part of the molecular structure of a nucleotide.

N-linked glycosylation The attachment of sugar units to an asparagine in a polypeptide.

Non-coding RNA An RNA molecule that does not code for a protein.

Non-metric trait Small, inherited variations in skeletal morphology.

Non-polar A hydrophobic (water-hating) chemical group.

Non-synonymous substitution A mutation that converts a codon for one amino acid into a codon for a second amino acid.

Non-template addition The tendency of some types of DNA polymerase to add an additional A to the 3′-end of a polynucleotide that it has just synthesized.

N-terminus The end of a polypeptide that has a free amino group.

Nuclear genome The DNA molecules present in the nucleus of a eukaryotic cell.

Nucleic acid The term first used to describe the acidic chemical compound isolated from the nuclei of eukaryotic cells. Now used specifically to describe a polymeric molecule comprising nucleotide monomers, such as DNA and RNA.

Nucleosome The complex of histones and DNA that is the basic structural unit in chromatin.

Nucleotide A purine or pyrimidine base attached to a five-carbon sugar, to which a monophosphate, diphosphate, or triphosphate is also attached. The monomeric unit of DNA and RNA.

Null allele An allele of an STR that is not amplified during PCR, probably due to a mutation in one of the priming sites.

O-glycosidic bond The link between the two monosaccharide units in a disaccharide.

Oligonucleotide A short synthetic single-stranded DNA molecule.

O-linked glycosylation The attachment of sugar units to a serine or threonine in a polypeptide.

Osteoarchaeology The study of the structure and function of archaeological skeletons.

Osteoblast The type of cell responsible for the synthesis of bone.

Osteoclast The type of cell responsible for the resorption of bone.

Osteocyte Osteoblasts that have become embedded in the bone that they have formed.

Osteon Cylindrical structural units in lamellar bone.

Ouchterlony technique An immunoassay in which the precipitin reaction is carried out in a thin slab of agar or agarose gel.

Out of Africa hypothesis A hypothesis which holds that modern humans evolved in Africa, moving to the rest of the Old World between 100 000 and 50 000 years ago, displacing the descendants of *Homo erectus* that they encountered.

Paleodietary studies The study of the diet and mode subsistence of past people.

Paleopathology The study of disease in the past.

Paleoproteomics The application of proteomics to the study of ancient proteins.

Parallel-fibered bone Bone with an intermediate arrangement between woven and lamellar.

Partition coefficient The ratio of the amounts of a compound in two immiscible phases, as occurs during chromatography.

Pathognomonic Refers to a skeletal lesion that is a definite indicator of the presence of a particular disease.

Patrilocal A marriage system in which the woman moves and the new couple live within or close to the man's family.

Pentose A sugar comprising five carbon atoms.

Peptide bond The chemical link between adjacent amino acids in a polypeptide.

Peptide mass fingerprinting A method based on mass spectrometry that can identify the peptides obtained by digestion of a protein.

Phenotype The observable characteristics displayed by a cell or organism.

Phosphodiester bond The chemical link between adjacent nucleotides in a polynucleotide.

Phylogenetic tree A tree depicting the evolutionary relationships between a set of DNA sequences, species, or other taxa.

Phytolith Microscopic inorganic deposit, usually of silica, sometimes formed in plant cells.

Plant microfossil A microscopic plant remain, primarily starch grains, phytoliths, and pollen.

Plasma ionization mass spectrometry (PIMS) A type of mass spectrometry in which the material to be studied is ionized by plasma exposure.

Plasmid A usually circular piece of DNA often found in bacteria and some other types of cell.

Polar A hydrophilic (water-loving) chemical group.

Polyacrylamide gel electrophoresis Electrophoresis carried out in a polyacrylamide gel and used to separate DNA molecules between 10 and 1500 bp in length.

Polyclonal antibody A collection of immunoglobulins recognizing different epitopes.

Polygenic Refers to a characteristic that is specified by a group of genes acting together.

Polymerase chain reaction (PCR) A technique that results in exponential amplification of a selected region of a DNA molecule.

Polymorphism Refers to a locus that is represented by a number of different alleles or haplotypes in the population as a whole.

Polynucleotide A single-stranded DNA or RNA molecule.

Polypeptide A polymer of amino acids.

Polysaccharide A polymeric carbohydrate such as starch.

Population genetics The area of genetics devoted to study of genes in populations.

Post-marital residence pattern Refers to where a married couple live, in relation to their parents, after marriage.

Potsherd A fragment of an archaeological pot.

Precipitin reaction Formation of an insoluble antibody–antigen binding complex.

Primary antibody The antibody that directly detects the antigen in an indirect immunoassay.

Primary structure The sequence of amino acids in a polypeptide.

Primer A short oligonucleotide that is attached to a single-stranded DNA molecule in order to provide a start point for strand synthesis.

Principal component analysis A procedure that attempts to identify patterns in a large dataset of variable character states.

Prokaryote An organism whose cells lack a distinct nucleus.

Protein profiling The methodology used to identify the proteins in a proteome.

Protein The polymeric compound made of amino acid monomers.

Proteome The collection of functioning proteins synthesized by a living cell.

Proteomics A variety of techniques used to study proteomes.

Pulp cavity The central cavity within a tooth.

Purifying selection A form of selection that gradually removes deleterious mutations from the genome.

Purine One of the two types of nitrogenous base found in nucleotides.

Pyranose The cyclic form of an aldohexose.

Pyrimidine One of the two types of nitrogenous base found in nucleotides.

Pyrolysis GC-MS A type of gas chromatography-mass spectrometry that uses a high temperature to volatilize the sample under study.

Pyrosequencing A novel DNA sequencing method in which addition of a nucleotide to the end of a growing polynucleotide is detected directly by conversion of the released pyrophosphate into a flash of chemiluminescence.

Quadrupole mass spectrometer A type of mass spectrometry that utilizes four magnetic rods placed parallel to one another, surrounding a central channel through which the ions must pass.

Quaternary structure The structure resulting from the association of two or more polypeptides.

Radioimmunoassay (RIA) A form of ELISA in which the detecting antibody is radioactively labelled.

Recessiveness Refers to the allele that is not expressed in a heterozygote.

Recombinant A vector molecule into which a new piece of DNA has been inserted.

Recombination A large-scale rearrangement of a DNA molecule.

Renaturation The return of a denatured molecule to its natural state.

Reporter enzyme An enzyme whose activity is easily monitored and that is therefore used as the detection process in, for example, an immunoassay.

Restriction endonuclease An enzyme that cuts DNA molecules at a limited number of specific nucleotide sequences.

Restriction fragment length polymorphism (RFLP) analysis A method that obtains information on a DNA sequence indirectly by comparing the sizes of fragments obtained after digestion with a restriction endonuclease.

Reverse phase chromatography Chromatography using a matrix whose surface is covered with non-polar chemical groups such as hydrocarbons.

Reverse transcriptase A polymerase that synthesizes DNA on an RNA template.

Reverse transcriptase PCR (RT-PCR) PCR in which the first step is carried out by reverse transcriptase, so RNA can be used as the starting material.

Ribonucleic acid One of the two forms of nucleic acid in living cells; the genetic material for some viruses.

Ribosomal RNA (rRNA) The RNA molecules that are components of ribosomes.

Ribosome One of the protein–RNA assemblies on which translation occurs.

RNA Ribonucleic acid, one of the two forms of nucleic acid in living cells; the genetic material for some viruses.

RNA polymerase An enzyme that synthesizes RNA on a DNA or RNA template.

RT-PCR PCR in which the first step is carried out by reverse transcriptase, so RNA can be used as the starting material.

Saponification The process leading to soap formation, involving heating a triglyceride with alkali.

Saturated Refers to a fatty acid in which every carbon carries two hydrogen atoms.

Scaffold A protein core within a chromosome to which DNA is attached.

Secondary antibody The antibody that carries the reporter enzyme in an indirect immunoassay.

Secondary product Animal products that can be acquired without killing the animal.

Secondary structure The conformations, such as α-helix and β-sheet, taken up by a polypeptide.

Sex cell A reproductive cell; a cell that divides by meiosis.

Sex identification Identification of the biological and/or genetic sex of an individual.

Sex reversal Refers to an individual whose genetic sex differs from their biological sex.

Sexually dimorphic Refers to a feature that is different in males and females.

Short tandem repeat (STR) A type of simple sequence length polymorphism comprising tandem copies of, usually, dinucleotide, trinucleotide, or tetranucleotide repeat units. Also called a microsatellite.

Siberian ice maiden A female frozen mummy from the 5th century BC, found in a tomb belonging to the Pazyryk culture.

Silica binding A method for DNA preparation involving binding to silica, usually in a chromatography column.

Simple triglyceride A triglyceride in which all three fatty acids are identical.

Single ion monitoring A mass spectrometry method in which a particular ion, or group of ions, is detected in the output from a chromatography system.

Single nucleotide polymorphism (SNP) A point mutation that is carried by some individuals of a population.

Somatic cell A non-reproductive cell; a cell that divides by mitosis.

Sphagnum bog A type of bog whose peat is derived largely from sphagnum moss.

Sphingolipid A type of membrane lipid comprising sphingosine, a hydrophilic head group, and a fatty acid.

Spongy bone A less dense type of bone permeated with channels.

Stable isotope An isotope that is not radioactive, or that decays with such a long half-life as to be considered stable.

Stable isotope analysis Measurement of the stable isotope ratios in a material, for example to study the diets of past populations.

Starch grain Stored energy reserves in plant cells, sometimes recovered from archaeological material.

Stationary phase The stationary component of a column chromatography system.

Steroid A sterol derivative with a different chemical group replacing the hydroxyl attached to the C_3 carbon.

Sterol A type of terpene formed by cyclization of squalene.

Stored starch synthesis Formation of long-lived starch reserves in amyloplasts.

Streptavidin A protein with a strong affinity for biotin, used to purify biotin-labelled molecules from a mixture.

STRUCTURE A software program that identifies relationships between genotypes in order to assign individuals to groups.

Stuttering An artifact arising during STR genotyping, which results in a series of linked peaks of varying heights in the electrophoretogram.

Subsistence The way in which people obtain their food.

Super-numerary bone A small component of the skeleton not possessed by all individuals.

Synonymous substitution A mutation that changes a codon into a second codon that specifies the same amino acid.

Tandem MS A type of mass spectrometry in which two or more mass analyzers are linked in series.

Tandemly repeated DNA DNA sequence motifs that are repeated head to tail.

***Taq* DNA polymerase** The thermostable DNA polymerase used in PCR.

Telomere The end of a eukaryotic chromosome.

Template The polynucleotide that is copied during a strand synthesis reaction catalyzed by a DNA or RNA polymerase.

Template switching An event that can lead to chimeric products when PCR is carried out with damaged DNA molecules.

Termination codon One of the three codons that mark the position where translation of an mRNA should stop.

Terpene A type of lipid made up of isoprene units, the largest class of natural products.

Terpenoid A terpene in which one or more isoprene units have undergone chemical modification.

Tertiary structure The level of protein structure resulting from folding the secondary structural components of the polypeptide into a three-dimensional configuration.

Thermal ionization mass spectrometry (TIMS) A type of mass spectrometry in which the material to be studied is ionized by heating on a metal filament.

Thymine One of the pyrimidine bases found in DNA.

T_m Melting temperature.

Total ion content A measure of all the ions produced from a particular gas chromatography fraction.

Trabecular bone A less dense type of bone permeated with channels.

Transcription The synthesis of an RNA copy of a gene.

Transcriptome The entire mRNA content of a cell.

Transcriptomics The use of high throughput techniques related to next generation sequencing to study the RNA content of a tissue.

Transfer RNA (tRNA) A small RNA molecule that acts as an adaptor during translation and is responsible for decoding the genetic code.

Transitory starch synthesis Formation of short-lived starch reserves in chloroplasts.

Translation The synthesis of a polypeptide, the amino acid sequence of which is determined by the nucleotide sequence of an mRNA in accordance with the rules of the genetic code.

Translocation The transfer of one part of chromosome to another chromosome.

Transposition The movement of a genetic element from one site to another in a DNA molecule.

Transposon A genetic element that can move from one position to another in a DNA molecule.

Triacylglycerol A lipid made up of three fatty acids that have each formed an ester linkage with a single molecule of glycerol.

Triglyceride A lipid made up of three fatty acids that have each formed an ester linkage with a single molecule of glycerol.

Tropocollagen Triple helical collagen structures.

Two-dimensional gel electrophoresis A method for separation of proteins used especially in studies of the proteome.

Tyrolean iceman A natural mummy found in the Ötstal region of the Alps on the border between Austria and Italy.

Unsaturated Refers to a fatty acid in which one or more carbon pairs are linked by double bonds.

Wave of advance A hypothesis which holds that the spread of agriculture into Europe was accompanied by a large-scale movement of human populations.

Woven bone Bone in which the collagen fibrils take up a random arrangement.

Index

Biomolecular Archaeology: An Introduction, by Terry Brown and Keri Brown © 2011 Terry Brown & Keri Brown

Printed and bound by CPI Group (UK) Ltd, Croydon, CR0 4YY

09/06/2025

14685997-0001